Catalytic Transformations of Sustainable and Versatile Furanic Chemicals

Diminishing confined fossil resources has spurred the scientific community to strive for alternative, sustainable resources, such as terrestrial biomass, which can potentially substitute fossil-based derivatives. Lignocellulosic biomass is deemed an indispensable carbon source for meeting industrial and social demands regarding energy/fuels and chemicals. Over the past decade, significant advances have been shown in developing a broad spectrum of high-value chemicals and functional materials derived from biomass-based substrates. In connection with this, furanic chemicals, such as 5-hydroxymethylfurfural (HMF) and furfural, have recently received considerable attention due to their potential applications.

Catalytic Transformations of Sustainable and Versatile Furanic Chemicals aims to explicitly display the latest technical developments on designing functional catalysts, optimizing process and reaction parameters, and improving the stability of the employed materials on the transformation of HMF and furfural to a wide range of value-added chemicals via oxidation, reduction, hydrodeoxygenation, and reductive amination processes.

Features:

- Highlighting the importance and significance of HMF and furfural.
- Evaluating the transformations of HMF and furfural with various catalytic processes.
- Designing robust and efficient catalytic systems.
- Understanding the stability of catalytic materials.
- Understanding the characteristic features of the catalytic materials.

This book aims to target not only researchers focusing on biomass valorization from academic to industrial sectors but also graduate students interested in pursuing their higher education in biochemicals (especially furanic compounds), covering various aspects in organic synthesis routes, preparation of inorganic materials and physicochemical properties, for both chemists and chemical engineers.

Catalytic Transformations of Sustainable and Versatile Furanic Chemicals

Edited by

Shunmugavel Saravanamurugan
Center of Innovative and Applied Bioprocessing
Mohali, Punjab, India

Hu Li
State-Local Joint Laboratory for Comprehensive Utilization of Biomass Center for R&D of Fine Chemicals Guizhou University
Guiyang, Guizhou, China

Ashok Pandey
Centre for Energy and Environmental Sustainability
Lucknow, Uttar Pradesh, India

Young-Woong Suh
Department of Chemical Engineering, Hanyang University
Seoul, Republic of Korea

CRC Press
Taylor & Francis Group
Boca Raton London New York

CRC Press is an imprint of the
Taylor & Francis Group, an **informa** business

Designed cover image: Shunmugavel Saravanamurugan

First edition published 2025
by CRC Press
2385 NW Executive Center Drive, Suite 320, Boca Raton FL 33431

and by CRC Press
4 Park Square, Milton Park, Abingdon, Oxon, OX14 4RN

CRC Press is an imprint of Taylor & Francis Group, LLC

ISBN: 9781032735672 (hbk)
ISBN: 9781032735696 (pbk)
ISBN: 9781003464839 (ebk)

DOI: 10.1201/9781003464839

Typeset in Times
by Newgen Publishing UK

Contents

Preface

Terrestrial biomass, such as lignocellulosic biomass (LCB) – not dependent on the food cycle – includes agricultural and forestry residues, grasses and shrubs, and are deemed as an alternative to the dwindling confined fossil-fuel reserves that are the main source of today's energy, fuels, and chemicals, which can provide a sophisticated life for human beings. However, the standard of life of human beings is being improved worldwide, it is therefore crucial to utilise the steadily declining fossil fuels effectively and to strive for efficient transformation protocols for the LCB to produce energy, fuels, and chemicals. The intrinsic nature of LCB, predominantly composed of homopolymer (cellulose) and heteropolymers (hemicellulose and lignin), makes it challenging to selectively undergo chemical and biological approaches to obtain target products. However, cellulose and hemicellulose have been widely explored to obtain intermediate products, such as furanics (e.g., 5-hydroxymethylfurfural (HMF) and furfural), rather than lignin (due to its intricate nature).

HMF and furfural are potential platform chemicals that can be obtained from corresponding hexoses (e.g., glucose) and pentoses (e.g., xylose), the monomer units of cellulose and hemicellulose, respectively. The book *Catalytic Transformations of Sustainable and Versatile Furanic Chemicals* exclusively focuses on transforming these two furanics into value-added products. The main goal of this book is to exhibit the potential applications of HMF and furfural via transformation into various products, such as 2,5- dimethyl furan (DMF), 2,5-difoymylfuran (DFF), 2,5- furandicarboxylic acid (FDCA) and 2-methylfuran (MF), based on multiple catalytic strategies. This book provides the recent update on various catalytic approaches for converting HMF and furfural to the target products, providing state-of-the-art perspectives.

The book comprises 11 chapters. Chapter 1 gives a brief overview of the importance of LCB, the market size, and the applications of HMF and furfural. Selective hydrodeoxygenation (HDO) of HMF can yield DMF, a promising liquid fuel with a high energy density compared to ethanol and n-butanol. Chapter 2 discusses the catalytic HDO of HMF to DMF with various metal-containing materials being discussed in detail based on how several parameters, such as the nature of the support, hydrogen donors and bimetallic synergy, influence the formation of intermediates and the target product DMF. When switching the reaction atmosphere from H_2 to O_2 in the transformation of HMF, the target products can be entirely different, and in this case, FDCA and DFF are focused and presented in Chapters 3 and 4. Chapter 3 describes various catalytic oxidation approaches, such as thermocatalysis, electrocatalysis, photocatalysis, and biocatalysis, of HMF to FDCA using a number of metal catalysts. The catalytic reaction mechanism of each approach covered is discussed for better understanding. Chapter 4 discusses the selective photocatalytic oxidation of HMF to DFF, describing the role of hydroxyl and superoxide radicals on the product selectivity. Like HDO of HMF, furfural can also undergo a deoxygenation reaction under similar reaction conditions to yield MF over supported metal catalysts, besides other hydrogenated products, such as furfuryl alcohol, tetrahydrofurfuryl alcohol, 2-methyltetrahydrofuran, 1,5-pentanediol and cyclopentanone, are delineated

(Chapters 5 and 6). Introducing ammonia into the furfural hydrogenation reaction conditions changes the product distribution, forming the corresponding furfuryl amine. Employing suitable metal(s) on a suitable support to influence the transformation of the Schiff base intermediate to primary amine is discussed (Chapter 7).

HMF is hygroscopic and tends to oligomerize at room temperature, which means it must be refrigerated constantly. Alkoxy HMF, for example, 5-ethxoymethylfurfural (EMF), is a stable molecule at room temperature and an alternative to HMF. In this regard, a chapter has been included discussing how acid-functionalized solid catalysts play a key role in transforming HMF to EMF, sugars (e.g., fructose) to EMF via HMF and 5-chloromethylfurfural to EMF via nucleophilic substitution (Chapter 8). Likewise, FDCA, a solid compound, has very limited solubility in water as a reaction solvent, which is considered an advantage and disadvantage. However, to address the issue of solubility, oxidative esterification of HMF in alcohol to produce a corresponding esterified product of FDCA (e.g., furan-2,5-dimethylcarboxylate (FDMC)) is also reported, and in line with this, how solid catalysts play a key role on the oxidative esterification of HMF/furfural to an ester of FDCA/furoic acid is included (Chapter 9). Carbon enhancement in the product is another strategic approach for producing biofuel/biofuel additives; about this, aldol addition of furfural with carbonyl compounds with acid and/or base functionalized heterogeneous catalysts has also been covered in this book (Chapter 10). Furfural, viz. tandem reaction, can also be converted to valuable solvents, such as γ-valerolactone (GVL), as a precursor fuel. How Lewis and Brønsted acids containing heterogeneous catalysts play a crucial role in the one-pot multi-step reaction of furfural to GVL has been elaborated in Chapter 11.

Overall, this book delineates the crucial role of metal-containing heterogeneous catalysts in shifting product selectivity during the transformation of furanics (namely, HMF and furfural) into valuable products. The book provides insight into the influence of factors such as the type of metal, the choice of support, and the reaction environment on the catalytic performance of the employed materials.

The editors express their sincere gratitude to all the authors who have contributed their chapters on time and cooperated in submitting and revising the manuscript within the given timeframe. Despite their busy schedules, the referees are acknowledged for their valuable suggestions in evaluating the submitted chapters and improving the content. The editors would like to thank CRC Press, Taylor & Francis for publishing this book. Our thanks are due to Ms. Renu Upadhyay, Publishing Editor and Ms Jyotsna Jangra, Editorial Assistant, and the CRC Press support team for completing this task.

We strongly believe that this book is an excellent reference for graduate students studying courses related to biomass (LCB) valorization and will also attract researchers from academia and industries who are engaged in R&D activities related to biomass valorization, designing heterogeneous catalytic material, furanics transformation, biochemicals and biofuels.

Editors

Shunmugavel Saravanamurugan

Hu Li

Ashok Pandey

Young-Woong Suh

Editor Biographies

Shunmugavel Saravanamurugan is Scientist-F at the Center of Innovative and Applied Bioprocessing, Mohali, India. He completed his PhD at Anna University, India, 2005 and worked as a post-doctoral fellow/senior researcher at KAIST, Inha University, and DTU. His research topic is catalytic biomass transformations. He has over 80 publications and co-edited two books with >6000 citations (*h*-index 35). He was elected as a FRSC in 2021.

Hu Li is Professor at the Center for R&D of Fine Chemicals, Guizhou University (GZU), China. Dr. Li stayed as a visiting PhD student researcher (CSC scholarship) at the Technical University of Denmark, and the University of Chinese Academy of Sciences prior to obtaining his PhD from GZU. He then worked as a postdoctoral fellow at Tohoku University, and Nanjing Agricultural University. His research focuses on the catalytic conversion of biomass and waste resources into biofuels and chemicals with functional materials. Dr. Li has published more than 20 patents, 8 books, and over 200 peer-reviewed papers (*h*-index 50, ~7000 citations) on sustainable catalysis and green chemistry. He was awarded the National Youth Top-notch Talent Support Program, the Guizhou Provincial Natural Science Award (first prize), the Guizhou Provincial Youth Medal, etc. Currently, he is Associate Editor of *Frontiers in Energy Research and Current Green Chemistry*, and Editorial Board Member of several international journals.

Ashok Pandey is currently Executive Director (Honorary) at the Centre for Energy and Environmental Sustainability – India. His major research and technological development interests are industrial & environmental biotechnology and energy biosciences, focusing on biomass to biofuels & chemicals, waste to wealth & energy, industrial enzymes, etc.

Young-Woong Suh is a full professor at the Department of Chemical Engineering, Hanyang University, Republic of Korea. He obtained his Ph.D. degree at Seoul National University, Republic of Korea, in 2003. He was a postdoctoral research associate at Northwestern University and then worked as a senior research scientist at the Korea Institute of Science and Technology (KIST) until 2011. His research expertise and interests include biomass conversion, chemical hydrogen storage, and environmentally-friendly processing based on heterogeneous catalysis.

Contributors

Ruijia Cao
State Key Laboratory of Green Pesticide, Key Laboratory of Green Pesticide and Agricultural Bioengineering, Ministry of Education, State-Local Joint Laboratory for Comprehensive Utilization of Biomass, Center for R&D of Fine Chemicals, Guizhou University, Guiyang, Guizhou 550025, China.

Jian He
Key Laboratory of Hunan Forest Products and Chemical Industry Engineering, National and Local United Engineering Laboratory of Integrative Utilization Technology of Eucommia Ulmoides, Jishou University, Zhangjiajie, 427000, PR China.

Jinshu Huang
State Key Laboratory of Green Pesticide, Key Laboratory of Green Pesticide and Agricultural Bioengineering, Ministry of Education, State-Local Joint Laboratory for Comprehensive Utilization of Biomass, Center for R&D of Fine Chemicals, Guizhou University, Guiyang, Guizhou 550025, China.

Yeongin Jo
Department of Chemical Engineering
Hanyang University
Seoul, Republic of Korea.

Sang Hyeok Ko
Department of Chemical Engineering
Hanyang University
Seoul, Republic of Korea.

Hu Li
State Key Laboratory of Green Pesticide, Key Laboratory of Green Pesticide and Agricultural Bioengineering, Ministry of Education, State-Local Joint Laboratory for Comprehensive Utilization of Biomass, Center for R&D of Fine Chemicals, Guizhou University, Guiyang, Guizhou 550025, China.

Jie Li
State Key Laboratory of Green Pesticide, Key Laboratory of Green Pesticide and Agricultural Bioengineering, Ministry of Education, State-Local Joint Laboratory for Comprehensive Utilization of Biomass, Center for R&D of Fine Chemicals, Guizhou University, Guiyang, Guizhou 550025, China.

Mingrui Li
State Key Laboratory of Green Pesticide, Key Laboratory of Green Pesticide and Agricultural Bioengineering, Ministry of Education, State-Local Joint Laboratory for Comprehensive Utilization of Biomass, Center for R&D of Fine Chemicals, Guizhou University, Guiyang, Guizhou 550025, China.

Huan Liu
State Key Laboratory of Green Pesticide, Key Laboratory of Green Pesticide and Agricultural Bioengineering, Ministry of Education, State-Local Joint Laboratory for Comprehensive Utilization of Biomass, Center for R&D of Fine Chemicals, Guizhou University, Guiyang, Guizhou 550025, China.

Xiaofang Liu
The Key Laboratory for Critical Degradation Technologies of Pesticide Residues in Superior Agricultural Products in Guizhou Ecological Environment, College of Biology and Environmental Engineering, Guiyang University, Guiyang 550005, China.

Yixuan Liu
State Key Laboratory of Green Pesticide, Key Laboratory of Green Pesticide and Agricultural Bioengineering, Ministry of Education, State-Local Joint Laboratory for Comprehensive Utilization of Biomass, Center for R&D of Fine Chemicals, Guizhou University, Guiyang, Guizhou 550025, China.

Pan Meng
The Key Laboratory for Critical Degradation Technologies of Pesticide Residues in Superior Agricultural Products in Guizhou Ecological Environment, College of Biology and Environmental Engineering, Guiyang University, Guiyang 550005, China.

Ye Meng
State Key Laboratory of Green Pesticide, Key Laboratory of Green Pesticide and Agricultural Bioengineering, Ministry of Education, State-Local Joint Laboratory for Comprehensive Utilization of Biomass, Center for R&D of Fine Chemicals, Guizhou University, Guiyang, Guizhou 550025, China.

Dinesh Kumar Mishra
Research Institute of Industrial Science (RIIS)
Department of Chemical Engineering
Hanyang University
Seoul, Republic of Korea.

Ashok Pandey
Centre for Energy and Environmental Sustainability
Lucknow, India.

Zhengfei Pei
The Key Laboratory for Critical Degradation Technologies of Pesticide Residues in Superior Agricultural Products in Guizhou Ecological Environment, College of Biology and Environmental Engineering, Guiyang University, Guiyang 550005, China.

Kanika Saini
Laboratory of Bioproduct Chemistry
Center of Innovative and Applied Bioprocessing
Mohali, Punjab, India.

Shunmugavel Saravanamurugan
Laboratory of Bioproduct Chemistry
Center of Innovative and Applied Bioprocessing
Mohali, Punjab, India.

Neeraj Sharma
Laboratory of Bioproduct Chemistry
Center of Innovative and Applied Bioprocessing
Mohali, Punjab, India.

Young-Woong Suh
Department of Chemical Engineering
Hanyang University
Seoul, Republic of Korea.

Cong Chien Truong
Research Institute of Industrial Science (RIIS)
Department of Chemical Engineering
Hanyang University
Seoul, Republic of Korea.

Hongguo Wu
State Key Laboratory of Green Pesticide, Key Laboratory of Green Pesticide and Agricultural Bioengineering, Ministry of Education, State-Local Joint Laboratory for Comprehensive Utilization of Biomass, Center for R&D of Fine Chemicals, Guizhou University, Guiyang, Guizhou 550025, China.

Mei Wu
College of Chemistry and Chemical Engineering, Jishou University, Jishou, 416000, PR China.

Lei Yu
The Key Laboratory for Critical Degradation Technologies of Pesticide Residues in Superior Agricultural Products in Guizhou Ecological Environment, College of Biology and Environmental Engineering, Guiyang University, Guiyang 550005, China.

Zhuangzhuang Zhang
The Key Laboratory for Critical Degradation Technologies of Pesticide Residues in Superior Agricultural Products in Guizhou Ecological Environment, College of Biology and Environmental Engineering, Guiyang University, Guiyang 550005, China.

1 Introduction to 5-Hydroxymethylfurfural and Furfural

Shunmugavel Saravanamurugan, Hu Li, Ashok Pandey, and Young-Woong Suh

1.1 LIGNOCELLULOSIC BIOMASS

The standard of human lives is improving globally day by day, depending largely on the energy and chemicals from fossil-based resources, simultaneously, population growth is predicted to significantly increase until the end of this century, reaching ~11 billion (Henrich et al. 2015), therefore, finding renewable and sustainable energy has become inevitable. Terrestrial biomass is such an indispensable carbon source that it can replace the diminishing and confined fossil fuels on which the whole world predominantly relies to run day-to-day life. The first-generation biomass, for example, corn and sugarcane, used for bioethanol production, is disrupting the food chain; thus, second-generation biomass, such as lignocellulosic biomass (LCB), becomes essential to substitute fossil-based chemicals and fuels (Corma et al. 2007). At the same time, renewable energy sources, such as wind and solar, cannot derive chemicals, and that is why LCB is a cornerstone of carbon for producing chemicals rather than fuels.

The major sources of LCB are forestry and agricultural residues and grasses, which account for the production of about 180 billion tons globally per year (Dahmen et al. 2019). Out of LCB, less than 5% is used as fodder, and 2.5% belongs to agricultural residues with an effective utilisation of 25%, indicating that the overall utilisation of LCB is very poor and significantly underutilised. These data suggest that valorising unutilised/underutilised LCB into valuable products/chemicals/fuels is highly essential to improve the bioeconomy. LCB is composed of a rigid homopolymer of polysaccharides (crystalline cellulose), a heteropolymer of polysaccharides (amorphous hemicellulose) and a complex aromatic heteropolymer (amorphous lignin); polysaccharides account for more than 75% in most biomass but energy content more in lignin than polysaccharides. Due to the intricate nature of lignin, the polysaccharide parts of LCB (cellulose and hemicellulose) have been widely explored for producing more value-added chemicals, as aromatic chemical products can rarely be derived from it in moderately high selectivity. These two polysaccharides (cellulose and hemicellulose) are composed of monomer units of glucose and xylose-arabinose-glucose-mannose-galactose with side branches (e.g., glucuronic acid), respectively. Cellulose and hemicellulose are feedstocks for the production of two central platform

DOI: 10.1201/9781003464839-1

chemicals, such as 5-hydroxymethylfurfural (HMF) and furfural, via triple dehydration of glucose and xylose – monomer unit of cellulose and hemicellulose – which can be further converted to a wide a range of chemicals. This book explores the significance of these two chemicals.

1.2 5-HYDROXYMETHYLFURFURAL

HMF, a dehydration product of C6 carbohydrates (e.g., the glucose-monomer unit of the cellulosic part of LCB), is a versatile chemical having two reactive functional groups, such as aldehyde and alcohol, which can undergo multiple reactions, including oxidation, hydrogenation, hydrodeoxygenation (HDO), reductive amination rehydration and ring-opening reactions to yield 2,5-diformylfuran (DFF), 5-formyl-2-furancarboxylic acid (FFCA), 2,5-furandicarboxylic acid (FDCA), 2,5-bihydroxymethylfuran (BHMF), 2,5-bihydroxytetrahydrofuran (BHTHF), 2,5-dimethylfuran (DMF), 2,5-dimethyltetrahydrofuran (DMTHF) furan-2,5-diyldimethanamine (FDDMA), (tetrahydrofuran-2,5-diyl)dimethanamine (THFDMA), levulinic acid, fumaric acid, and maleic acid (Pal and Saravanamurugan 2019) (Figure 1.1). The IMARC group reports that the global market size of HMF in 2023 is US$ 62.2 million and is expected to reach US$ 73.1 million by 2032 with a compound annual growth rate (CAGR) of 1.8% (IMARC 2024a). Asia Pacific is a major player compared to other parts of the world: North America, Europe, Latin America, the Middle East, and Africa. The competitive players for the production of HMF are AVA Biochem AG, Merck

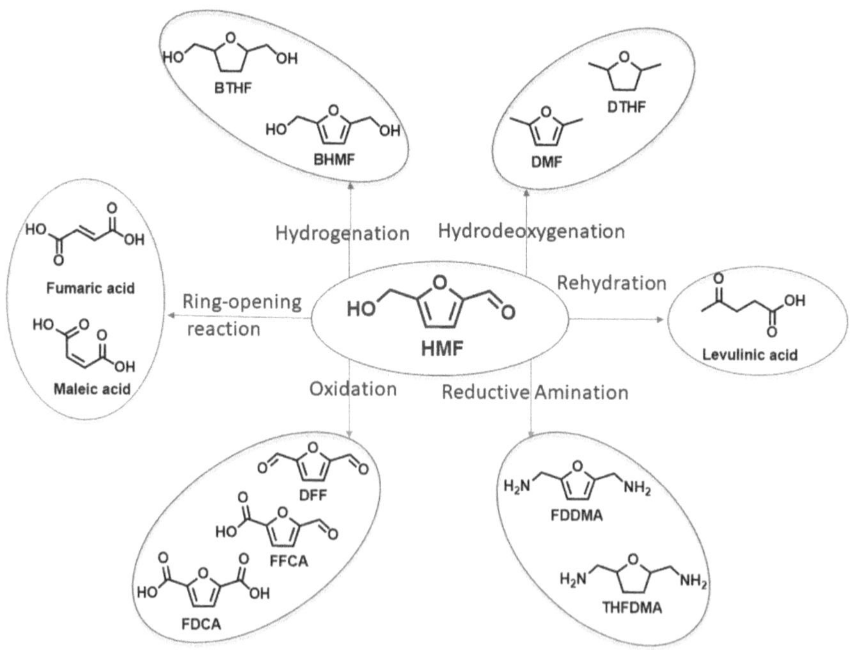

FIGURE 1.1 Transformation of HMF to various products.

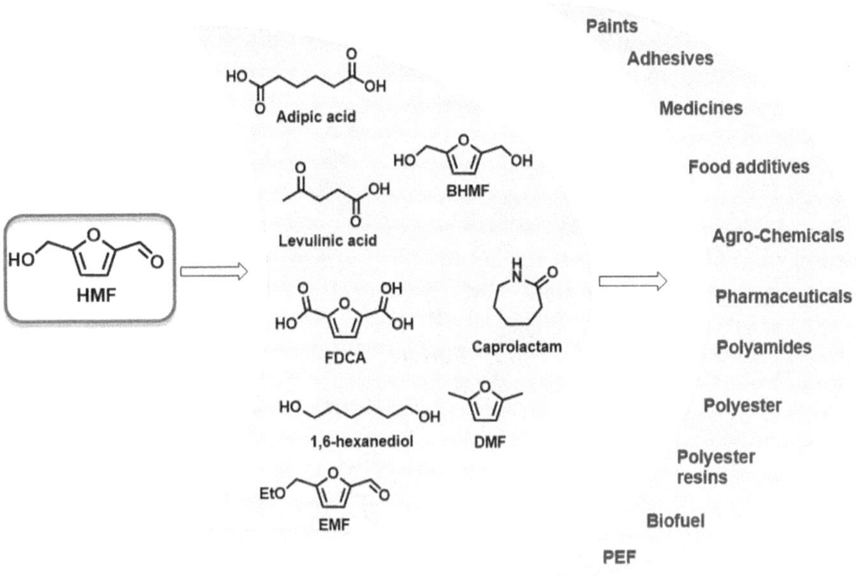

FIGURE 1.2 Application of HMF.

KGaA, Robinson Brothers Limited, Xuzhou Ruisai Technology Industrial Co Ltd., and Zhejiang Sugar Energy Technology Co. Ltd.

As AVALON Industries AG mentioned, HMF has potential in the production of a wide range of valuable chemicals, with applications in agrochemicals, solvents, and polymers (AVALON 2016) (Figure 1.2). One of the potential applications of HMF is to produce FDCA, which can substitute terephthalic acid in polyethylene terephthalate (PET) bottles to produce polyethylene furanoate (PEF). FDCA held a market size of US$ 481 million in 2022 and is expected to grow to US$ 980 million by 2031 at a CAGR of 8.2%, indicating the importance of producing HMF from biomass-based substrates to obtain FDCA (FDCA outlook 2023).

1.3 FURFURAL

Like HMF, furfural is a triple dehydration product of xylose/pentoses having one reactive functional group of aldehyde, which can undergo similar reactions, such as oxidation, hydrogenation, HDO, reductive amination and ring-opening reactions to produce furfuryl alcohol (FA), tetrahydrofurfuryl alcohol (THFA), furoic acid, 2-methylfuran (MF), tetrahydromethylfuran (THMF), furfuryl amine (FFA), tetrahydrofurfuryl amine (THFFA), levulinic acid, maleic acid and succinic acid (Figure 1.3). Furfural is made from bagasse/corncob via acid-catalysed dehydration of xylose, about 300 k tons/year (Kamm et al. 2007, Jong et al. 2010). The global market size of furfural (US$ 770 million in 2023) is several-fold higher than HMF and is expected to increase to US$ 1,290 million in the next ten years with a CAGR of 5.7% as reported by IMARC (IMARC 2024b).

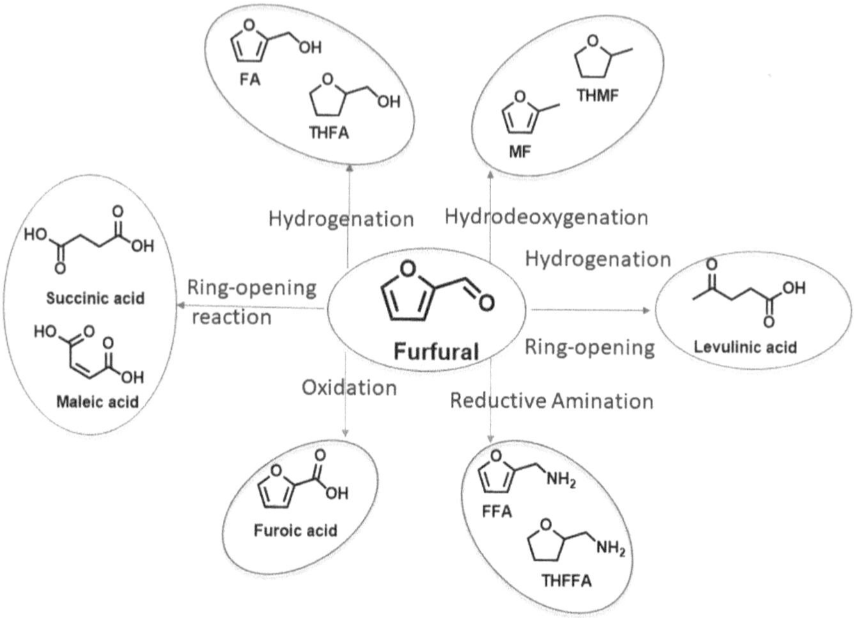

FIGURE 1.3 Transformation of furfural to various products.

As reported by IMARC, the key market players of furfural are Behran Oil Company, Central Hongye Holding Group Corporation Limited, Illovo Sugar Ltd. (Associated British Foods plc), KRBL Limited, Linzi Organic Chemical Inc. Ltd., Lenzing Aktiengesellschaft, Pennakem LLC, Romana Corporation Ltd., Shandong Crownchem Industries Co. Ltd, Silvateam S.p.a., Tanin d.d., Sevnica, TransFurans Chemicals bvba and Xian Welldon Trading Co. Ltd. The major raw materials for the production of furful are rice husk and corn cob. The major application of furfural is in the production of furfuryl alcohol, which is the main component as a monomer in producing furan-based resins (e.g., foundry resins). Besides, furfural is an intermediate for producing value-added chemicals with huge applications in the chemical industry, pharmaceuticals and transportation fuels (Mariscal et al. 2016).

1.4 CONCLUSIONS

While the steady diminishing of confined fossil fuels is alarming, the annual production of LCB in hundreds of billion of tonnes is unutilised/underutilised, is a promising substitute for fossil fuels, and can be converted into valuable products. The data relating to the market size of HMF and furfural – which can be derived from the two polysaccharide parts (cellulose and hemicellulose) of LCB – indicate promising prospects for these two furanic chemicals being transformed into value-added products with potential applications, as mentioned in this chapter and discussed in the following chapters of the book.

REFERENCES

AVALON. 2016. www.bioplasticsmagazine.com/en/news/meldungen/20161213-AVALO N-Industries-takes-over-operations-from-AVA-CO2.php

Corma, Avelino, Sara Iborra, and Alexandra Velty. 2007. "Chemical Routes for the Transformation of Biomass into Chemicals." *Chem. Rev.* 107. American Chemical Society: 2411–2502. doi: 10.1021/cr050989d.

Dahmen, Nicolaus, Iris Lewandowski, Susanne Zibek, and Annette Weidtmann. 2019. "Integrated Lignocellulosic Value Chains in a Growing Bioeconomy: Status Quo and Perspectives." *GCB Bioenergy.* 11. John Wiley & Sons Ltd: 107–117. doi: 10.1111/gcbb.12586.

FDCA outlook. 2023. www.transparencymarketresearch.com/furandicarboxylic-acid-market.html

Henrich, Edmund, Nicolaus Dahmen, Eckhard Dinjus, and Jorg Sauer. 2015. "The Role of Biomass in a Future World without Fossil Fuels." *Chem. Ing. Tech.* 87 (12). Wiley-VCH Verlag: 1667–85. doi: 10.1002/cite.201500056.

IMARC. 2024a. www.imarcgroup.com/5-hydroxymethylfurfural-market#:~:text=Market%20Overview%3A,1.8%25% 20during%202024%2D2032.

IMARC. 2024b. www.imarcgroup.com/furfural-market#:~:text=The%20global%20furfural%20market%20was,5.7%25%20during%202024%2D2032.

Jong, Wiebren De and Gianluca Marcotullio. 2010. "Overview of Biorefineries based on Co-Production of Furfural, Existing Concepts and Novel Developments." *Int. J. Chem. React. Eng.* 8. Walter de Gruyter GmbH & Co. KG. doi: 10.2202/1542-6580.2174.

Kamm, Birgit, Patrick R. Gruber and Michael Kamm. 2007. "Biorefineries—Industrial Processes and Products, ed." Wiley-VCH, Weinheim. doi: 10.1002/14356007.l04_l01.

Mariscal, R, P. Maireles-Torres, M. Ojeda, I. Sádaba and M. López Granados. 2016. "Furfural: A Renewable and Versatile Platform Molecule for the Synthesis of Chemicals and Fuels." *Energy Environ. Sci. 9*. The Royal Society of Chemistry: 1144–1189. doi: 10.1039/C5EE02666K

Pal Priyanka and Shunmugavel Saravanamurugan. 2019. "Recent Advances in the Development of 5-Hydroxymethylfurfural Oxidation with Base (Nonprecious)-Metal-Containing Catalysts." *ChemSusChem* 12. Wiley-VCH Verlag: 145–163. doi:10.1002/cssc.201801744.

2 Catalytic Hydrogenolysis 5-Hydroxymethylfurfural to 2,5-Dimethylfuran with Non-Noble Metals

Ruijia Cao, Hu Li, and Jinshu Huang

2.1 REACTION PATHWAYS AND HYDROGEN DONORS FOR HYDROGENOLYSIS OF HMF TO DMF

As technology progresses, studies about the hydrogenolysis of HMF towards DMF have developed further. DMF has emerged as a promising fuel candidate due to its unique physicochemical properties. Its energy density stands at 31.5 MJ/L, surpassing ethanol's 23.0 MJ/L and approaching gasoline's 35.0 MJ/L. Furthermore, 2,5-DMF boasts a boiling temperature range of 365–367 K, rendering it less volatile than ethanol, which boils at 351 K, and is insoluble in water. These attributes collectively position 2,5-DMF as a potential alternative fuel (Castillo et al. 2023). DMF can be utilized to produce p-xylene (PX) through the Diels-Alder reaction. The global PX market, valued at 33.03 billion USD in 2015, is anticipated to surge to 60.04 billion USD by 2024, with a projected compound annual growth rate (CAGR) of 7%. This growth is largely attributed to the expanding PET market. Additionally, PX, as a key component of xylenes, finds diverse applications in various industries, including solvents, pesticides, and pharmaceuticals (Dutta et al. 2023). In this section, we made a comprehensive overview of the selectivity of reaction pathways, the types of hydrogen donors, and the advantages of different hydrogen donors in HMF hydrogenolysis.

2.1.1 Reaction Pathways

The -CHO group and C-OH group of HMF tend to react using tailored catalysts and hydrogen donors due to their different reactivity. Therefore, there are usually two pathways for the hydrogenolysis of HMF (Scheme 2.1). Pathway 1 involves the hydrogenation of the C=O bond to form the C-OH group, resulting in the outcome of 2,5-dihydroxymethylfuran (DHMF). Subsequently, the hydrogenolysis of two C-OH groups one by one leads to methyl furfuryl alcohol (MFA), and ultimately to DMF (Pisal et al. 2021; Talpade et al. 2019). In pathway 2, the C-O bond undergoes hydrogenolysis to produce 5-methylfurfural (MF), then the C=O bond

 DOI: 10.1201/9781003464839-2

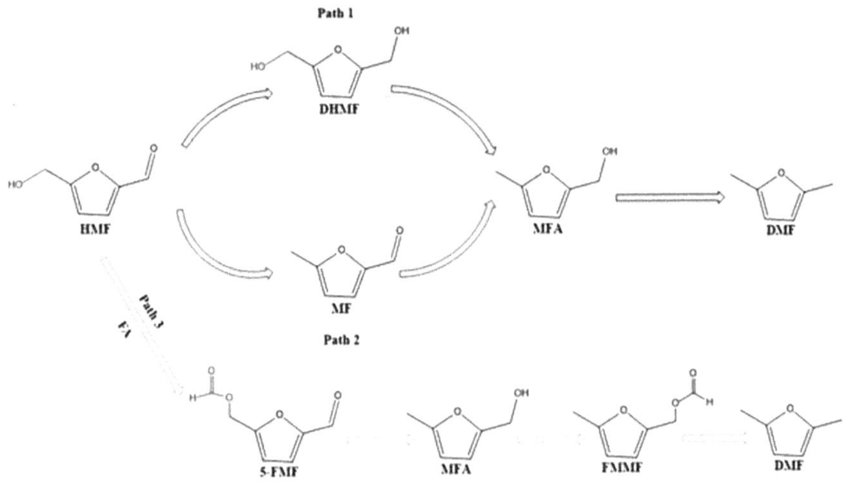

SCHEME 2.1 Pathways of HMF-to-DMF conversion.

is hydrogenated to generate MFA. Finally, the C-OH group on MFA is hydrolysed to yield DMF (Guo et al. 2020; Xia et al. 2022). Additionally, when formic acid (FA) is applied as a hydrogen donor, an alternative pathway for DMF emerges. This pathway initiates with an esterification reaction involving the C-OH group, leading to the production of 5-formyloxymethylfurfural (5-FMF). Subsequently, both the -CHO group and C-O ester-bond occur in the hydrogenolysis simultaneously, leading to the intermediate MFA. Eventually, MFA repeats the previous reaction to form the intermediate FMMF to DMF (Thananatthanachon et al. 2010). There will certainly be side reactions and by-products in the process of hydrogenolysis of HMF, which will affect the yield of DMF. Therefore, we must reduce or even eliminate the side reactions in this conversion and improve the conversion rate.

2.1.2 Hydrogen Donor

H_2 is most widely used as a hydrogen source for hydrogenolysis of HMF. Zu et al. utilized a straightforward coprecipitation technique to synthesize the Ru/Co_3O_4 catalyst with different contents of Ru. Under specific reaction conditions specifically, at the temperature of 403 K and a pressure of 0.7 MPa, with H_2 as the hydrogen donor, they achieved a remarkable yield of DMF, reaching 93.4% (Zu et al. 2014). Zhang et al. constructed a core-shell $Co@CuCo/Al_2O_3$ catalyst to convert HMF, while the yield of DMF can achieve more than 91% under 130 °C and 1 MPa H_2 (Zhang et al. 2024). Although hydrogen is good at hydrogenation and hydrogenolysis of HMF with catalysts, there are some intrinsic drawbacks in hydrogen, such as the cost of transportation being high, combustible, and almost insoluble in organic reagents, which will take more risk to use it (Li et al. 2017; Li et al. 2017; Hao et al. 2016). Compared to hydrogen, alcohols, and formic acid are much safer and cheaper, and the two have also been widely used as hydrogen donors at present (Hansen et al. 2012).

Formic acid, like HMF, can be derived from lignocellulosic biomass and can also be obtained by CO_2 reduction, thus offering an environmentally friendly and renewable alternative as a hydrogen donor in HMF-to-DMF conversion reactions (Voss et al. 2019; Wei et al. 2021; Zhang et al. 2021; Sun et al. 2021). Zhao et al. successfully constructed a CoFe bimetallic catalyst with structural advantages to selectively hydrogenated HMF to DMF (Zhao et al. 2021). Utilizing FA as the source of hydrogen, they achieved a DMF yield exceeding 92% under reaction conditions of 513 K for 2 h. While FA can exhibit several distinct advantages over H_2, we need to focus on how to improve the stability of catalysts under acidic conditions.

Apart from hydrogen and FA, alcohols such as ethanol, isopropyl alcohol, and 1-butanol have also been explored as hydrogen donors for hydrogenolysis of HMF. Zhang et al. fabricated a copper-based catalyst with an Al/Cu molar ratio of 1:3 by utilizing hydrotalcite as the precursor (Zhang et al. 2017). To evaluate its catalytic activity, methanol was utilized not only as a solvent but also as a hydrogen donor for the hydrogenolysis reaction of HMF. Under these conditions, the reaction was conducted at 240 °C for 1.5 h, and the yield of DMF could reach almost 97%. However, recent studies have highlighted a challenge that applying alcohol as a hydrogen source and organic solvent can lead to solvent conversion, which makes it hard to recycle and reuse alcohol, and leads to alcohol waste.

From what has been explained above, we may draw a conclusion that H_2, FA, and many kinds of alcohols are three of the most commonly used hydrogen donors in the hydrogenolysis of HMF-to-DMF conversion. Additionally, H_2O activated through photo- and electrocatalysis has also been explored (Zhang et al. 2019). Furthermore, polymethylhydrosiloxane (PHMS) can also be used as a hydrogen source. Zhang et al. applied $PdCl_2$ as a reaction catalyst and the yield of DMF could reach 89.7% even at 25 °C (Zhang et al. 2019).

2.2 NON-NOBLE MONOMETALLIC CATALYSTS

Non-noble metals have been widely and extensively employed as catalysts for HMF hydrogenolysis. In particular, Ni, Co, and Cu, compared to other non-noble metals, have shown superior catalytic properties for HMF-to-DMF conversion from previous to present studies. Therefore, these three kinds of metal catalysts are mainly introduced here.

2.2.1 Ni-Based Catalysts

Ni exhibits excellent catalytic efficiency, which makes it a common choice for use in the catalytic hydrogenation reaction. These kinds of unique properties make it prefer to break the C=O bond under relatively mild reaction conditions. Therefore, a large amount of different Ni-based catalysts have been prepared to enhance the hydrogenolysis reaction of HMF to form DMF. Some of the Ni-based catalysts are listed in Table 2.1.

Instead of loading nickel with a carrier, Kong et al. applied a Raney Ni catalyst for the hydrogenation and hydrogenolysis of HMF (Kong et al. 2014)

TABLE 2.1
Representative Results of the Hydrogenolysis of HMF to DMF over Different Single Metal Ni Catalysts

Entry	Catalyst	Solvent	H-donor	*T* (°C)	*t* (h)	HMF Conversion (%)	DMF Yield (%)	Ref.
1	Raney Ni	1,4-dioxane	H_2 (1.5 Mpa)	180	15	100	88.5	(Kong et al. 2014)
2	5%-Ni/HC	1,4-dioxane	H_2 (1.5 Mpa)	220	25	100	70.4	(Zhang et al. 2020)
3	Ni@SAZn-PC	ethanol	H_2 (0.6 Mpa)	150	-	79	68	(Mani et al. 2017)
4	35Ni/NDC	-	H_2 (8.0 Mpa)	150	-	99	80	(Brandi et al. 2020)
5	Ni/Co_3O_4	THF	H_2 (1.0 Mpa)	130		>99	76	(Yang et al. 2015)
6	$Ni^0NiO/ATZ3_{WI}$		Formic acid	210	24	99	46	(Cortez-Elizalde et al. 2022)
7	Ni/ZSM-5	THF	H_2 (0.25 Mpa)	180	7	91	96	(Guo et al. 2020)
8	Ni/C	THF	H_2 (4.5 Mpa)	180	2	100	75	(Gyngazova et al. 2017)
9	Ni/CN	H_2O	H_2 (3.0 Mpa)	200	6	100	98.7	(Goyal et al. 2016)
10	$5\%Ni/WO_3$	H_2O	H_2 (1.0 Mpa)	180	6	>99	>95	(Siddiqui et al. 2018)
11	Ni/SBA-15	1,4-dioxane	H_2 (3.0 Mpa)	180	0.5	93	71	(Chen et al. 2020)
12	Ni/TiO_2	1,4-dioxane	H_2 (3.0 Mpa)	220	2	-	85	(Przydacz et al. 2020)
13	Ni/Al_2O_3	1,4-dioxane	H_2 (1.2 Mpa)	180	4	100	92	(Yang et al. 2019)

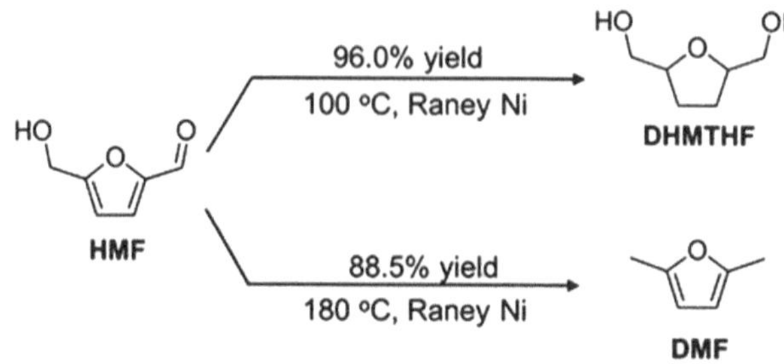

FIGURE 2.1 Hydrogenation and hydrogenolysis of HMF over high-performance Raney Ni.

(Figure 2.1). It is worth noting that Raney Ni is highly sensitive to temperature, and by adjusting the temperature, we can get different products for reducing HMF. Under the same conditions, when adjusting the temperature to 100 °C, the conversion rate of HMF can reach 100%, and the selectivity of HMF hydrogenation to DHMTHF can reach 96%. As the temperature increases, the conversion rate of HMF remains at 100%, and the selectivity of DHMTHF gradually decreases. However, the selectivity of DMF continues to increase, reaching 88.5% at 180 °C. The above experimental results indicate that at a relatively low temperature (100 °C), Raney Ni mainly hydrogenates the C=C bond in the furan ring. At relatively high temperatures (180 °C), Raney Ni tends to primarily hydrogenate the hydroxyl and aldehyde groups of the furan ring.

Carbon materials are frequently chosen as carriers for supported metal catalysts, attributed to their desirable combination of properties. These include a high surface area, large pore volume, chemical inertness, etc. These properties make carbon material ideal for use in a wide range of catalytic applications (Luo et al. 2021; Li et al. 2023; Zhang et al. 2021). For example, Zhang et al. prepared hydrothermal carbon (HC) by sucrose through the hydrothermal method (Zhang et al. 2020), which has a large surface area, and a series of Ni/HC catalysts (with Ni content of 2%, 5%, 8%, and 10%, respectively) can be prepared by accurately adjusting the amount of nickel nitrate added. In this research, it is found that under the same conditions, different loading amounts of Ni lead to different sizes of Ni NPs on HC, which also results in different dispersibility of Ni NPs. When using 10% Ni/HC catalysis, it promotes the hydrogenation reaction of HMF to DHMF, but when 5% Ni/HC was used to catalytic hydrogenolysis HMF, the highest yield of DMF could reach 85.1% and HMF conversion of 100% could be obtained at 220 °C for 25 h. Based on the previously developed three-dimensional carbon tetragonal structure with high specific surface area, multilayer, and porous. Mani et al. prepared new nickel nanoparticles loaded onto the material, which is referred to as Ni@SAZn-PC used for the hydrogenolysis of HMF (Mani et al. 2017). As a result, this composite catalyst can achieve a conversion rate of 79% for HMF and a yield of 68% for DMF at a hydrogen pressure of 0.6 MPa and a temperature of 150 °C. Nitrogen-doped carbon materials, as common carriers of catalysts, are able to effectively exert synergistic effects with transition metals. Brandi et al. prepared a catalyst with a small spherical structure, noted as 35Ni/NDC, which contains Ni with a weight percentage of 35% (Brandi et al. 2020). Using this catalyst in a continuous flow system, under optimized conditions of 8.0 MPa hydrogen pressure and 423 K, achieved remarkable results. Specifically, the conversion of HMF reached a high of 99%, resulting in a DMF yield of 80%. By monitoring the conditions to track the intermediate products, it can be found that under low space-time temperature, the reaction path follows HMF→DHMF→MFA→DMF.

Other than carbon and nitrogen-doped carbon materials, metal oxides and mesoporous molecular sieves can also serve as carriers for Ni and can produce excellent catalytic effects. For instance, Yang et al. prepared the Ni/Co_3O_4 catalyst, which was the first non-noble metal catalyst to be utilized for the hydrogenolysis of HMF towards DMF (Yang et al. 2015). In comparison with other pure Co_3O_4, under the same reaction conditions and reaction time (130 °C, 1 MPa H_2, and 24 h), the conversion rate of HMF over the two catalysts is very high, more than 99%, but the selectivity of Ni/Co_3O_4 to DMF is 76%, which is relatively high by 6%. It may be concluded that Ni loaded on Co_3O_4 by hydrothermal method can enhance the ability to crack the C-O bond. Cortez-Elizalde et al. synthesized a Ni supported on the oxides of Al, Ti, and Zr via a sol-gel method (Cortez-Elizalde et al. 2022), which can be labeled as Ni/ATZ. The authors conducted a detailed investigation to understand the influence of the interaction between metal and acid-base active sites present on the surface of the catalyst on the hydrogenolysis of HMF. By adjusting the ratio of the three oxides of ATZ in the catalyst, the loading amount of Ni and various reaction conditions, using $Ni^0NiO/ATZ3_{WI}$ (the composition ratio of ATZ support is Al_2O_3: TiO_2: ZrO_2= 20:40:40 (weight ratio) and the loading amount of Ni is 15 wt%) as the catalyst, the conversion of HMF is close to 100%, and the yield of DMF is 46%. For molecular sieve, Guo et al. utilized ZSM-5 as the carrier, which is a kind of zeolite material with a unique pore structure and high specific surface area, and nickel is loaded on it in contents ranging from 10% to 50%, with each 10% increment representing a different catalyst (Guo et al. 2020). Under the conditions of 180 °C, 0.25 MPa H_2, and 7 h reaction time, the conversion of HMF and DMF increased with the increasing nickel content. However, when the catalyst was 40% Ni/ZSM-5, it exhibited the best catalytic effect, achieving a conversion of HMF of 91.2% and a selectivity of DMF of 96.2%. By tracking the experimental process, it was observed that the hydrogenolysis of HMF leads to the formation of DMF, with MF serving as the intermediate, and the synergistic effect between these two components results in the catalyst's remarkable performance.

In general, Ni is high-performance in the hydrogenation of the C=O bond and promotes the breaking of the C-O bond. However, the narrow d-band width of Ni makes it tend to hydrogenate the C=C bond, and this characteristic also leads to a high possibility of side reactions, which can hinder the production of DMF from HMF. Consequently, it is necessary to utilize a carrier or introduce another with even more metals to enhance the selectivity of DMF.

2.2.2 Co-Based Catalysts

Up to now, there have been many studies reporting that Co-based catalysts have special catalytic activity for the catalytic hydrogenolysis of HMF, especially for the hydrogenation of the C=O bond. And cobalt is often added as a promoter of different catalysts to improve their activity. Some examples of Co-based catalysts, which enhance the conversion of HMF to DMF, are summarized in Table 2.2.

TABLE 2.2
Representative Results of the Hydrogenolysis of HMF to DMF over Different Single Metal Co Catalysts

Entry	Catalyst	Solvent	H-donor	*T* (ºC)	*t* (h)	HMF Conversion (%)	DMF Yield (%)	Ref.
1	Co@CoO	THF	H_2 (1.0 Mpa)	130	2	-	89.0	(Xiang et al. 2022)
2	Co/CeO_2	THF	H_2 (2.0 Mpa)	160	5	100	99	(Wang et al. 2023)
3	$Co_{0.25}$/N–C-500	THF	H_2 (0.25 Mpa)	150	4	99.9	98.5	(Zeng et al. 2023)
4	$CoNC_x$/NiFeO	THF	H_2 (1.0 Mpa)	180	6	99.8	94.3	(Xia et al. 2022)
5	$Co_{1.0}$/rGO	ethanol	H_2 (2.0 Mpa)	200	1	100	94.1	(Yang et al. 2019)
6	Co@NGs	ethanol	H_2 (2.0 Mpa)	200	6	99.9	94.7	(Wang et al. 2020)
7	Co–N-C/NiAl-MMO	THF	H_2 (1.5 Mpa)	170	6	100	99.9	(Ma et al. 2020)
8	CoZnAl-LDH	THF	H_2 (0.7 Mpa)	130	24	>99.9	74.2	(An et al. 2019)
9	$Co–CoO_x$	1,4-dioxane	H_2 (1.0 Mpa)	170	12	-	83.3	(Li et al. 2019)

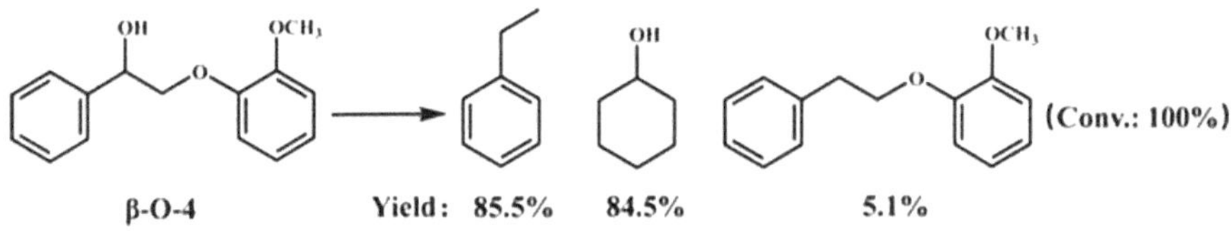

FIGURE 2.2 Hydrogenolysis of lignin β-O-4 model compound with Co_3O_4-250.

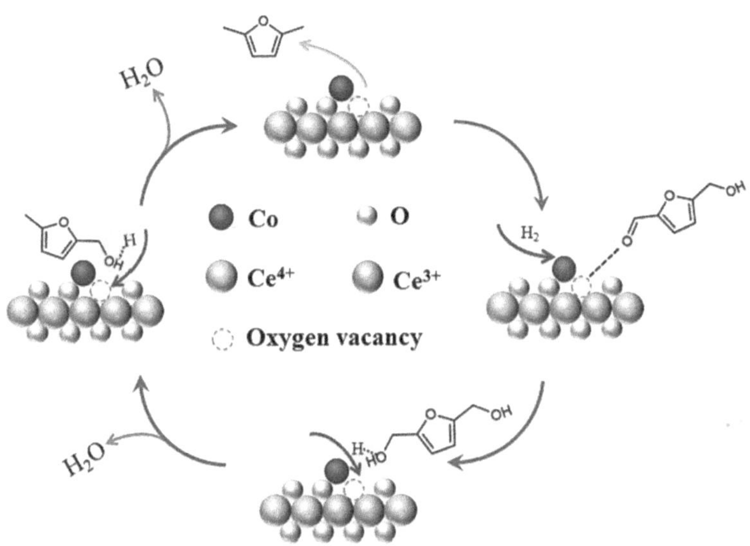

FIGURE 2.3 The possible reaction mechanism for hydrogenolysis of HMF over Co/CeO_2-NF.

Xiang et al. successfully prepared a core-shell structured catalyst using a straightforward precipitation method (Xiang et al. 2022) (Figure 2.2). By adjusting the reduction temperature, they fabricated a series of temperature-specific Co_3O_4 catalysts, among which Co_3O_4-250 (reduced at 250 °C) exhibited the optimal catalytic performance due to numerous oxygen vacancies. When Co_3O_4-250 was utilized as the reaction catalyst, the maximum yield of DMF could reach 89.2% under optimized reaction conditions of 130 °C and 1 MPa hydrogen pressure for a duration of 2 h. In addition to the hydrogenolysis of HMF, Co_3O_4-250 also showed excellent activity on lignin β-O-4, especially for hydrogenolysis of C=O and C=C bonds.

Wang et al. employed CeO_2 as the substrate and further loaded Ni onto it in various forms, including nanoflower (NF), nanorod (NR), nanocube (NC), and nanosheet (NS) (Wang et al. 2023) (Figure 2.3). It was discovered that the unique nanoflower structure, enriched with oxygen vacancies, possessing optimal acidic properties, and characterized by a high dispersion of cobalt species, exhibited superior catalytic activity in comparison to other catalyst morphologies. Under the reaction conditions of 160 °C and 2 MPa H_2, this nanoflower-structured catalyst achieved an impressive conversion rate of HMF and a corresponding yield of DMF of 99% within 5 h. To

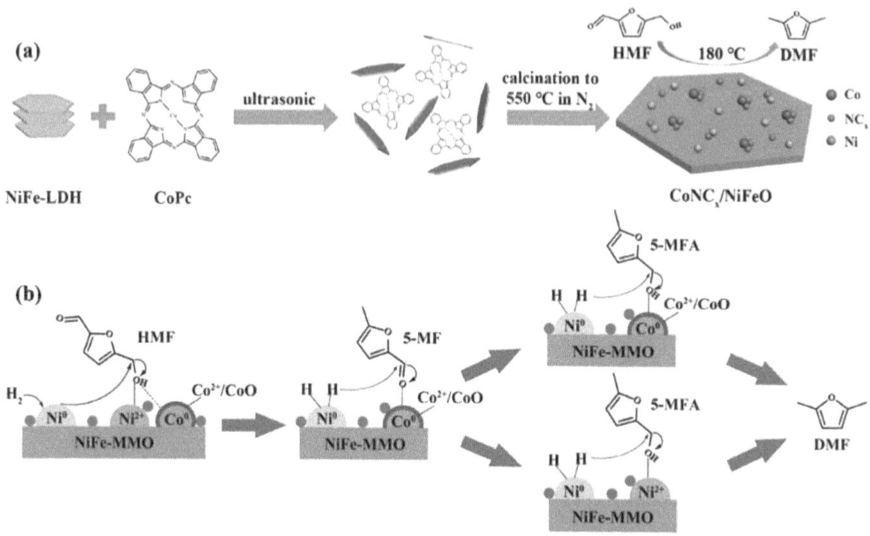

FIGURE 2.4 (a) Synthesis route of $CoNC_x/NiFeO$ catalyst. (b) Proposed mechanism for the hydrogenation of HMF to DMF over the $CoNC_x/NiFeO$ catalyst.

come to the conclusion systematically by the analysis above, the possible reaction mechanism is shown in Figure 2.3.

Nitrogen-doped carbon materials have attracted increasing attention due to their excellent properties arising from their special nanostructures (Jeon et al. 2020; Ruan et al. 2021). The doping of nitrogen not only creates basic sites but also serves as anchor points for metal nanoparticles in metal single-atom catalysts, effectively suppressing metal aggregation. Zeng et al. successfully synthesized the precursor Co_x-ZIF-9 for catalysts with varying cobalt content by precisely adjusting the molar ratio of cobalt nitrate and benzimidazole (Zeng et al. 2023). Subsequently, they calcined this precursor at different temperatures (*T* represented 450 °C, 500 °C, and 600 °C respectively) to obtain the final three catalysts designated as Co_x/N-C-450, Co_x/N-C-500, and Co_x/N-C-600. The calcination temperature played a crucial role in regulating the catalytic activity. At lower temperatures, the cobalt oxide may not be fully reduced, and the formation of cobalt nanoparticles may be incomplete. Conversely, at higher temperatures, cobalt nanoparticles tend to agglomerate, leading to a decrease in activity. Through comparative analysis, it was found that when $Co_{0.25}$/N-C-500 was used as the catalyst, a remarkable HMF conversion of 99.9% and DMF selectivity of 98.8% could be achieved under optimized reaction conditions of 150 °C and 0.25 MPa hydrogen pressure. Xia et al. initially synthesized NiFe-LDH and subsequently integrated it with CoPc through a series of processes including ultrasonication and heating (Xia et al. 2022) (Figure 2.4). Consequently, they successfully fabricated the desired catalyst, $CoNC_x/NiFeO$. It was observed that the catalyst exhibited its maximum BET specific surface area at 550 °C compared to $CoNC_x$ and NiFe-MMO, thereby enhancing the availability of active sites and ultimately boosting its catalytic efficiency. Under specific reaction conditions of 180 °C, 1.0 MPa H_2, and a duration of 6 h, the conversion rate of HMF soared to 99.8%, while the selectivity towards DMF attained a remarkable level of 94.3%. After conducting an in-depth analysis of the catalyst's active sites and carefully tracing the reaction process, it is hypothesized that the C-O bond is preferentially cleaved, leading to the formation of MF. Subsequently, MF undergoes hydrogenation and hydrogenolysis, ultimately resulting in the production of DMF. Consequently, it is deduced that the entire reaction proceeds according to path 1 (Scheme 2.1).

In summary, Co metal is seldom employed solely as a catalyst for the hydrogenolysis of HMF. Instead, it typically serves as a promoter in bimetallic or polymetallic catalyst systems, effectively enhancing their catalytic activity.

2.2.3 Cu-Based Catalysts

Cu metal has garnered significant attention for the hydrogenolysis of HMF due to their low cost and exceptional hydrogenation activity, and they offer vast potential for wide-scale applications (Table 2.3).

Wang et al. successfully synthesized a range of catalysts, designated as $Cu_xCo_yCe_z$-*T*, where *x*, *y*, and *z* represent the molar ratios of Cu, Co, and Ce respectively, and *T* indicates the reduction temperature (Wang et al. 2022). These catalysts included variations such as $Cu_6Co_6Ce_4$-300, $Cu_4Co_8Ce_4$-300, $Cu_4Co_{10}Ce_2$-300, and $Cu_4Co_{10}Ce_2$-600. Notably, $Cu_4Co_{10}Ce_2$-600 exhibited remarkable catalytic performance when using 2-propanol as the hydrogen donor. At 170 °C, it achieved a complete conversion of HMF and a selectivity of 95.4% towards DMF, making it the most effective catalyst for the hydrogenolysis of HMF to DMF. Among the catalysts synthesized, the Cu species play a pivotal role in facilitating the liberation of hydrogen atoms from 2-propanol, then efficiently transferred to the HMF molecule, promoting its hydrogenation reaction. Furthermore, the $CoCeO_x$ site enhances the conversion of aldehyde groups on HMF to DHMF through the hydrogenation of active hydrogen atoms. Lastly, the CuCo bimetallic nanoparticles play a crucial role in promoting the hydrogenolysis of the two hydroxyl groups in DHMF, ultimately leading to the production of DMF. Hsiao et al. employed PBSAC as a carrier to support Cu nanoparticles through a wet impregnation method (Hsiao et al. 2022). This approach resulted in a high dispersion of Cu, significantly enhancing the hydrogenolysis efficiency of HMF. Notably, isopropanol (IPA) functioned not only as a solvent but also as an endogenous hydrogen donor. When using this catalyst in the reaction at 190 °C, the conversion of 5-hydroxymethyl furfural (HMF) reached 91.9% within 6 h, with a DMF selectivity of 71.7%. The Cu nanoparticles effectively promoted the dehydrogenation of IPA, reducing the energy barrier in the reaction process and thereby efficiently catalyzing the hydrogenolysis of HMF to DMF.

Silicon-based materials can serve as excellent catalyst support materials due to their extensive surface area and effective dispersion of metal active sites. Lakshmi et al. synthesized a range of catalysts by loading various amounts of Cu onto the silica-based mesoporous material SBA-15 (Lakshmi et al. 2021) (Figure 2.5). These catalysts were categorized as 5%-Cu/SBA-15, 10%-Cu/SBA-15, 15%-Cu/SBA-15,

TABLE 2.3
Representative Results of the Hydrogenolysis of HMF to DMF over Different Single Metal Cu Catalysts

Entry	Catalyst	Solvent	H-donor	*T* (ºC)	*t* (h)	HMF Conversion (%)	DMF Yield (%)	Ref.
1	$Cu_4Co_{10}Ce_2$-600	2-propanol	2-propanol	170	12	100	95.4	(Wang et al. 2022)
2	Cu/PBSAC	isopropanol	isopropanol	190	6	91.9	71.7	(Hsiao et al. 2022)
3	15%Cu/SBA-15	THF	H_2 (2.0 Mpa)	180	8	100	90.0	(Lakshmi et al. 2021)
4	CuSi-PS-30	1,4-dioxane	H_2 (15 bar)	180	2	-	93.4	(Zhu et al. 2020)
5	Cu_3Al	methanol	methanol	240	1.5	99	96.7	(Zhang et al. 2017)
6	Cu/ZnO	1,4-dioxane	H_2 (1.5 Mpa)	220	5	100	91.8	(Zhu et al. 2015)
7	Cu/ZnO-Al_2O_3	1,4-dioxane	H_2 (1.2 Mpa)	180	5	100	90.1	(Wang et al. 2020)
8	NC-Cu/MgAlO	cyclohexanol	cyclohexanol	220	0.5	100	96.1	(Gao et al. 2018)

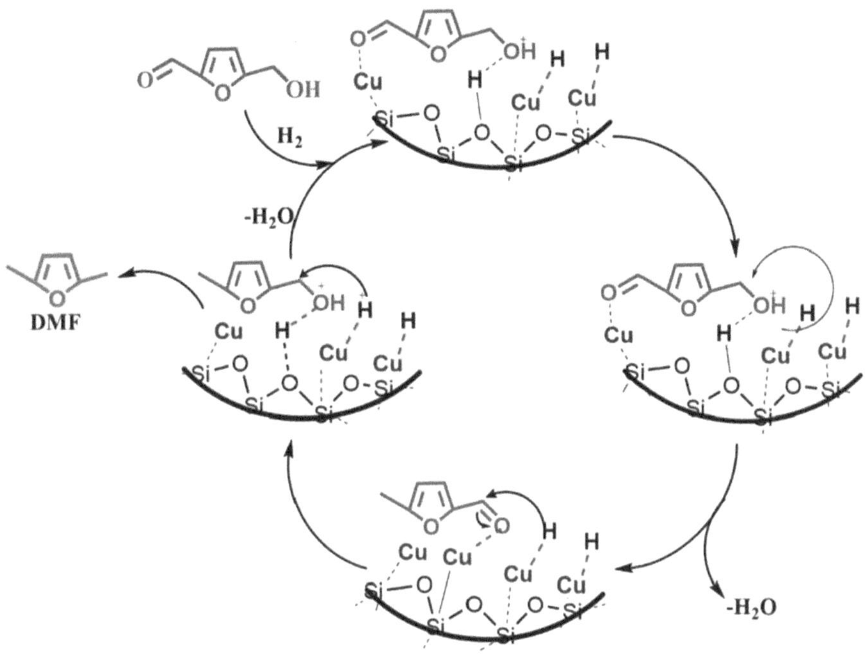

FIGURE 2.5 The possible reaction mechanism for hydrogenolysis of HMF over Cu/SBA-15.

and 20%-Cu/SBA-15, based on the weight percentage of Cu. By adjusting the relevant reaction conditions, 15% Cu/SBA-15 emerged as the optimal catalyst, exhibiting peak performance at 180 °C, with H_2 as the hydrogen donor at a pressure of 2.0 MPa, and a reaction duration of 8 h. Under these conditions, the conversion of HMF achieved a remarkable 100%, with a corresponding DMF yield of 90%. It can be analyzed that the highly dispersed Cu particles effectively enhance the adsorption and activation of H_2, while the acidic sites facilitate the hydrogenation and hydrogenolysis of hydroxyl and aldehyde groups, but a higher acidity can trigger undesired side reactions on the furan ring. Therefore, the appropriate acidity is the key to the efficient and selective generation of DMF. And the possible reaction mechanism is shown in Figure 2.5.

Zhu et al. synthesized a series of CuSi-PS-*x* catalysts, where *x* represents the weight percentage of Cu content (Zhu et al. 2020). Due to the formation of a covalent bond structure between Cu and Si (Cu-O-Si), the Cu^0 and $Cu^{\delta+}$ sites were formed, in which the Cu^0 site is conducive to the activation of hydrogen, while the Lewis acidic $Cu^{\delta+}$ site facilitates the hydrogenolysis of C-O bond. The synergistic effect of the two sites greatly improves the activity of the catalyst. Under optimal reaction conditions of 180 °C and 15 bar H_2 pressure, the CuSi-PS-30 catalyst achieved an excellent DMF yield of approximately 93.4% within 2 h, significantly surpassing the performance of other catalysts.

Although Cu catalysts demonstrate superior activity in the hydrogenolysis of the C-O bond on the furan ring, they often require more severe reaction conditions and the assistance of an appropriate support material to achieve optimal performance. Therefore, researchers have focused on integrating Cu with other metals to prompt its activity and improve its reaction conditions to be more moderate.

2.3 Non-Noble Bimetallic Catalysts

The monometallic catalysts mentioned previously all have unique characteristics and exhibit specific catalytic activity, but they often encounter challenges such as stringent reaction conditions and the formation of numerous by-products due to excessive hydrogenation and hydrogenolysis of DMF. In order to tackle and address the above challenges, researchers have turned to bimetallic catalysts, such as Ni-Cu and Cu-Co, leveraging the synergistic interactions between different metal species to enhance catalytic performance.

2.3.1 Ni-Cu Bimetallic Catalysts

Both Ni and Cu exhibit good performance in the hydrogenolysis of HMF, leading to a growing trend among researchers to jointly utilize Ni and Cu in the preparation of catalysts (Table 2.4).

Zhang et al. successfully fabricated a Cu-Ni bimetallic electrode through the electrodeposition of Ni onto a Cu substrate (Zhang et al. 2019). The electrode's high surface area facilitated the dispersion of Ni, enhancing its activity. The CuNi alloy effectively suppressed the hydrogen evolution reaction on the Ni surface, thereby promoting the selectivity towards DMF. Notably, this catalytic reaction employed water as the hydrogen donor, making it environmentally friendly compared to other similar reactions. At a potential of –0.9 V vs. Ag/AgCl, the CuNi electrode exhibited the Faraday efficiency of 88.0% and a selectivity of 91.1% towards DMF. As mentioned above, mesoporous molecular sieve materials are widely used because of their high specific surface area, ordered and regular pores, and tunable pore size. Umasankar et al. fabricated a Ni-Cu bimetallic catalyst on SBA-16 as support via the impregnation approach (Umasankar et al. 2020). They evaluated its performance under a temperature of 210 °C and a hydrogen pressure of 2 MPa for 4 h, which led to a 100% conversion rate of HMF and 60.7% yield of DMF. Similarly, Sun et al. prepared a Ni-Cu supported on SBA-15, employing the identical SBA-15 material as a support structure (Sun et al. 2019). They utilized formic acid as a hydrogen donor to facilitate the transformation of FMF into DMF, achieving a 71% conversion efficiency. Moreover, Guo et al. initially synthesized porous TS-1 and then prepared a series of Cu and Ni single metal catalysts 40%M/TS-1 (M=Cu, Ni) and bimetallic catalysts 40%Ni_xCu/TS-1 (x = 1, 3, 5, 7%, respectively) using different ratios of Cu and Ni loaded on TS-1 (Guo et al. 2021) (Figure 2.6). Among them, 40%Ni/TS-1 demonstrated a complete conversion rate of HMF, but it also led to the production of a high proportion of excessive hydrogenation products, thus resulting in unsatisfactory DMF selectivity. Compared to 40%Ni/TS-1, 40%Cu/TS-1 exhibited a milder hydrogenation activity, but the conversion and selectivity are still at a low level. Among the catalysts tested, 40%Ni-5%Cu/TS-1 exhibited the most promising catalytic activity.

TABLE 2.4
Representative Results of the Hydrogenolysis of HMF to DMF over Bimetallic Ni-Cu and Co-Cu Catalysts

Entry	Catalyst	Solvent	H-donor	*T* (ºC)	*t* (h)	HMF Conversion (%)	DMF Yield (%)	Ref.
1	CuNi bimetallic electrode	H_2O	H_2O	-	-	-	91.1	(Zhang et al. 2019)
2	5Ni–8Cu/SBA-16	THF	H_2 (2.0 Mpa)	210	4	100	60.7	(Umasankar et al. 2020)
3	36Ni-12Cu/SBA-15	THF	formic acid	220	5	100	60.5	(Sun et al. 2019)
4	40%Ni-5%Cu/TS-1	THF	H_2 (0.5 Mpa)	180	7	100	97.3	(Guo et al. 2021)
5	Cu–Ni/γ–Al_2O_3	THF	H_2 (3.0 Mpa)	200	6	100	56	(Srivastava et al. 2017)
6	Cu–Ni/TiO_2	1,4-dioxane	H_2 (2.5 Mpa)	200	8	100	84.3	(Seemala et al. 2017)
7	Ni-Cu/ZrO_2	1-butanol	H_2 (1.5 Mpa)	275	8	-	70	(Viar et al. 2020)
8	Cu–Ni/BC	THF	H_2 (4.0 Mpa)	220	12	100	93.5	(Zhu et al. 2019)
9	Ni-Cu/HT	formic acid	formic acid	90	15	99	67	(Gupta et al. 2020)
10	$NiCu_3$/C	-	H_2 (3.3 Mpa)	180	-	100	98.7	(Luo et al. 2017)
11	Ni–Cu	cyclohexanol	cyclohexanol	240	-	99	99	(Li et al. 2017)
12	CuCo@C-TsOH-IM	THF	H_2 (3.0 Mpa)	220	10	-	71.1	(Li et al. 2019)
13	2CuCo–IG	2-methylimidazolem	H_2 (2.0 Mpa)	180	4	100	93.7	(Hao et al. 2022)
14	Cu-Co/Al_2O_3	THF	H_2 (3.0 Mpa)	220	8	100	78	(Srivastava et al. 2017)
15	Cu-Co/Al_2O_3	-	H_2 (3.0 Mpa)	220	6	-	87	(Srivastava et al. 2018)
16	Cu_1-Co_8/CeOx	THF	H_2 (1.5 Mpa)	180	8	100	96.5	(Wang et al. 2022)
17	CuCo®/NGr/α-Al_2O_3	THF	H_2 (2.0 Mpa)	180	16	>99	>99	(Guo et al. 2016)
18	Cu-Co@C	ethanol	H_2 (5.0 Mpa)	180	8	100	99.4	(Chen et al. 2017)
19	Cox-Cu@C	isopropanol	H_2 (2.5 Mpa)	160	3	100	>85	(Zhang et al. 2021)
20	Co@Cu/CoAlOx	1,4-dioxane	H_2 (1.5 Mpa)	180	5	100	98.5	(Wang et al. 2020)

Specifically, under the conditions of 180 °C, 0.5 MPa H_2, and a reaction time of 7 h, this catalyst achieved a 100% HMF conversion rate while maintaining a high DMF selectivity of 97.3%, effectively suppressing side reactions. TS-1 plays a vital role as acid sites during this reaction process, which can effectively facilitate the cleavage of the C-O bond, and metal sites present contribute to the reduction of the -CHO group. The roles of two sites in two pathways during the catalytic process are shown in Figure 2.6.

Metal oxides serve as carriers, exhibiting excellent support and thermal stability. They are capable of modifying the electronic structure of metal sites present on the catalyst surface, thereby influencing the catalytic activity. Srivastava et al. utilized γ-Al_2O_3 as a carrier due to its excellent support and thermal stability (Srivastava et al. 2017). The Cu-Ni/γ-Al_2O_3 catalyst, with Ni and Cu contents both set at 10 wt%, was synthesized through the impregnation method and demonstrated the highest reaction activity. Under conditions of 200 °C and a hydrogen pressure of 3.0 MPa, this catalyst achieved a DMF selectivity of up to 56%. Seemala et al. prepared two bimetallic catalysts, Cu-Ni/Al_2O_3 and Cu–Ni/TiO_2, utilizing Al_2O_3 and TiO_2 as supports (Seemala et al. 2017). Compared to Cu-Ni/Al_2O_3, the presence of TiO_2 promoted the

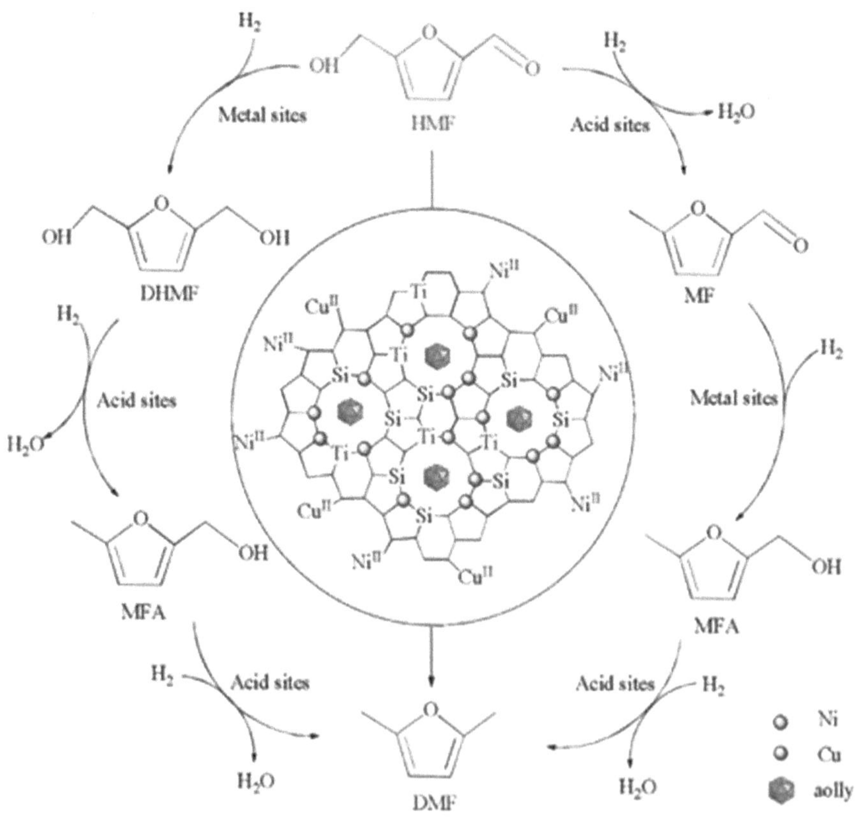

FIGURE 2.6 Proposed reaction pathways for the hydrogenolysis of HMF.

formation of a unique core-shell structure between Ni and Cu metals. Additionally, the presence of Cu metal not only weakens the effect in the Ni-hydrogenated furan ring, but also improves the hydrogenolysis of the aldehyde group. The conversion of HMF could reach 95.9% and the yield of DMF was 84.3% (after 8 h) with the optimal conditions. Viar et al. prepared a series of Ni-Cu/ZrO_2 bimetallic catalysts using ZrO_2 as a carrier through the impregnation method (one-step and two-step), and also adjusted the impregnation order of Ni and Cu, as well as the different contents of the two metals (Viar et al. 2020). The catalyst obtained by the two-step impregnation method has the smallest metal particle size and a wider distribution of metal sites. Among the catalysts tested, 30Ni-15CuZr stands out with its relatively low Cu content, effectively preventing the further hydrogenation of DMF into DMTHF. And it also exhibits the most superior catalytic performance, achieving a yield of 70% for DMF.

As the combination of Ni and Cu metals in different ways, Ni-Cu bimetallic catalysts contain the advantages of two metals. Compared to the monometallic catalysts of Ni and Cu, we can flexibly adjust the content of Cu and Ni to improve their activity of hydrogenolysis and prevent the over hydrogenation, reducing the generation of excessive hydrogenation by-products. However, the conditions for this reaction process are still relatively strict. Therefore, we need to further explore Ni-Cu bimetallic catalysts based on the previous work.

2.3.2 Cu-Co Bimetallic Catalysts

In addition to Ni-Cu bimetallic catalysts, Cu-Co bimetallic catalysts have also attracted the attention of researchers and have been studied for the hydrogenation of HMF to DMF, owing to their remarkable selectivity towards DMF (Table 2.4).

Li et al. initially prepared CuCo@C as a precursor, in which the Cu-Co bimetallic core is wrapped in carbon, then CuCo@C-TsOH-IM was successfully prepared by mixing CuCo@C with p-toluenesulfonic acid (TsOH) via impregnation method (Li et al. 2019). This catalyst exhibits remarkable efficiency in directly converting fructose to HMF, and further facilitating the synthesis of DMF from HMF without the need for separation from the reaction solution. After 10 h of reaction at 220 °C and 3 MPa H_2, the yield of DMF can reach 71.1%. Nitrogen-doped carbon materials have been widely used as catalyst supports, and modifying the morphology of catalysts to enhance their functionality is a commonly employed strategy. Hao et al. loaded nitrogen-doped carbon nanotubes onto carbon sphere precursors to form a unique popcorn structure, with Cu and Co mainly distributed at the tip of the nanotubes (noted as 2CuCo–IG) (Hao et al. 2022). And it was found that this structure can only be formed when 2-methylimidazolem is used as the nitrogen source and glucose is used as the carbon source. This popcorn structure offers a significantly enlarged surface area, with the tip region potentially serving as a highly active catalytic site. Furthermore, nitrogen doping effectively modifies the electronic structure of Cu and Co metals, thereby significantly enhancing the catalyst's activity. Under optimized conditions of 180 °C and 2 MPa hydrogen pressure, the conversion rate of HMF reaches 100% after 4 h of reaction, while the selectivity towards DMF remains as high as 93.7%.

Metal oxides are also used as carriers of Cu-Co bimetallic catalysts to adjust their catalytic activity. Srivastava et al. conducted a comprehensive study on the effects of different metal oxides, including CeO_2, ZrO_2, and Al_2O_3, as supports for Cu-Co bimetallic catalysts (Srivastava et al. 2017). Among them, Al_2O_3 is beneficial for the dispersion of Cu and promotes the introduction of appropriate acidic sites, thereby enhancing the activity of the catalyst, resulting in a yield of 78% for DMF. In the next year, the authors delved deeper into the reaction mechanism of Cu-Co/Al_2O_3 and the influence of various parameters such as temperature, pressure, catalyst loading, and reaction time on the catalytic activity. A detailed analysis revealed that Cu sites primarily facilitate the hydrogenation of aldehyde groups, while CoO_x sites function as acidic centers, further promoting the hydrogenation of the intermediate DHMF to DMF (Srivastava et al. 2018). Wang et al. developed a series of Cu-Co/CeO_x catalysts by reducing CuCoCe layered double hydroxides (LDHs) with varying Cu and Co molar ratios (Wang et al. 2022). The successfully constructed catalysts contain a large number of oxygen vacancies, especially at the interface structure of CoCe-Vö (where Vö represents oxygen vacancies). The presence of these oxygen vacancies significantly enhances the catalyst's adsorption capacity for HMF, enabling almost complete conversion of HMF across all Cu-Co/CeO_x catalysts. The Cu sites within the catalyst play a crucial role in promoting the dissociation of hydrogen, thereby enhancing the ability of hydrogen transfer to the CoCe-Vö interface due to hydrogen spillover effects. The HMF adsorbed on the oxygen vacancies are then easily hydrogenated and converted into DMF. Under identical conditions of 180 °C and a hydrogen pressure of 1.5 MPa, Cu_1-Co_8/CeO_x stands out as the most effective catalyst, achieving the highest DMF yield of 96.5%.

To sum up, the roles of Cu and Co metals are distinct in Cu-Co bimetallic catalysts. The Cu site is more conducive to activating H_2 and the -CHO group on the furan ring, while the Co site excels at hydrogenolysis of the C-O bond. It can be found from previous work that Cu-Co bimetallic catalysts have better selectivity to DMF, but most of them require a pressure greater than 2 MPa when H_2 is used as a hydrogen source.

2.3.3 More Non-Noble Bimetallic Catalysts

Over and above the most widely utilized Ni-Cu and Cu-Co bimetallic catalysts, many other bimetallic catalysts have been explored and have shown good activity in the hydrogenolysis of HMF to DMF.

It was found that Fe metal has good adsorption properties for oxygen-containing species, such as -OH, -CHO, and the furan ring. Therefore, Fe-doped bimetals can also serve as catalysts for the hydrogenation of HMF. Shao et al. initially prepared Ni-Al LDH as a starting material, subsequently incorporating Fe metal into it to synthesize FeNi alloys (Shao et al. 2023). This alloy formation not only minimizes the interaction between Ni and Al species but also inhibits the generation of Lewis acidic sites. These advantages effectively reduce the adsorption and activation of hydrogen, thereby controlling excessive hydrogenation reactions such as the C=C bond on the furan ring. Additionally, the FeNi alloy exhibits a certain level of hydrophilicity, enhancing the adsorption of reaction substrates and intermediates, and thereby boosting the catalytic activity for HMF hydrogenation. Under the conditions of 200 °C, 4.0 MPa H_2, and a reaction duration of 6 h, it was observed that among all various catalysts with distinct metal molar ratios, and the 1.5Ni-1.5Fe-1.0Al composition exhibited the most superior catalytic activity, achieving a complete conversion of HMF and a yield of 93.4% for DMF. Yu et al. successfully synthesized amorphous Ni-Fe borides through a reduction process (Yu et al. 2023). The monometallic borides of Ni ($NiBO_x$) and Fe ($FeBO_x$) exhibited either low or non-existent catalytic activity during the hydrogenation of HMF. However, the bimetallic Ni-Fe boride, specifically $Ni_{1.52}Fe_{0.36}BO_x$, demonstrated the most prominent catalytic activity among all catalysts tested with varying Ni and Fe molar ratios. Under optimal reaction conditions, the formation rate of DMF was close to quantification, reaching 16.5 $mmol_{DMF}$ $gcat^{-1}$ h^{-1}, while the yield of DMF was also impressively high, reaching 98.9%. Due to the incorporation of Fe metal, electrons are more inclined to transfer to Ni sites, which also promotes the generation of more acidic sites. The Ni sites were primarily responsible for promoting hydrogenation and the adsorption of aldehyde groups, whereas the acidic sites played a crucial role in enhancing the hydrogenation process. The combined effect of these two components synergistically contributed to the efficient production of DMF.

Single metal catalysts of Ni and Co have been widely used, and Ni-Co bimetallic catalysts have also been studied in related fields. Zhao et al. successfully developed a range of magnetic Ni_xCo_y catalysts, whose magnetic characteristics are conducive to the recovery and reuse of the catalysts (Zhao et al. 2022). Among them, $Ni_{1.5}Co_1$ has the best catalytic activity. Under very gentle reaction conditions, including a hydrogen pressure of 0.50 MPa at 100 °C and a reaction duration of 4 h, this catalyst achieved 100% conversion of HMF with a selectivity of 80.1% towards DMF. The metal sites of Ni and Co can promote hydrogenation reactions, while the oxide species of Ni and Co act as acidic sites, which can enhance the adsorption of C=O bonds and thus enhance the activity of hydrogen hydrolysis of HMF.

In general, the synergistic interaction between bimetals, in contrast to single metals, enhances the hydrogenolysis of HMF and makes the reaction conditions milder, which is beneficial for reducing the risk and cost of the reaction process.

2.4 CONCLUSION

This chapter delves into the utilization of diverse non-precious metal catalysts, both mono- and bimetallic species, in the hydrogenolysis of HMF to produce DMF. Currently, the catalytic conversion of HMF to DMF primarily follows two distinct routes, either via DHMF or MF as an intermediate. Moreover, the main active sites and mechanisms of the conversion are also listed in some examples. Nevertheless, further exploration in this research domain remains imperative.

REFERENCES

An, Zhe, Wenlong Wang, Suhua Dong, Jing He. 2019. "Well-distributed cobalt-based catalysts derived from layered double hydroxides for efficient selective hydrogenation of 5-hydroxymethyfurfural to 2,5-methylfuran." *Catal. Today*. 319. ELSEVIER SCIENCE BV: 128–138. doi: 10.1016/j.cattod.2018.03.052.

Brandi, Francesco, Marius Bäumel, Irina Shekova, Valerio Molinari, Majd Al-Naji. 2020. "5-hydroxymethylfurfural hydrodeoxygenation to 2,5-dimethylfuran in continuous-flow system over Ni on nitrogen-doped carbon." *Sustainable Chemistry*. 1(2). MDPI: 106–115. doi: 10.3390/suschem1020009.

Castillo, Cristian, Maria H. Brijaldo, Ludmila P. C. Silva, Fabio B. Passos. 2023. "Recent advances in catalytic conversion of 5-hydroxymethylfurfural (5-HMF) to 2,5-dimethylfuran (2,5-DMF): Mechanistic insights and optimization strategies." *Chemcatchem*. 15(15). WILEY-V C H VERLAG GMBH. doi: 10.1002/cctc.202300480.

Chen, Bingfeng, Fengbo Li, Zhijun Huang, Guoqing Yuan. 2017. "Carbon-coated Cu-Co bimetallic nanoparticles as selective and recyclable catalysts for production of biofuel 2,5-dimethylfuran." *Appl. Catal. B-Environ*. 200. ELSEVIER: 192–199. doi: 10.1016/j.apcatb.2016.07.004.

Chen, Shuo, Carmen Ciotonea, Karine De Oliveira Vigier, Francois Jerome, Robert Wojcieszak, Franck Dumeignil, Eric Marceau, Sebastien Royer. 2020. "Hydroconversion of 5-hydroxymethylfurfural to 2,5-dimethylfuran and 2,5-dimethyltetrahydrofuran over non-promoted Ni/SBA-15." *ChemcatChem*. 12(7). WILEY-V C H VERLAG GMBH: 2050–2059. doi: 10.1002/cctc.201902028.

Cortez-Elizalde, Jorge, Gerardo E. Cordova-Perez, Adib Abiu Silahua-Pavon, Hermicenda Perez-Vidal, Adrian Cervantes-Uribe, Adrian Cordero-Garcia, Juan Carlos Arevalo-Perez, Norma Leticia Becerril-Altamirano, Nayi Cristel Castillo-Gallegos, Maria Antonia Lunagomez-Rocha, Jorge Noe Diaz de Leon, Zenaida Guerra-Que, Alejandra E. Espinosa de Los Monteros, Jose Gilberto Torres-Torres. 2022. "2,5-dimethylfuran production by catalytic hydrogenation of 5-hydroxymethylfurfural using Ni supported on Al_2O_3-TiO_2-ZrO_2 prepared by Sol-Gel method: The effect of hydrogen donors." *Molecules*. 27(13). MDPI: 4187. doi: 10.3390/molecules27134187.

Dutta, Saikat, Navya Subray Bhat. 2023. "Catalytic synthesis of renewable p-xylene from biomass-derived 2,5-dimethylfuran: A mini review." *Biomass Convers. Bior*. 13(1). SPRINGER HEIDELBERG: 541–554. doi: 10.1007/s13399-020-01042-z.

Gao, Zhi, Chenyue Li, Guoli Fan, Lan Yang, Feng Li. 2018. "Nitrogen-doped carbon-decorated copper catalyst for highly efficient transfer hydrogenolysis of 5-hydroxymethylfurfural to convertibly produce 2,5-dimethylfuran or 2,5-dimethyltetrahydrofuran." *Appl. Catal. B-Environ*. 226. ELSEVIER SCIENCE BV: 523–533. doi: 10.1016/j.apcatb.2018.01.006.

Goyal, Reena, Bipul Sarkar, Arijit Bag, Nazia Siddiqui, Deepa Dumbre, Nishita Lucas, Suresh Kumar Bhargava, Ankur Bordoloi. 2016. "Studies of synergy between metal – Support interfaces and selective hydrogenation of HMF to DMF in water." *J. Catal*. 340. ACADEMIC PRESS INC ELSEVIER SCIENCE: 248–260. doi: 10.1016/j.jcat.2016.05.012.

Guo, Dongwen, Xianxiang Liu, Feng Cheng, Wenguang Zhao, Sha Wen, Yanping Xiang, Qiong Xu, Ningya Yu, Dulin Yin. 2020. "Selective hydrogenolysis of 5-hydroxymethylfurfural to produce biofuel 2,5-dimethylfuran over Ni/ZSM-5 catalysts." *Fuel*. 274. ELSEVIER SCI LTD: 117853. doi: 10.1016/j.fuel.2020.117853.

Guo, Dongwen, Jinhua Lai, Feng Cheng, Wenguang Zhao, Han Chen, Hui Li, Xianxiang Liu, Dulin Yin, Ningya Yu. 2021. "Titanium silicalite-1 supported bimetallic catalysts for selective hydrogenolysis of 5-hydroxymethylfurfural to biofuel 2,5-dimethylfuran." *Chemical Engineering Journal Advances* 5. Elsevier B.V.: 100081. doi: 10.1016/j.ceja.2020.100081.

Guo, Weiwei, Hangyu Liu, Suqi Zhang, Hongling Han, Huizhen Liu, Tao Jiang, Buxing Han, Tianbin Wu. 2016. "Efficient hydrogenolysis of 5-hydroxymethylfurfural to 2,5-dimethylfuran over a cobalt and copper bimetallic catalyst on N-graphene-modified Al_2O_3." *Green Chem*. 18(23). ROYAL SOC CHEMISTRY: 6222–6228. doi: 10.1039/c6gc02630c.

Gupta, Dinesh, Rohit Kumar, Kamal K. Pant. 2020. "Hydrotalcite supported bimetallic (Ni-Cu) catalyst: A smart choice for one-pot conversion of biomass-derived platform chemicals to hydrogenated biofuels." *Fuel.* 277. ELSEVIER SCI LTD: 118111. doi: 10.1016/j.fuel.2020.118111.

Gyngazova, Maria S., Leila Negahdar, Lena C. Blumenthal, Regina Palkovits. 2017. "Experimental and kinetic analysis of the liquid phase hydrodeoxygenation of 5-hydroxymethylfurfural to 2,5-dimethylfuran over carbon-supported nickel catalysts." *Chem. Eng. Sci.* 173. PERGAMON-ELSEVIER SCIENCE LTD: 455–464. doi: 10.1016/j.ces.2017.07.045.

Hao, Peng, Jianliang Zuo, Wurong Tong, Jing Lin, Qiying Wang, Zili Liu. 2022. "Selective hydrogenation of 5-hydroxymethylfurfural to 2,5-dimethylfuran over popcorn-like nitrogen-doped carbon-confined CuCo bimetallic catalyst." *Front. Chem.* 10. FRONTIERS MEDIA SA: 882670. doi: 10.3389/fchem.2022.882670.

Hao, Weiwei, Weifeng Li, Xing Tang, Xianhai Zeng, Yong Sun, Shijie Liu, Lu Lin. 2016. "Catalytic transfer hydrogenation of biomass-derived 5-hydroxymethyl furfural to the building block 2,5-bishydroxymethyl furan." *Green Chem.* 18(4). ROYAL SOC CHEMISTRY: 1080–1088. doi: 10.1039/c5gc01221j.

Hansen, Thomas S., Katalin Barta, Paul T. Anastas, Peter C. Ford, Anders Riisager. 2012. "One-pot reduction of 5-hydroxymethylfurfural via hydrogen transfer from supercritical methanol." *Green Chem.* 14(9). ROYAL SOC CHEMISTRY: 2457–2461. doi: 10.1039/c2gc35667h.

Jeon, In-Yup, Hyuk-Jun Noh, Jong-Beom Baek. 2020. "Nitrogen-doped carbon nanomaterials: Synthesis, characteristics and applications." *Chem.-Asian J.* 15(15). WILEY-V C H VERLAG GMBH: 2282–2293. doi: 10.1002/asia.201901318.

Kong, Xiao, Yifeng Zhu, Hongyan Zheng, Fang Dong, Yulei Zhu, Yong-Wang Li. 2014. "Switchable synthesis of 2,5-dimethylfuran and 2,5-dihydroxymethyltetrahydrofuran from 5-hydroxymethylfurfural over Raney Ni catalyst." *RSC Adv.* 4(105). ROYAL SOC CHEMISTRY: 60467–60472. doi: 10.1039/c4ra09550b.

Lakshmi, D. Dhana, B. Srinivasa Rao, Yogita, N. Lingaiah. 2021. "Selective hydrodeoxygenation of 5-hydroxymethylfurfural to 2,5-dimethylfuran over mesoporous silica supported copper catalysts." *Mater. Sci. Energy Technol.* 4. Elsevier B.V.: 357–366. doi: 10.1016/j.mset.2021.08.012.

Li, Dan, Qiying Liu, Changhui Zhu, Haiyong Wang, Chunhua Cui, Chenguang Wang, Longlong Ma. 2019. "Selective hydrogenolysis of 5-hydroxymethylfurfural to 2,5-dimethylfuran over Co_3O_4 catalyst by controlled reduction." *J. Energy Chem.* 30. ELSEVIER SCIENCE B.V.: 34–41. doi: 10.1016/j.jechem.2018.03.008.

Li, Hu, Wenfeng Zhao, Anders Riisager, Shunmugavel Saravanamurugan, Zhongwei Wang, Zhen Fang, Song Yang. 2017. "A Pd-catalyzed in situ domino process for mild and quantitative production of 2,5-dimethylfuran directly from carbohydrates." *Green Chem.* 19(9). ROYAL SOC CHEMISTRY: 2101–2106. doi: 10.1039/c7gc00580f.

Li, Hu, Wenfeng Zhao, Zhen Fang. 2017. "Hydrophobic Pd nanocatalysts for one-pot and high-yield production of liquid furanic biofuels at low temperatures." *Appl. Catal. B-Environ.* 215. ELSEVIER SCIENCE BV: 18–27. doi: 10.1016/j.apcatb.2017.05.039.

Li, Jifan, Zhe Song, Yifeng Hou, Ziyi Li, Chunli Xu, Chun-Ling Liu, Wen-Sheng Dong. 2019. "Direct production of 2,5-dimethylfuran with high yield from fructose over a carbon-based solid acid-coated CuCo bimetallic catalyst." *ACS Appl. Mater. Interfaces.* 11(13). AMER CHEMICAL SOC: 12481–12491. doi: 10.1021/acsami.8b22183.

Li, Jiang, Heyang Liu, Zhidong An, Yue Kong, Lei Huang, Delong Duan, Ran Long, Piaoping Yang, Yuan-Ye Jiang, Jiaxing Liu, Junjie Zhang, Tong Wan, Jiayi Fu, Runze Pan, Xintian Wang, Dionisios G. Vlachos. 2023. "Nitrogen-doped carbon for selective pseudo-metal-free hydrodeoxygenation of 5-hydroxymethylfurfural to 2,5-dimethylfuran: Importance of trace iron impurity." *J. Catal.* 417. ACADEMIC PRESS INC ELSEVIER SCIENCE: 396–407. doi: 10.1016/j.jcat.2022.12.021.

Li, Wei, Guoli Fan, Lan Yang, Feng Li. 2017. "Highly efficient synchronized production of phenol and 2,5-dimethylfuran through a bimetallic Ni-Cu catalyzed dehydrogenation-hydrogenation coupling process without any external hydrogen and oxygen supply." *Green Chem.* 19(18). ROYAL SOC CHEMISTRY: 4353–4363. doi: 10.1039/c7gc01387f.

Luo, Jing, Matteo Monai, Cong Wang, Jennifer D. Lee, Tomas Duchon, Filip Dvorak, Vladimir Matolin, Christopher B. Murray, Paolo Fornasiero, Raymond J. Gorte. 2017. "Unraveling the surface state and composition of highly selective nanocrystalline Ni-Cu alloy catalysts for hydrodeoxygenation of HMF." *Catal. Sci. Technol.* 7(8). ROYAL SOC CHEMISTRY: 1735–1743. doi: 10.1039/c6cy02647h.

Ma, Ning, Yong Song, Feng Han, Geoffrey I. N. Waterhouse, Yan Li, Shiyun Ai. 2020. "Highly selective hydrogenation of 5-hydroxymethylfurfural to 2,5-dimethylfuran at low temperature over a Co-N-C/NiAl-MMO catalyst." *Catal. Sci. Technol.* 10(12). ROYAL SOC CHEMISTRY: 4010–4018. doi: 10.1039/d0cy00363h.

Mani, Christian Mbaya, Max Braun, Valerio Molinari, Markus Antonietti, Nina Fechler. 2017. "A high-throughput composite catalyst based on nickel carbon cubes for the hydrogenation of 5-hydroxymethylfurfural to 2,5-dimethylfuran." *Chemcatchem.* 9(17). WILEY-V C H VERLAG GMBH: 3388–3394. doi: 10.1002/cctc.201700506.

Luo, Mingsheng, Shuo Li, Zuoxing Di, He Li, Qinglong Liu, Baozhong Lü, Aimei Wang, Buchang Shi, Iltaf Khan. 2021. "Fischer–tropsch synthesis: Study of different carbon materials as cobalt catalyst support." *Reactions.* 2(1). MDPI AG: 43–61. doi: 10.3390/reactions2010005.

Pisal, Devendra S., Ganapati D. Yadav. 2021. "Production of biofuel 2,5-dimethylfuran using highly efficient single-step selective hydrogenation of 5-hydroxymethylfurfural over novel Pd-Co/Al-Zr mixed oxide catalyst." *Fuel.* 290. ELSEVIER SCI LTD: 119947. doi: 10.1016/j.fuel.2020.119947.

Przydacz, Martyna, Marcin Jedrzejczyk, Magdalena Brzezinska, Jacek Rogowski, Nicolas Keller, Agnieszka M. Ruppert. 2020. "Solvothermal hydrodeoxygenation of hydroxymethylfurfural derived from biomass towards added value chemicals on Ni/TiO_2 catalysts." *J. Supercrit. Fluids.* 163. ELSEVIER: 104827. doi: 10.1016/j.supflu.2020.104827.

Ruan, Luna, An Pei, Jianhua Liao, Li Zeng, Guangrong Guo, Kai Yang, Qin Zhou, Ning Zhao, Lihua Zhu, Bing Hui Chen. 2021. "Nitrogen-doped carbon nanotubes-supported PdNiCo nanoparticles as a highly efficient catalyst for selective hydrogenation of furfural." *Fuel.* 284. ELSEVIER SCI LTD: 119015. doi: 10.1016/j.fuel.2020.119015.

Shao, Yuewen, Mingzhu Guo, Mengjiao Fan, Kai Sun, Guoming Gao, Chao Li, Felix Merime Bkangmo Kontchouo, Lijun Zhang, Shu Zhang, Xun Hu. 2023. "Importance of oxyphilic FeNi alloy in NiFeAl catalysts for selective conversion of biomass-derived 5-hydroxymethylfurfural to 2,5-dimethylfuran." *Renew. Energy.* 208. PERGAMON-ELSEVIER SCIENCE LTD: 105–118. doi: 10.1016/j.renene.2023.03.026.

Seemala, Bhogeswararao, Charles M. Cai, Charles E. Wyman, Phillip Christopher. 2017. "Support induced control of surface composition in Cu-Ni/TiO_2 catalysts enables high yield co-conversion of HMF and furfural to methylated furans." *ACS Catal.* 7(6). AMER CHEMICAL SOC: 4070–4082. doi: 10.1021/acscatal.7b01095.

Siddiqui, Nazia, Anupam Singha Roy, Reena Goyal, Rubina Khatun, Chandrashekar Pendem, Appala Naidu Chokkapu, Ankur Bordoloi, Rajaram Bal. 2018. "Hydrogenation of 5-hydroxymethylfurfural to 2,5 dimethylfuran over nickel supported tungsten oxide nanostructured catalyst." *Sustain. Energ. Fuels.* 2(1). ROYAL SOC CHEMISTRY: 191–198. doi: 10.1039/c7se00363c.

Srivastava, Sanjay, G. C. Jadeja, Jigisha Parikh. 2017. "Influence of supports for selective production of 2,5-dimethylfuran via bimetallic copper-cobalt catalyzed 5-hydroxymethylfurfural hydrogenolysis." *Chinese J, Catal.* 38(4). SCIENCE PRESS: 699–709. doi: 10.1016/S1872-2067(17)62789-X.

Srivastava, Sanjay, G. C. Jadeja, Jigisha K. Parikh. 2018. "Optimization and reaction kinetics studies on copper-cobalt catalyzed liquid phase hydrogenation of 5-hydroxymethylfurfural to 2,5-dimethylfuran." *Int. J. Chem. React. Eng.* 16(9). WALTER DE GRUYTER GMBH: 20170197. doi: 10.1515/ijcre-2017-0197.

Srivastava, Sanjay, G. C. Jadeja, Jigisha Parikh. 2017. "Synergism studies on alumina-supported copper-nickel catalysts towards furfural and 5-hydroxymethylfurfural hydrogenation." *J. Mol. Catal. A Chem.* 426. ELSEVIER SCIENCE BV: 244–256. doi: 10.1016/j.molcata.2016.11.023.

Sun, Ruiyan, Yuhe Liao, Shao-Tao Bai, Mingyuan Zheng, Cheng Zhou, Tao Zhang, Bert F. Sels. 2021. "Heterogeneous catalysts for CO_2 hydrogenation to formic acid/formate: From nanoscale to single atom." *Energy Environ. Sci.* 14(3). ROYAL SOC CHEMISTRY: 1247–1285. doi: 10.1039/d0ee03575k.

Sun, Yong, Caixia Xiong, Quanchang Liu, Jiaren Zhang, Xing Tang, Xianhai Zeng, Shijie Liu, Lu Lin. 2019. "Catalytic transfer hydrogenolysis/hydrogenation of biomass-derived 5-formyloxymethylfurfural to 2,5-dimethylfuran over Ni-Cu bimetallic catalyst with formic acid as a hydrogen donor." *Ind. Eng. Chem. Res.* 58(14). AMER CHEMICAL SOC: 5414–5422. doi: 10.1021/acs.iecr.8b05960.

Talpade, Abhijit D., Manishkumar S. Tiwari, Ganapati D. Yadav. 2019. "Selective hydrogenation of bio-based 5-hydroxymethyl furfural to 2,5-dimethylfuran over magnetically separable Fe-Pd / C bimetallic nanocatalyst." *Mol. Catal.* 465. ELSEVIER SCIENCE BV: 1–15. doi: 10.1016/j.mcat.2018.12.009.

Thananatthanachon, Todsapon, Thomas B. Rauchfuss. 2010. "Efficient production of the liquid fuel 2,5-dimethylfuran from fructose using formic acid as a reagent." *Angew. Chem. Int. Ed.* 49(37). WILEY-V C H VERLAG GMBH: 6616–6618. doi: 10.1002/anie.201002267.

Umasankar, S., P. Tamizhdurai, P. Santhana Krishnan, S. Narayanan, V. L. Mangesh, K. Shanthi. 2020. "Effect of copper on NiCu bimetallic catalyst supported on SBA-16 for the catalytic hydrogenation of 5-hydroxymethylfurfural to 2,5-dimethylfuran." *Biomass Bioenerg.* 143. PERGAMON-ELSEVIER SCIENCE LTD: 105868. doi: 10.1016/j.biombioe.2020.105868.

Viar, Nerea, Jesus M. Requies, Ion Agirre, Aitziber Iriondo, Miryam Gil-Calvo, Pedro L. Arias. 2020. "Ni-Cu bimetallic catalytic system for producing 5-hydroxymethylfurfural-derived value-added biofuels." *ACS Sustain. Chem. Eng.* 8(30). AMER CHEMICAL SOC: 11183–11193. doi: 10.1021/acssuschemeng.0c02433.

Voss, Dorothea, Heike Pickel, Jakob Albert. 2019. "Improving the fractionated catalytic oxidation of lignocellulosic biomass to formic acid and cellulose by using design of experiments." *ACS Sustain. Chem. Eng.* 7(11). AMER CHEMICAL SOC: 9754–9762. doi: 10.1021/acssuschemeng.8b05095.

Wang, Jiashi, Qinhong Wei, Qingxiang Ma, Zhongya Guo, Fangfang Qin, Zinfer R. Ismagilov, Wenzhong Shen. 2020. "Constructing Co@N-doped graphene shell catalyst via Mott-Schottky effect for selective hydrogenation of 5-hydroxylmethylfurfural." *Appl. Catal. B-Environ.* 263. ELSEVIER: 118339. doi: 10.1016/j.apcatb.2019.118339.

Wang, Mengying, Ningzhao Shang, Wei Gao, Xiang Cheng, Shutao Gao, Chun Wang. 2023. "Anchoring Co on CeO2 nanoflower as an efficient catalyst for hydrogenolysis of 5-hydroxymethylfurfural." *Fuel*. 354. ELSEVIER SCI LTD: 129433. doi: 10.1016/j.fuel.2023.129433.

Wang, Qian, Junting Feng, Lirong Zheng, Bin Wang, Ruxia Bi, Yufei He, Haichao Liu, Dianqing Li. 2020. "Interfacial structure-determined reaction pathway and selectivity for 5-(Hydroxymethyl)furfural hydrogenation over Cu-based catalysts." *ACS Catal.* 10(2). AMER CHEMICAL SOC: 1353–1365. doi: 10.1021/acscatal.9b03630.

Wang, Qian, Zhi Yu, Junting Feng, Paolo Fornasiero, Yufei He, Dianqing Li. 2020. "Insight into the effect of dual active Cu^0/Cu^+ sites in a Cu/ZnO-Al_2O_3 catalyst on 5-hydroxylmethylfurfural hydrodeoxygenation." *ACS Sustain. Chem. Eng.* 8(40). AMER CHEMICAL SOC: 15288–15298. doi: 10.1021/acssuschemeng.0c05235.

Wang, Xiaofeng, Chengcheng Zhang, Zuyi Zhang, Yaoming Gai, Qingbo Li. 2022. "Insights into the interfacial effects in Cu-Co/CeO_x catalysts on hydrogenolysis of 5-hydroxymethylfurfural to biofuel 2,5-dimethylfuran." *J. Colloid. Interface. Sci.* 615. ACADEMIC PRESS INC ELSEVIER SCIENCE: 19–29. doi: 10.1016/j.jcis.2022.01.168.

Wang, Xiaofeng, Chengcheng Zhang, Zuyi Zhang, Qingbo Li. 2022. "Tuning dual active sites of Cu/$CoCeO_x$ catalysts for efficient catalytic transfer hydrogenation of 5-hydroxymethylfurfural to biofuel 2,5-dimethylfuran." *Fuel*. 320. ELSEVIER SCI LTD: 123996. doi: 10.1016/j.fuel.2022.123996.

Wei, Xinfa, Yan Li, Lisong Chen, Jianlin Shi. 2021. "Formic acid electro-synthesis by concurrent cathodic CO_2 reduction and anodic CH_3OH oxidation." *Angew. Chem. Int. Ed.* 60(6). WILEY-V C H VERLAG GMBH: 3148–3155. doi: 10.1002/anie.202012066.

Xia, Jing, De Gao, Feng Han, Ruifu Lv, Geoffrey I. N. Waterhouse, Yan Li. 2022. "Hydrogenolysis of 5-hydroxymethylfurfural to 2,5-dimethylfuran over a modified CoAl-hydrotalcite catalyst." *Front. Chem.* 10. FRONTIERS MEDIA SA: 907649. doi: 10.3389/fchem.2022.907649.

Xia, Jing, De Gao, Feng Han, Yan Li, Geoffrey I. N. Waterhouse. 2022. "Efficient and selective hydrogenation of 5 – hydroxymethylfurfural to 2,5 – dimethylfuran over a non – noble $CoNC_x$/NiFeO catalyst." *Catal. Lett.* 152(11). SPRINGER: 3400–3413. doi: 10.1007/s10562-022-03919-2.

Xiang, Shuang, Lin Dong, Zhi-Qiang Wang, Xue Han, Luke L. Daemen, Jiong Li, Yongqiang Cheng, Yong Guo, Xiaohui Liu, Yongfeng Hu, Anibal J. Ramirez-Cuesta, Sihai Yang, Xue-Qing Gong, Yanqin Wang. 2022. "A unique Co@CoO catalyst for hydrogenolysis of biomass-derived 5-hydroxymethylfurfural to 2,5-dimethylfuran." *Nat. Commun.* 13(1). NATURE PORTFOLIO: 3657. doi: 10.1038/s41467-022-31362-9.

Yang, Fan, Jingbo Mao, Shenmin Li, Jingmei Yin, Jinxia Zhou, Wei Liu. 2019. "Cobalt – graphene nanomaterial as an efficient catalyst for selective hydrogenation of 5-hydroxymethylfurfural into 2,5-dimethylfuran." *Catal. Sci. Technol.* 9(6). ROYAL SOC CHEMISTRY: 1329–1333. doi: 10.1039/c9cy00330d.

Yang, Panpan, Qiqi Cui, Yanhong Zu, Xiaohui Liu, Guanzhong Lu, Yanqin Wang. 2015. "Catalytic production of 2,5-dimethylfuran from 5-hydroxymethylfurfural over Ni /Co 3 O 4 catalyst." *Catal. Commun.* 66. ELSEVIER SCIENCE BV: 55–59. doi: 10.1016/j.catcom.2015.02.014.

Yu, Yixuan, Huai Liu, Junhua Zhang, Heng Zhang, Yong Sun, Lincai Peng. 2023. "Highly efficient, amorphous bimetal Ni-Fe borides-catalyzed hydrogenolysis of 5-hydroxymethylfurfural into 2,5-dimethylfuran." *Renew. Energy*. 209. PERGAMON-ELSEVIER SCIENCE LTD: 453–461. doi: 10.1016/j.renene.2023.04.023.

Zeng, Zhijuan, Liu Yang, Xiaoting Zhu, Wenguang Zhao, Xianxiang Liu, Zexing Huang, Qiong Xu, Wenzhou Zhong. 2023. "Highly selective production of the biofuel 2,5-dimethylfuran from 5-hydroxymethylfurfural over Co/N-C catalysts." *React. Chem. Eng.* 8(2). ROYAL SOC CHEMISTRY: 455–464. doi: 10.1039/d2re00339b.

Zhang, Fengyu, Huifang Wu, Qian Wang, Dianqing Li, Junting Feng. 2024. "Construction of core-shell Co@CuCo/Al_2O_3 catalyst to boost the low-temperature hydrogenation of 5-hydroxymethylfurfural." *Aiche J.* 70(2). WILEY: e18289. doi: 10.1002/aic.18289.

Zhang, Jun, Jinzhu Chen. 2017. "Selective transfer hydrogenation of biomass-based furfural and 5-hydroxymethylfurfural over hydrotalcite-derived copper catalysts using methanol as a hydrogen donor." *ACS Sustain. Chem. Eng.* 5(7). AMER CHEMICAL SOC: 5982–5993. doi: 10.1021/acssuschemeng.7b00778.

Zhang, Jun, Kaijun Dong, Weimin Luo. 2019. "PdCl 2 -catalyzed hydrodeoxygenation of 5-hydroxymethylfurfural into 2,5-dimethylfuran at room-temperature using polymethylhydrosiloxane as the hydrogen donor." *Chem. Eng. Sci.* 201. PERGAMON-ELSEVIER SCIENCE LTD: 467–474. doi: 10.1016/j.ces.2019.03.011.

Zhang, Qingtu, Jianliang Zuo, Lu Wang, Feng Peng, Shengzhou Chen, Zili Liu. 2021. "Non noble-metal copper-cobalt bimetallic catalyst for efficient catalysis of the hydrogenolysis of 5-hydroxymethylfurfural to 2,5-dimethylfuran under mild conditions." *ACS Omega.* 6(16). AMER CHEMICAL SOC: 10910–10920. doi: 10.1021/acsomega.1c00676.

Zhao, Wenguang, Zexing Huang, Liu Yang, Xianxiang Liu, Hongye Xie, Zixuan Liu. 2022. "Highly efficient syntheses of 2,5-bis(hydroxymethyl)furan and 2,5-dimethylfuran via the hydrogenation of biomass-derived 5-hydroxymethylfurfural over a nickel-cobalt bimetallic catalyst." *Appl. Surf. Sci.* 577. ELSEVIER: 151869. doi: 10.1016/j.apsusc.2021.151869.

Zhang, Yan-Ru, Bing-Xin Wang, Lei Qin, Qiang Li, Yong-Ming Fan. 2019. "A non-noble bimetallic alloy in the highly selective electrochemical synthesis of the biofuel 2,5-dimethylfuran from 5-hydroxymethylfurfural." *Green Chem.* 21(5). ROYAL SOC CHEMISTRY: 1108–1113. doi: 10.1039/c8gc03689f.

Zhang, Hongming, Lu Zhang, Ismael A. Rodriguez-Perez, Wenkang Miao, Kangli Chen, Wenfeng Wang, Yuan Li, Shumin Han. 2021. "Carbon nanospheres supported bimetallic Pt-Co as an efficient catalyst for $NaBH_4$ hydrolysis." *Appl. Surf. Sci.* 540. ELSEVIER: 148296. doi: 10.1016/j.apsusc.2020.148296.

Zhang, Zhaonan, Chengwei Liu, Dan Liu, Yaning Shang, Xianqiang Yin, Peng Zhang, Bhekie B. Mamba, Alex T. Kuvarega, Jianzhou Gui. 2020. "Hydrothermal carbon-supported Ni catalysts for selective hydrogenation of 5-hydroxymethylfurfural toward tunable products." *J. Mater. Sci.* 55(29). SPRINGER: 14179–14196. doi: 10.1007/s10853-020-05052-0.

Zhang, Zhaofu, Shuaishuai Liu, Minqiang Hou, Guangying Yang, Buxing Han. 2021. "Continuous-flow formic acid production from the hydrogenation of CO_2 without any base." *Green Chem.* 23(5). ROYAL SOC CHEMISTRY: 1978–1982. doi: 10.1039/d0gc04233a.

Zhao, Jingwen, Mengran Liu, Guoli Fan, Lan Yang, Feng Li. 2021. "Efficient transfer hydrogenolysis of 5-hydromethylfurfural to 2,5-dimethylfuran over CoFe bimetallic catalysts using formic acid as a sustainable hydrogen donor." *Ind. Eng. Chem. Res.* 60(16). AMER CHEMICAL SOC: 5826–5837. doi: 10.1021/acs.iecr.1c01029.

Zu, Yanhong, Panpan Yang, Jianjian Wang, Xiaohui Liu, Jiawen Ren, Guanzhong Lu, Yanqin Wang. 2014. "Efficient production of the liquid fuel 2,5-dimethylfuran from 5-hydroxymethylfurfural over Ru/Co_3O_4 catalyst." *Appl. Catal. B-Environ.* 146. ELSEVIER SCIENCE BV: 244–248. doi: 10.1016/j.apcatb.2013.04.026.

Zhu, Changhui, Haiyong Wang, Hu Li, Bijuan Cai, Wei Lv, Chiliu Cai, Chenguang Wang, Long Yan, Qiying Liu, Longlong Ma. 2019. "Selective hydrodeoxygenation of 5-hydroxymethylfurfural to 2,5-dimethylfuran over alloyed Cu-Ni encapsulated in biochar catalysts." *ACS Sustain. Chem. Eng.* 7(24). AMER CHEMICAL SOC: 19556–19569. doi: 10.1021/acssuschemeng.9b04645.

Zhu, Yifeng, Xiao Kong, Bo Peng, Luping Li, Zhen Fang, Yulei Zhu. 2020. "Efficient Cu catalyst for 5-hydroxymethylfurfural hydrogenolysis by forming Cu-O-Si bonds." *Catal. Sci. Technol.* 10(21). ROYAL SOC CHEMISTRY: 7323–7330. doi: 10.1039/d0cy01032d.

Zhu, Yifeng, Xiao Kong, Hongyan Zheng, Guoqiang Ding, Yulei Zhu, Yong-Wang Li. 2015. "Efficient synthesis of 2,5-dihydroxymethylfuran and 2,5-dimethylfuran from 5-hydroxymethylfurfural using mineral-derived Cu catalysts as versatile catalysts." *Catal. Sci. Technol.* 5(8). ROYAL SOC CHEMISTRY: 4208–4217. doi: 10.1039/c5cy00700c.

3 Catalytic Oxidation of 5-Hydroxymethylfurfural to 2,5-Furandicarboxylic Acid

Ye Meng, Huan Liu, Jie Li, and Hu Li

3.1 THERMOCATALYTIC HMF-TO-FDCA OXIDATION

Biomass-based feedstock as a renewable alternative to fossil fuels can help reduce the consumption of crude oil in the production of petrochemical products. The related derivatives of the furfural series are important biomass-based raw materials. HMF, a prominent biomass derivative, is recognized as a potential key compound. It originates from the dehydration of hexose carbohydrates and serves as a versatile chemical sourced from biomass. As presented in Figure 3.1a, the preparation of HMF from cellulose is generally divided into three reaction steps: (1) hydrolysation of cellulose into glucose monomer; (2) glucose isomerization to fructose; (3) HMF production from fructose after dehydration three times (Kunnikuruvan et al. 2019; Meng et al. 2022). The structure of HMF contains -OH, -CHO, and a furan ring, which has very active chemical activity.

HMF, as a key molecule that connects biomass chemicals and petroleum-based chemicals with a market value exceeding USD 60 million, plays a crucial role in the production of valuable chemicals used across various industries such as medicine, the polyester sector, and functional materials (Zhang et al. 2024). FDCA, a biomass furan-based monomer with a rigid planar structure, is receiving attention for its potential to serve as a substitute for the petroleum-based material, terephthalic acid, in the production of polymer synthetic materials such as polyethylene terephthalate (PET) and polybutylene terephthalate (PBT). The Transparency Market Research indicates that the global FDCA market was valued at USD 481.4 million in 2022 and is expected to grow at a compound annual growth rate (CAGR) of 8.2% from 2023 to 2031, reaching USD 980.3 million by the end of 2031. FDCA offers a sustainable and renewable alternative to traditional petrochemicals, making it an attractive option for reducing the environmental impact of polymer production. Additionally, the unique properties of FDCA, such as its high thermal stability and mechanical strength, make it a suitable candidate for a wide range of applications in industries such as packaging, textiles, and automotive. The transition from petroleum-based materials to biomass-derived alternatives like FDCA not only leads to a decrease in greenhouse

DOI: 10.1201/9781003464839-3

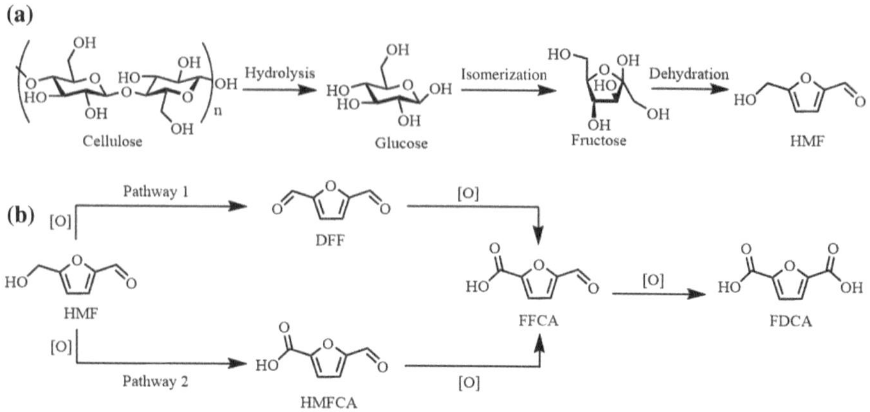

FIGURE 3.1 Paths for the synthesis of HMF from cellulose (a); a duo of potential routes for the conversion of HMF to FDCA (b). 5-hydroxymethylfurfural (HMF); 5-hydroxymethyl-2-furan-carboxylic acid (HMFCA); 2,5-diformylfuran (DFF); 5-formyl-2-furan-carboxylic acid (FFCA); 2,5-furandicarboxylic acid (FDCA).

gas emissions but also contributes to the development of a more circular and sustainable economy. By utilizing renewable resources for the production of polymers, we can reduce our reliance on finite fossil fuels and move towards a more environmentally friendly and economically viable future. Continued research and investment in the development of FDCA and other biomass-derived monomers will be crucial in accelerating the transition towards a greener and more sustainable polymer industry. Additionally, FDCA holds significant potential in the commercial fabrication of polyurethane materials. The catalytic HMF-to-FDCA oxidation depicted in Figure 3.1b exhibits two potential pathways facilitated by diverse catalytic approaches such as thermocatalysis, electrocatalysis, photocatalysis, and biocatalysis. The paths comprise the oxidation of the formyl group of HMF to create HMFCA, then the oxidation of HMFCA's hydroxymethyl group to produce FFCA, and lastly, the oxidation of FFCA's formyl group to generate FDCA (HMF → HMFCA → FFCA → FDCA). Another possible pathway includes oxidizing the hydroxymethyl group in HMF to form DFF, then further oxidizing it to produce FFCA, and finally leading to the creation of FDCA (HMF → DFF → FFCA → FDCA).

3.1.1 Homogeneous Catalysis

Ribeiro et al. used cobalt acetylacetonate $Co(acac)_3$ as a homogeneous catalyst and reacted at 80 °C for 4 h with the participation of oxidizing agent methylisobutylketone for catalytic HMF-to-FDCA oxidation (19% yield) (Ribeiro et al. 2003). Hansen et al. developed CuCl and t-Buooh hydroperoxide as a catalytic oxidation system (Hansen et al. 2013). At room temperature, acetonitrile, as the reaction solvent, reacts in CuCl and t-Buooh for 48 h, and the yield of FDCA can reach 50%. Cu(II) is identified as the critical element in the catalytic oxidation mechanism. During this process, t-Buooh triggers the transformation of Cu(I) in CuCl to Cu(II) and facilitates the selective oxidation of HMF. The effectiveness of Cu(II) was validated, and employing $CuCl_2$ as a catalyst proved to be equally effective in the conversion of HMF to FDCA. Incorporating stoichiometric thermal oxidants enhances the corrosiveness of the reaction medium. Hence, opting for molecular oxygen O_2 as an oxidant is preferred, promoting a more environmentally friendly and sustainable reaction pathway.

3.1.2 Heterogeneous Catalysis

The oxidation of HMF to FDCA through catalytic means is a complex process that requires a properly constructed catalyst to achieve selective conversion. The efficiency and selectivity of the conversion process heavily depend on the design and composition of the catalyst utilized. Several factors, such as the type of metal, support material, and reaction conditions, play a crucial role in determining the effectiveness of the catalyst in facilitating the conversion of HMF to FDCA. The careful selection and preparation of the catalyst are essential for optimizing the oxidation reaction and maximizing the yield of FDCA. Overall, the success of the HMF-to-FDCA oxidation reaction hinges on the careful design and construction of the catalyst, which is pivotal in enabling the selective conversion of HMF to FDCA. Homogeneous catalysts are difficult to recover, and easily cause side reactions, pollution, and so on, which makes them less studied. However, heterogeneous catalysts can overcome these shortcomings perfectly and become the research focus and main research direction for catalytic HMF-to-FDCA oxidation.

(1) Noble Metal Catalysis

The basic auxiliary agent and noble-metal-supported catalyst make up the catalytic system that is commonly employed in the oxidation of HMF to FDCA. Due to the good oxygen dissociation ability of precious metals, related catalyst types have been continuously developed, including Au, Pd, Pt, Ru, etc. (Casanova et al. 2009; Siankevich et al. 2014; Zhang et al. 2015; Gorbanev et al. 2011; Nishimura et al. 2013; Luo et al. 2008; Donoeva et al. 2017). Au-supported catalysts have been shown to exhibit superior stability in aqueous solutions when compared to catalysts supported by Pt, Pd, and Ru. However, there are still challenges associated with Au-supported catalysts, including deactivation due to the deposition of by-products or intermediates on the catalyst surface during the reaction. This can hinder the overall efficiency and effectiveness of the catalyst, ultimately impacting the success of the reaction. In aqueous environments, Au-supported catalysts have shown promise due to their enhanced stability. This characteristic is crucial for ensuring the catalyst can withstand the harsh conditions of the reaction and continue to perform optimally. However, the issue of deactivation looms large, as by-products or intermediates can accumulate on the catalyst surface over time, leading to decreased activity and efficiency. While Au-supported catalysts may hold certain advantages over Pt-, Pd-, and Ru-supported catalysts, the challenge of deactivation must be addressed to fully capitalize on their potential in aqueous solutions. Researchers and scientists must continue to explore ways to mitigate the effects of deposition on the catalyst surface, ultimately maximizing the overall performance and longevity of Au-supported catalysts in various reactions. Therefore, the addition of some alkaline additives can effectively improve the solubility of FDCA in a water solution, and avoid the deactivation of the catalyst active site due to product adsorption. The most widely used types of alkaline additives include NaOH, Na_2CO_3, and $NaHCO_3$.

Under relatively mild reaction conditions, H_2O as the green solvent of the reaction and O_2 as the oxidation agent of the reaction conform to the concept of "green sustainable chemistry", while avoiding some disadvantages brought by homogeneous catalysts or stoichiometric thermal oxidants. For example, Suib's group reported an Au-supported high-specific surface graphitic carbon carrier (HSAG) catalyst for HMF-to-FDCA oxidation (Donoeva et al. 2017). The characteristics of the HSAG support surface are utilized to control the size and dispersion of Au nanoparticles, consequently impacting the performance of the catalyst. The Au particles' spherical size ranges from 2 to 4 nm post-loading. The catalytic efficacy of these particles for HMF oxidation was influenced by specific acid-base sites on HSAG carbon materials following diverse pretreatments. Alkaline carbon materials tend to attract OH^- ions to the catalyst surface, which promotes the formation of intermediate HMFCA and ultimately accelerates the production of FFCA and FDCA. However, the acidic side of the HSAG carbon carrier is negatively charged in the reaction, which will reduce the rate of H atom transfer during the reaction. The synergy between the support material and the supported gold particles played a critical role in enhancing the catalyst's effectiveness. When $NaHCO_3$ was used as the reaction agent and water as the reaction solvent, the conversion rate of HMF was able to reach 99%, with an FDCA selectivity of 75% achieved at a temperature of 90 °C and pressure of 0.1 MPa in an atmosphere containing O_2 for 12 h.

The performance of catalysts supported by noble metals is largely influenced by the size of the metal particles being supported. The active-site Au species has unsaturated valence electrons and therefore can bind to adsorptive species during oxidation. When the nanometer size of the Au atoms decreases, more nanoparticles with unsaturated electrons will be distributed on the catalyst support, resulting in some specific chemical properties. The noble-metal catalyst demonstrated significant catalytic activity in the production of FDCA from HMF oxidation, with the reactions being completed in a short amount of time and at low temperatures under mild reaction conditions (Table 3.1). This indicates its potential for efficient and environmentally friendly FDCA synthesis. For example, an Au/CeO_2 catalyst, supported by CeO_2 is a good oxygen storage material because of oxygen vacancy and other defects. The loading of Au particles will lead to a further increase of defect sites, and the O_2 will dissociate at the oxygen vacancy point of the catalyst. Therefore, the abundant oxygen vacancy on the surface gives the catalytic material an advantage in oxygen dissociation and thus promotes the oxidation reaction. Furthermore, the metal atoms in an unsaturated state exhibit Lewis acid characteristics, thereby aiding in the oxidation of hydroxyl groups in HMF. Transformation of HMF was nearly total, achieving an FDCA selectivity of over 85% with CeO_2, TiO_2, C, and hydrotalcite HT as chosen supports (Table 3.1, Entry 1–4).

Pd and Pt catalysts supported by metals are commonly utilized in the process of oxidizing HMF to produce FDCA (Table 3.1, Entries 5–9). Bara and colleagues

TABLE 3.1
The Catalytic MF-to-FDCA Oxidation Over Noble-Metal-Supported Catalysts

Entry	Catalyst	Additive	Reaction condition	Con. (%)	Sel. (%)	Ref.
1	Au/CeO_2	NaOH	130 °C, 5 h, 0.1 MPa O_2	100	96	(Casanova et al. 2009)
2	Au/TiO_2	NaOH	60 °C, 6 h, 0.3 MPa O_2	>99	85	(Cai et al. 2013)
3	Au_8-Pd_2/C	NaOH	60 °C, 2 h, 3 MPa O_2	>99	>99	(Pasini et al. 2011)
4	Au/HT	No base	95 °C, 7 h, 1 MPa O_2	>99	>99	(Gupta et al. 2011)
9	Pt/PVP	No base	80 °C, 24 h, 0.1 MPa O_2	100	94	(Siankevich et al. 2014)
5	Pd/C@Fe_3O_4	K_2CO_3	80 °C, 6 h, 1 MPa O_2	98.4	86.7	(Mei et al. 2015)
8	Pt/γ-Al_2O_3	Na_2CO_3	75 °C, 12 h, 0.1 MPa O_2	96	96	(Sahu et al. 2014)
6	Pd/PVP	K_2CO_3	90 °C, 6 h, 1 MPa O_2	>99	90	(Siyo et al. 2014)
7	Pt-Bi/C	Na_2CO_3	100 °C, 6 h, 4 MPa Air	100	>99	(Rass et al. 2013)

studied the synthesis of Pd-based catalysts by carefully adjusting the size of the Pd particles (Siyo et al. 2014). Under mild reaction conditions, the O_2 flow rate was 35 mL/min at 90 °C, NaOH was used as the reaction assistant, and H_2O was the reaction solvent. The yield of FDCA reached a maximum of 90% when the particle size of Pd was 1.8 nm for 6 h. When the particle size of Pd is reduced to 2 nm, the yield of FDCA decreases to 81%. The Pd/PVP catalysts exhibit enhanced catalytic activity when particle size is reduced, attributed to the increased surface atom density and greater coordination of unsaturated metal sites. It was also found that the activity of the catalyst depended on the O_2 flow rate. The Pd nanoparticles will be quickly deactivated when the O_2 flow rate is far from the optimal value. When the oxygen flow rate is too low, an excess of by-products will gather on the catalyst's surface, obstructing the surface-active site. The Pd surface will interact with oxygen and consume a certain active site when the oxygen flow rate is too high. The reaction temperature had a certain effect on the oxidation process of HMF on the Pd/PVP catalyst. At low reaction temperatures below 70 °C, it was observed that the transformation of intermediate product HMFCA to FFCA occurred at a slower rate compared to the conversion of FFCA to FDCA. Additionally, the predominant product under these conditions was found to be HMFCA. However, when the reaction temperature was raised to 90 °C, the oxidation rate of HMFCA to FFCA was comparable to that of FFCA to FDCA. This suggests that the temperature plays a crucial role in determining the reaction kinetics and product distribution in this process. Further research may be needed to fully understand the mechanisms underlying these temperature-dependent effects.

(2) Non-Noble Metal Catalysis

The supported catalyst comprised of precious metals exhibited remarkable catalytic performance during the HMF-to-FDCA oxidation. However, the noble-metal catalyst has a high cost, and the catalytic stability of the supported catalyst was reduced due to the active phase leaching, which affected the industrialization of FDCA research.

Studies have recently shown that acetic acid has been identified as a solvent in reactions, or t-BuOOH has been used as an oxidant in the oxidation of HMF. Using organic reagents as reaction solvents and potent oxidants can lead to environmental issues, contradicting the principles of "green chemistry". Therefore, it is still the focus of research to develop a reaction system with transition metal as catalyst, H_2O as solvent, and O_2 as oxidant. Among them, Co, Mn, Fe, and other catalyst types have been continuously developed for catalytic HMF-to-FDCA oxidation.

Spinel composite oxides are commonly utilized in the realm of catalytic materials due to their steadfast crystal structure, expansive specific surface area, and elevated surface activity. The stable crystal structure of spinel composite oxides ensures their durability and long-term performance in catalytic processes. Additionally, the large specific surface area of these materials allows for more active sites to be exposed, enhancing their catalytic efficiency. Moreover, the high surface activity of spinel composite oxides results in improved reactivity, making them ideal for a variety of catalytic applications. Mugweru et al. synthesized a spinel composite oxide catalyst $Li_2CoMn_3O_8$ and chose NaBr as the co-catalyst. The catalyst showed promising results in catalyzing a specific reaction, demonstrating its potential for application in various chemical processes. The incorporation of NaBr as a co-catalyst further enhanced the catalytic activity of the spinel composite oxide, showcasing the importance of co-catalysts in fine-tuning the performance of catalysts. These findings contribute to the ongoing research efforts aimed at developing efficient catalysts for industrial and environmental applications (Jain et al. 2015). At 150 °C for 8 hours, utilizing acetic acid as the reaction solvent under 800 psi of O_2 pressure, the conversion of HMF was able to achieve 100%, resulting in an FDCA yield of 80%. In their study, Martin and colleagues created a magnetic nano-Fe_3O_4-COO_x catalyst by utilizing methyl isobutyl ketone as the solvent for the reaction (Kroger et al. 2000). The conversion of HMF reached 97.2% and the yield of FDCA reached 68.6% in an O_2 atmosphere at 80 °C for 12 h.

The interaction between transition metals and polymers is also an effective means of preparing highly efficient catalysts. In their study, Saha and colleagues created Fe^{3+} -POP^{-1}, a thermally durable poly-porous organic polymer based on porphyrin, to catalyze the conversion of HMF to FDCA (Saha et al. 2013). Solid-state ^{13}C-NMR analysis demonstrated the successful polymerization of pyrrole with an aldehyde linker. Additionally, high-resolution transmission electron microscopy (HRTEM) revealed that the nanospheres contained micropores measuring approximately 1.0 nm in size. The oxidation of HMF follows the free radical mechanism of peroxides and results in the uniform cleavage of RO-OH bonds at the iron-active sites via the Fenton reaction. During a 10-hour reaction period using O_2 as the oxidant and H_2O as the solvent, the conversion rate of HMF reached 99% with an 85% selectivity for FDCA. The catalyst was characterized and it was found that the active-site Fe still maintained an oxidation state, reflecting good stability after the reaction.

As the precursor of carbon materials, lignin, and transition metal coordination can improve the dispersion of active metals and thus improve the catalytic performance. In their study, Zhou and colleagues described a Co-Mn/N@C catalyst with Co and Mn nanoparticles coordinated by lignin (Zhou et al. 2019). The results showed that Mn/ N-doped carbon materials can enhance the activity of metal nanoparticles Co NPs embedded in the lignin structure during HMF oxidation. Simultaneously, it was discovered that nitrogen doping not only enhanced the reactivity of inert cobalt nanoparticles but also influenced the pathway of HMF selective oxidation. The catalyst's activity was enhanced by the dispersion of manganese atoms within, which adjusted the electronic structure of the nanoparticles. Through isotope labeling experiments, it was revealed that the oxygen atom from H_2O was incorporated into FDCA, while O_2 acted as an electron scavenger to eliminate the protons produced during HMF adsorption onto the catalyst surface. This resulted in the breaking of the C-OH bond and facilitated catalyst regeneration. Under the conditions of 85 °C for 12 hours, with water as the solvent and Co-Mn/N@C serving as the catalyst, the HMF conversion rate was 99.7% and the FDCA selectivity was 96.1%.

Liu et al. developed a mesoporous carbon material catalyst supported by CoO_x nanoparticles (Liu et al. 2020). CoO_x mainly exists in the forms of Co, CoO, and Co_3O_4 on graphitized carbon materials. The CoO_x nanoparticles, with an average size of approximately 25 nm, were evenly dispersed throughout the mesoporous carbon MC after undergoing the calcination process from the Co polymer precursor. It was found that the active-phase CoO_x accelerant was uniformly dispersed on MC carbon material, and the adsorption site on the surface of the support was combined with OH^- in an aqueous solution, which made the oxidation path of HMF to HMFCA more favorable, and promoted the desorption of the intermediate product and the catalyst, and improved the activity of the catalyst. CoO_x-MC showed excellent performance in HMF-to-FDCA oxidation. Utilizing O_2 as the oxidizing agent, the experiment was carried out in water at a temperature of 80 °C for 30 hours. The HMF conversion rate reached 98.3%, with a corresponding FDCA yield of 95.3%. The turnover frequency (TOF) for FDCA on the CoO_x-MC catalyst was measured at 2.22 h^{-1}, demonstrating superior performance compared to other non-precious metal catalysts. Following three cycles, the HMF conversion rate remained high at 91.2%, resulting in an FDCA yield of 88.6%.

For the efficient conversion of HMF to FDCA through selective oxidation, further improvement of the catalyst's stability is necessary. Ensuring a strong bond between metal nanoparticles and the support is essential to avoid undesirable outcomes such as agglomeration and leaching of active catalyst components. This interaction plays a critical role in maintaining the stability and effectiveness of the catalyst. Without a strong bond, the nanoparticles may clump together, reducing the surface area available for catalysis. Additionally, if the nanoparticles detach from the support, they may become ineffective or even toxic. Therefore, efforts must be made to enhance the interaction between metal nanoparticles and the support to ensure the optimal performance of the catalyst in various applications. To minimize plugging and deactivation of the active site caused by the adsorption of by-product properties, the design of bimetallic or trimetallic catalysts can be explored to improve catalytic activity. Transition metal catalysts offer stability and cost benefits over precious metal catalysts, highlighting the importance of developing efficient transition metal catalytic materials like Co-, Fe-, and Mn-based catalysts through various strategies for future industrial application of HMF. The current catalytic system involves most of the reaction process in an aqueous solution with alkaline additives, complicating product separation and purification, and resulting in insufficient greenness due to the corrosion of the additives. The focal point of ongoing research remains on developing eco-friendly catalytic systems exhibiting high catalytic activity under weak alkali or alkali-free conditions.

3.2 ELECTROCATALYTIC HMF-TO-FDCA OXIDATION

The process of electrochemical conversion differs from traditional heating, relying on electrochemical potential and Faraday current as the driving forces. External control parameters play a crucial role in adjusting the activation energy and thermodynamic driving force of a chemical reaction. By modifying these parameters, such as temperature or pressure, it is possible to influence the reaction rate, conversion, and selectivity of the process. This ability to adjust external control parameters can significantly impact the overall outcome of the reaction, allowing for improved efficiency and desired product formation. Ultimately, understanding how to manipulate these parameters effectively is essential in optimizing the performance of chemical processes. Additionally, electrochemical oxidation offers the benefits of current control and reaction monitoring. These advantages, such as operating at room temperature and atmospheric pressure, have garnered increasing interest. The utilization of electrochemical strategies in catalytic oxidation, starting with HMF oxidation in 2012, has since emerged as a promising alternative approach (Vuyyuru et al. 2012). This method has been extensively explored in multidisciplinary research for the valorization of biomass-derived compounds within the realm of green chemistry initiatives. The advancements in electrochemical techniques for HMF oxidation and their potential mechanisms in electrochemical catalysis are elaborated upon. In a water-based solution, the catalytic potential for oxidizing HMF to FDCA was found to be lower compared to the reaction for generating oxygen. Within the latest technology of a membrane electrolyzer, both the direct catalytic oxidation of HMF to FDCA at the anode and the production of H_2 at the cathode occurred simultaneously through a multi-step electron transfer process as outlined in Figure 3.2. Hence, the key to the successful implementation of the approach lies in the development of highly efficient

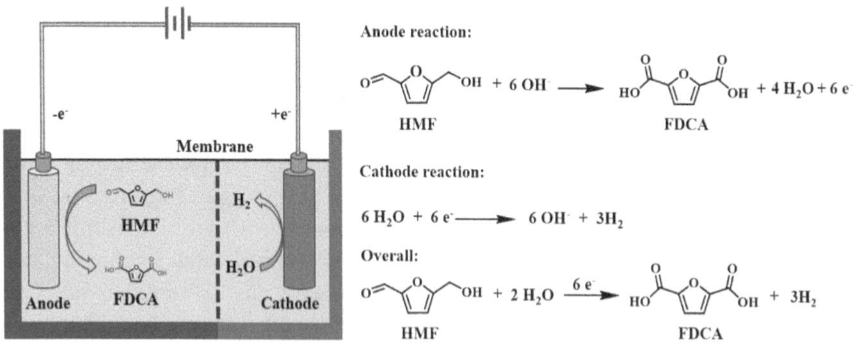

FIGURE 3.2 Electrocatalytic HMF-to-FDCA oxidation coupled with HER.

TABLE 3.2
Electrocatalytic HMF-to-FDCA Oxidation Over Various Catalysts

Entry	Catalyst	Reaction conditions	Potential (V_{RHE})	Con. (%)	(Yield)/ [Sel.] (%)	Ref.
1	Pd_1Au_2/C	0.1 M KOH, 25 °C, 1 h	0.90	100	(83.0)	(Chadderdon et al. 2014)
2	$CuCo_2O_4$	1.0 M KOH 25 °C	1.45	100	(93.7)	(Lu et al. 2020)
3	$NiCo_2O_4$	1.0 M KOH, 25 °C, 1 h	1.50	99.6	[90.8]	(Kang et al. 2019)
4	om-Co_3O_4/NF	1 M KOH, 25 °C	1.457	100	(99.8)	(Wang et al. 2021)
5	Co-P/DES	0.5 M $NaHCO_3$, 25 °C	1.45	99.0	(85.3)	(Kang et al. 2020)
6	Co-$CoSe_2$	1 M KOH, 25 °C, 1 h	1.43	100	(99.0)	(Huang et al. 2020)
7	NiCoMn-LDHs	1 M KOH, 35 °C, 2.5 h,	1.50	100	(91.7)	(Liu et al. 2021)
8	Ni_xB	1 M KOH, 25 °C, 0.5 h	1.45	100	(98.5)	(Barwe et al. 2018)
9	NiOOH	0.1 M KOH, 25 °C, 4.7 h	1.47	99.8	(96.0)	(Holzhäuser et al. 2020)
10	WO_3/Ni	1 M KOH, 25 °C, 5.85 h	1.57	99.4	(88.3)	(Hu et al. 2021)
11	$CuNi(OH)_2$	1 M KOH, 25 °C, 5 h	1.45	100	(93.3)	(Chen et al. 2019)
12	Cu nanocrystalline	0.1 M KOH, 25 °C	1.62	99.9	(96.4)	(Nam et al. 2018)
13	CuO	0.1 M KOH, 25 °C, 1.5 h	1.64	99.4	(90.09)	(Pham et al 2020)
14	TpBpy-Ni	0.1 M $LiClO_4$, 25 °C, 40 min	1. 55	96.0	(58)	(Cai et al. 2020)

catalysts, with the structure and makeup of these materials being crucial determinants of their catalytic effectiveness (Table 3.2).

(1) Noble Metal Catalysis

Metals like ruthenium, silver, platinum, gold, and palladium have long been important in the study of catalytic conversion because of their distinct electronic properties on the surface. Despite their high cost and limited availability, these metal materials are preferred by scientists for their outstanding catalytic capabilities. In 2012, Strasser et al. published the first study on electrocatalytic HMF oxidation using Pt/C as a catalyst with a high surface area (Vuyyuru et al. 2012). The resulting products showed a noteworthy 80% yield towards FDCA in an alkaline-aqueous solution with pH 13. Additionally, a high selectivity towards DFF was observed with an 18% yield in an alkaline-aqueous solution with a pH of 10. This highlights the significant impact of solvent effects on product distribution. At a lower pH level, the catalyst effectively captures the hydroxyl group from the substrate, but at a higher pH in the surrounding environment, it shows a preference for bonding with the aldehyde group of the substrate.

The production of valuable FDCA became achievable with the employment of a PdAu bimetallic catalyst, which consumes lower energy compared to conventional single-metal catalysts (Chadderdon et al. 2014). Product distribution was influenced by the surface alloy composition and electrode potential. The benefits of the bimetallic catalysts PdAu in a 1:2 ratio (Pd_1Au_2/C) and 2:1 ratio (Pd_2Au_1/C) included improved oxidation of aldehyde and alcohol, leading to substantial enhancements in HMF conversion and selectivity towards FDCA, with an 83% yield. Furthermore, Au-transition metal oxide nanostructures with a core-shell design were developed, with Au serving as the plasma surface. Surface-enhanced Raman spectroscopy (SERS) was utilized in the study to observe the progression of surface reaction intermediates and catalyst substances when subjected to specific voltage conditions (Heidary et al. 2020). The impact of surface interactions and adsorption on product selectivity and the evolution of surface intermediates under various potentials were effectively illustrated through operando Raman spectroscopy. The findings in this study offered important insights into the surface chemistry of HMF oxidation and the performance of electrocatalysts. It was observed that precious metals played a dual role as both active co-catalysts and platforms for analyzing catalytic materials. By developing a nanocatalyst known as $PtNiS_x/CB$, Wang et al. were able to establish a connection between the Pt co-catalyst and NiSx on carbon black. This novel approach proved to be effective in enhancing the catalytic properties of the electrocatalyst (Wang et al. 2020). This catalyst exhibited excellent performance in selectively oxidizing HMF to FDCA at low starting potentials.

(2) Transition-Metal Catalysis

Significant progress has been made in the creation and advancement of Co-containing electrocatalysts for the oxidation of HMF. Spinel oxides that contain Co in both the Co^{2+} and Co^{3+} states are viewed as potential catalysts for this process. In a study by Lu et al., the electrochemical efficiency of Co_3O_4, $ZnCo_2O_4$, and $CoAl_2O_4$ in catalyzing HMF oxidation was examined by manipulating specific blocks in the tetrahedral (Co^{2+}_{Td}) and octahedral (Co^{3+}_{Oh}) sites (Lu et al. 2020). It was found that Co_3O_4 exhibited superior performance compared to $ZnCo_2O_4$ and $CoAl_2O_4$, highlighting the catalytic activity of the octahedral CoO_6 sites. Additionally, the existence of tetrahedral and octahedral sites in Co_3O_4 significantly influenced the overall reaction rate. By considering the geometry of these sites and the introduction of Cu^{2+} to interact with octahedral Co^{3+} ions within the spinel oxides, the catalytic activity for the oxidation of HMF to FDCA was greatly improved. Consequently, an exceptionally efficient electrocatalyst ($CuCo_2O_4$) was developed through strategic adjustments to the geometric sites. The results showed that using $CuCo_2O_6$ as a catalyst for electrocatalytic oxidation of HMF to FDCA at 1.45 V led to a 93.7% yield of FDCA, with a Faraday efficiency of 94%.

The scientists described a new approach for creating efficient catalysts for electrocatalytic oxidation by manipulating the geometric structure of spinel Co_3O_4. They were able to create electrodes made of Co_3O_4 and $NiCo_2O_4$ using the hydrothermal synthesis method, with a filamentous nanostructure on nickel foam. These electrodes were designed to oxidize HMF to FDCA (Kang et al. 2019). The $NiCo_2O_4$ electrode showed a higher reduction in the Co^{3+} to Co^{2+} ratio compared to the Co_3O_4 electrode, indicating it is the primary active species for HMF oxidation. The film-like nanostructure of $NiCo_2O_4$ demonstrated superior efficiency in HMF oxidation compared to Co_3O_4. The rate of HMF conversion and the selectivity of FDCA were documented as 99.6% and 90.8%, respectively. In their study, Wang et al. introduced an innovative electrode made of 3D Co_3O_4/Ni foam (om-Co_3O_4/NF). This particular electrode successfully achieved full conversion of HMF and a 99.8% yield of FDCA at a potential of 1.457 V_{RHE} (Wang et al. 2021). Electrocatalytic oxidation of HMF took place on a Co-P-coated Cu foam, using either a deep eutectic solvent (DES) or water as the reaction medium (Kang et al. 2020). In DES, the Co-P material demonstrated outstanding performance with a 99% conversion rate for HMF and an 85.3% yield for FDCA. This version of Co-P displayed enhanced performance and long-term stability compared to its counterpart synthesized in H_2O.

The findings suggest that DES, possessing unique characteristics such as high ionic conductivity, compatibility with metal salts, and long-lasting product stability, holds promise as a potential substitute for water in electrode preparation processes utilizing electrodeposition. This is due to its robust sustainability and enhanced electrode efficiency. Huang et al. researched the synthesis of a Co-$CoSe_2$ catalyst containing oxygen vacancies through the addition of selenium to cobalt oxide (CoO) (Huang et al. 2020). The composite of CoO/$CoSe_2$ showed great effectiveness and longevity in the electrochemical conversion of HMF to FDCA. Utilizing a molar ratio of 23:1 for CoO to $CoSe_2$, FDCA was produced with a 99% efficiency and 97.9% Faraday efficiency at 1.43 V_{RHE}. This achievement was credited to the existence of oxygen vacancies, enhancing catalytic efficiency and specificity through the augmentation of the electrochemical surface area and reduction of charge-transfer resistance. In another study by Luo et al., it was discovered that Co^{3+} and Co^{4+} species could selectively oxidize functional groups of HMF (Deng et al. 2021). By electrodeposition, a CoO_xH_y catalyst was synthesized, and under varying voltage conditions, Co^{3+} and Co^{4+} species were formed. The oxidation of HMF to HMFCA was facilitated by Co^{3+}, while the generation of FDCA from HMFCA was favored by Co^{4+}. This study showcases the manipulation of voltage levels for the first time to achieve precise control over product distribution.

Catalytic materials based on nickel with strong redox characteristics are well-suited for improving the electrocatalytic transformation of HMF. In their study, Lu and colleagues amplified the number of damaged locations at the boundary of the NiO-Co_3O_4 electrocatalyst with a three-dimensional hierarchical nanostructure and explored its catalytic performance in the electrochemical oxidation of HMF (Lu et al.

2020). The impact of the interface led to an excess of positively charged vacancies, influencing the electronic characteristics of the Co and Ni elements and increasing the oxidation capability of the Ni species. The catalytic performance of the NiO-Co_3O_4 catalyst for HMF oxidation was exceptional, demonstrating a starting potential of 1.28 V_{RHE}. Additionally, an in-depth analysis was carried out to examine the oxidation process of the electrocatalyst. This investigation involved performing in-situ surface-specific vibrational spectroscopy using sum-frequency generation (SFG) to track the transformation of 5-hydroxymethylfurfural (HMF) into 2,5-furandicarboxylic acid (FDCA) via intermediate stages such as 5-hydroxymethyl-2-furancarboxylic acid (HMFCA) and 2,5-furandicarboxylic acid (FFCA).

The results revealed valuable insights into the mechanism of oxidation and provided a better understanding of the reaction pathway. Furthermore, the study highlighted the importance of understanding the oxidation route of electrocatalysts to improve their efficiency and performance in various applications. By gaining knowledge of the intermediate stages and products formed during oxidation, researchers can optimize the synthesis and design of electrocatalysts for enhanced catalytic activity and stability. This research contributes to the advancement of electrocatalysis and paves the way for developing more effective and sustainable catalysts for future use in renewable energy technologies. This study introduced a successful approach for generating positive charge gaps and highlighted the critical function of these gaps in the electrochemical conversion of HMF. Furthermore, researchers developed nanomaterials made up of ternary NiCoMn-layered double hydroxides (NiCoMn-LDHs) containing plentiful oxygen vacancies. These materials were designed for the targeted electrochemical conversion of HMF to FDCA, achieving an impressive FDCA yield of 91.7% (Liu et al. 2021).

The elevated amount of reactive Co-NiOOH locations on the ultrathin nanoparticle sheets, along with oxygen gaps, enhances the collaborative oxidation of HMF in both the Co-Ni and Mn^{2+} states. The nickel foam electrode was enhanced with high specific surface area nickel boride (Ni_xB) to improve the effective electrocatalytic oxidation of HMF to FDCA, achieving a close to 100% Faraday efficiency and a 98.5% yield (Barwe et al. 2018). A comparison was made between the reactivity of commercial nickel oxide (NiO) and mesoporous nickel oxide derived from the CMK-1 template (NiO-CMK-1) in alkaline solutions (Holzhäuser et al. 2020). The study found that NiO-CMK-1 exhibited higher reactivity and stability compared to commercial NiO in alkaline environments. This suggests that the mesoporous structure of NiO-CMK-1 plays a crucial role in enhancing its reactivity. The findings of this study have important implications for the development of more efficient and stable catalysts for alkaline solution reactions. Further research in this area could help in the design of better materials for various industrial applications. NiO-CMK-1 demonstrated an impressive FDCA selectivity of over 80%, surpassing the 30% selectivity of commercial NiO. The findings suggest that NiO-CMK-1 shows great potential for use in catalytic applications. Furthermore, NiOOH, CoOOH, and FeOOH were investigated for their catalytic properties in oxidizing HMF in alkaline solutions, particularly at pH 13 (Taitt et al. 2019).

The study involved a comprehensive analysis and comparison of the catalytic activity of these metal hydroxides. These results provide valuable insights into the potential applications of NiOOH, CoOOH, and FeOOH in the field of catalysis, particularly in the oxidation of biomass-derived compounds such as HMF. The NiOOH showed the highest efficiency as a catalyst, producing a remarkable 96.0% FDCA at a moderate potential of 1.47 V_{RHE}. Furthermore, when comparing FeOOH to other catalysts, such as Fe_3O_4 and Fe_2O_3, it was found to be less effective in catalyzing the oxidation of HMF. This lack of catalytic activity was particularly notable at potentials below the threshold for water oxidation. FeOOH did not exhibit any catalytic activity in this range, highlighting its limitations compared to other iron-based catalysts. These findings suggest that FeOOH may not be a suitable catalyst for the oxidation of HMF at lower potentials, calling into question its potential application in related reactions. In the study, it was found that NiOOH outperformed other catalysts in the conversion of HMF to FDCA, showcasing its superior catalytic properties.

In contrast, FeOOH failed to exhibit any significant activity in the oxidation process, highlighting the importance of selecting the appropriate catalyst for a desired reaction. When subjected to a decreased positive potential, CoOOH demonstrates greater efficiency in initiating the oxidation of HMF than NiOOH as a result of the transformation from $Co(OH)_2$ to CoOOH. Despite the promising potential of $Co(OH)_2$/CoOOH as a catalyst for HMF oxidation, it falls short in terms of achieving the necessary current density for sustained HMF oxidation at the desired potential. This highlights a limitation in the performance of $Co(OH)_2$/CoOOH in catalyzing the oxidation of HMF efficiently. Additional research and development may be necessary to enhance the catalytic activity of $Co(OH)_2$/CoOOH to meet the current density requirements for consistent HMF oxidation. Furthermore, this finding underscores the importance of continually optimizing and refining catalyst systems for electrochemical reactions. By addressing the challenges and limitations in catalytic performance, researchers can develop more effective catalysts for a wide range of applications, including the oxidation of HMF. This research also emphasizes the need for ongoing investigation into alternative catalyst materials and configurations to improve the efficiency and effectiveness of electrochemical processes.

Ultimately, by striving to overcome the current limitations in catalyst performance, researchers can unlock new opportunities for advancing the field of electrochemistry and enhancing the sustainability of chemical transformations. To comprehend the underlying factors, how the activity and alkalinity impacted the oxidation of HMF with Ni_3N acting as the catalyst. Zhou and colleagues employed a range of methodologies including in situ analysis with Raman spectroscopy, quasi-in-situ examination with X-ray photoelectron spectroscopy (XPS), operando observation with X-ray absorption spectroscopy (XAS), and operando assessment with electrochemical impedance spectroscopy (EIS) to acquire a deeper understanding of the phenomenon (Zhou et al. 2021). Through their comprehensive study, they were able to uncover the impact of activity and alkalinity on the oxidation of HMF. Their findings shed light on the differences in catalytic behavior between $Co(OH)_2$/CoOOH and NiOOH, highlighting the importance of understanding the mechanisms behind these processes for further optimization of HMF oxidation reactions. By utilizing advanced techniques such as in situ Raman spectroscopy and operando XAS, they were able to capture real-time data and gain a deeper understanding of the catalytic mechanisms at play. This research contributes valuable insights to the field of catalysis, paving the way for more efficient and sustainable oxidation reactions in the future.

The findings revealed that the catalytic activity of Ni_3N stemmed from the absorption of hydroxyl groups to create $Ni^{2+\delta}N(OH)_{ads}$. Furthermore, a two-stage process was proposed for the electrochemical conversion of 5-hydroxymethylfurfural (HMF) on Ni_3N. Initially, after reaching a potential of 1.35 V_{RHE}, the Ni atom experienced electron loss and attracted hydroxyl groups, resulting in the formation of positively charged $Ni^{2+\delta}N(OH)_{ads}$ species with electrophilic oxygen. This intermediate species then interacted with HMF, gaining protons and electrons and releasing HMF as water, while simultaneously converting HMF into FDCA. The speed of this electrocatalytic oxidation process was found to be highly efficient in strongly alkaline conditions, but significantly slower in less alkaline electrolytes. In a study, it was proposed that raising the alkalinity of the electrolyte may boost the creation of $Ni^{2+\delta}N(OH)ads$ species, leading to improved HMF adsorption on the Ni_3N electrode. Another research by Hu et al. introduced a new three-dimensional porous WO_3/Ni electrode produced through a simple one-step hydrothermal method at 180 °C, with the support of a polyethylene glycol template (Hu et al. 2021). This innovative electrode structure holds promise for improving the efficiency of electrochemical reactions involving HMF conversion, showcasing the potential for further advancements in catalytic materials for sustainable chemical transformations. Through these research findings, novel insights have been provided into the mechanisms underlying the electrochemical conversion of biomass-derived compounds, paving the way for the development of more effective and sustainable catalytic processes in the future.

This unique electrode architecture was utilized for the first time to produce FDCA with exceptional selectivity. Notably, in a 1.0 M KOH electrolyte at room temperature, the conversion rate of HMF and the yield of FDCA reached impressive values of 99.4% and 88.3%, respectively. This exceptional performance can be credited to various crucial factors: (i) Creating a three-dimensional porous WO_3/Ni electrode through careful construction with polyethylene glycol enhances the electrochemically active surface area and increases the exposure of active sites; (ii) the existence of nanorod-shaped WO_3 structures on the nickel foam surface, which greatly facilitated ion and mass transport; (iii) the participation of W^{6+} species within WO_3 in the electrocatalytic oxidation of HMF, with W^{6+} being transformed to W^{5+}. The introduction of Ni atoms into WO_3 has been shown to enhance the stability and effectiveness of the oxide. While WO_3 is inherently acidic and susceptible to instability in alkaline environments, the presence of Ni atoms serves to strengthen the overall performance of the material. This augmentation not only increases the durability of WO_3 but also improves its functionality in various applications. Additionally, the combination of WO_3 and Ni can lead to synergistic effects that further enhance the properties of the oxide. Overall, the incorporation of Ni atoms into WO_3 presents a promising opportunity to optimize the performance of the material for a range of uses. This study offers valuable insights into enhancing the catalytic efficiency and robustness of WO_3 in highly alkaline environments.

The electrochemical oxidation of 5-hydroxymethylfurfural (HMF) has been successfully carried out using copper (Cu) or Cu-based catalysts in previous studies.

One such study by Chen et al. involved the development of a Cu-Ni bimetallic catalyst with a specific $CuNi(OH)_2$ structure through incipient wetness impregnation (Chen et al. 2019). This Cu-Ni bimetallic catalyst exhibited promising results in the oxidation of HMF, showcasing its effectiveness in catalyzing this reaction. The structure of $CuNi(OH)_2$ likely plays a critical role in facilitating the electrochemical oxidation of HMF, demonstrating the importance of catalyst design in enhancing catalytic performance. Overall, the successful development and utilization of Cu-Ni bimetallic catalysts for the electrochemical oxidation of HMF provide valuable insights into the application of such catalysts in green chemistry processes. Further research in this area could lead to the optimization of catalyst properties and reaction conditions, ultimately improving the overall efficiency and sustainability of HMF oxidation reactions. At 1.45 V_{RHE}, the $NiCu(OH)_2$ catalyst achieved a current density of 9.2 mA cm^{-2}, leading to 100% conversion of HMF and a yield of 93.3% FDCA. Electrodeposition was employed to create both nanocrystalline and bulk Cu electrodes to examine variations in electrochemical oxidation and surface oxidation of HMF (Nam et al. 2018).

It was observed that the catalytic efficiencies of the oxidation and hydroxide layers developed on the Cu nanocrystalline electrode were notably distinct from those on the Cu bulk electrode regarding the efficiency of HMF oxidation. With a larger specific surface area and a higher geometric area of catalytically active Cu ion units, the Cu nanocrystalline electrode demonstrated enhanced FDCA yield and Faraday efficiency in a shorter timeframe. During the process of oxidizing HMF, the Cu nanocrystalline electrodes showed a clear correlation between the conversion of HMF and the production of FDCA, directly influenced by the quantity of charge transferred. No accumulation of intermediate oxidation by-products was observed. The FDCA yield on the Cu nanocrystalline electrode was significantly higher at 96.4% compared to the Cu bulk electrode, which only yielded 80.8% FDCA. Additionally, the Cu bulk electrode accumulated a notable amount of FFCA. The results demonstrate that the use of a Cu nanocrystalline electrode can lead to a higher yield of FDCA in comparison to a Cu bulk electrode. The significant difference in FDCA yield between the two electrodes highlights the efficiency and potential of nanocrystalline materials in electrochemical processes.

Furthermore, the accumulation of FFCA on the Cu bulk electrode further emphasizes the advantages of using nanocrystalline electrodes for improved performance in catalytic reactions. High electrical conductivity was exhibited by $Cu(OH)_2$ and CuO nanowires that were produced on the substrate using inorganic polymerization and calcination techniques. The origin of copper used in this process was copper foam in three dimensions, exhibiting excellent porosity (Pham et al 2020). Using nanowires as electrodes, the study analyzed the influence of copper oxide's crystal structure and chemical composition on the electrocatalytic oxidation of HMF-to-FDCA in a 0.1 M KOH solution. $Cu(OH)_2$ nanowires were initially grown on the substrate by stacking together through hydrogen bonding. This growth was facilitated by the square-planar coordination of Cu^{2+} ions on the crystal surfaces. The transition of these nanowires to CuO nanowires maintained the same morphology but increased the number of active sites on the surface. As a result, higher yields of FDCA were produced. The formation of $Cu(OH)_2$ nanowires on the substrate was driven by the hydrogen bonding between the stacked layers, which was made possible by the square-planar coordination of Cu^{2+} ions on the crystal surfaces. This transition to CuO nanowires did not alter the overall structure, but it did lead to an enhancement in the number of active sites on the surface. This increase in active sites ultimately resulted in an improvement in FDCA yields. In summary, the growth of $Cu(OH)_2$ nanowires on the substrate through stacking via hydrogen bonding was crucial for the subsequent transition to CuO nanowires.

This transition maintained the morphology while increasing the number of active sites on the surface, ultimately leading to higher FDCA yields. In another study, Lei et al. synthesized a bimetallic $CuMn_2O_4$ catalyst with oxygen vacancies, achieving a Faraday efficiency of 96% for FDCA production (Zhu et al. 2021). Transition metal-based electrocatalysts rich in earth elements show promise for selective oxidation of biomass-derived HMF, offering improved energy conversion efficiency and the potential for high-value product creation compared to water splitting.

The covalent organic framework (COF) is a unique porous polymer composed of elements such as carbon, hydrogen, oxygen, nitrogen, and boron, known for its exceptional thermal and chemical stability. This feature enables the efficient movement of reactants and products within the framework, making it a highly suitable material for supporting catalysts in various reactions (Kandambeth et al. 2018). Cai and his team effectively produced a TpBpy-Ni film catalyst by utilizing a TpBpy-based two-dimensional COF alongside Ni(II) acetate as a starting material (Cai et al. 2020). The TpBpy-Ni@FTO electrode showed excellent catalytic activity and stability for the electrochemical oxidation of HMF. The high selectivity towards 5-(hydroxymethyl) furfural oxide (5-HMFO) was obtained at a relatively high current density, indicating efficient conversion of HMF. The TpBpy-Ni@FTO electrode also exhibited good recyclability, with only a slight decrease in current density after multiple cycles. These results demonstrate the potential of the TpBpy-Ni@FTO electrode as a promising catalyst for the electrochemical oxidation of HMF. A steady applied potential was crucial in achieving a high conversion rate of HMF, with a remarkable rate of 96%. This highlights the importance of maintaining consistency in the experimental conditions to optimize the reaction process. Furthermore, the yields of FFCA and FDCA were also significant at 34% and 58%, respectively. Despite the marginally lower FDCA yield, the findings suggest that TpBpy-Ni@FTO offers a solid platform for advancing COF catalysts for the conversion of HMF derived from biomass. These results demonstrate the effectiveness of the applied potential in promoting the conversion of HMF to valuable products. It emphasizes the potential for further research and development in the field of green chemistry and renewable energy.

3.3 PHOTOCATALYTIC HMF-TO-FDCA OXIDATION

Photocatalysis is a promising technique for converting solar energy into chemical energy. This method has the potential to significantly impact various industries by providing a more sustainable and efficient way to produce chemical compounds. By harnessing the power of sunlight, photocatalysis offers a clean and renewable source of energy that can be used to drive chemical reactions. This innovative process holds great promise for advancing research in fields such as environmental remediation, energy conversion, and pharmaceutical synthesis. Moreover, the ability of photocatalysis to directly convert solar energy into chemical energy sets it apart as a highly promising solution for reducing our reliance on traditional fossil fuels and mitigating the impacts of climate change. As we continue to explore and refine the capabilities of photocatalysis, we can look forward to a future where this technology plays a key role in driving sustainable development and addressing global challenges. Utilizing the inexhaustible source of solar energy, the photocatalytic HMF oxidation holds significant value in various applications. A recent breakthrough in this field was the successful conversion of HMF to FDCA through photocatalysis using CoPz/g-C_3N_4 as a catalyst under artificial sunlight (Xu et al. 2017). The CoPz effectively triggers O_2 to produce singlet oxygen ($^1O_2$), which is crucial in the oxidation process. However, the hydrophobic nature of CoPz can lead to its aggregation, reducing the number of active sites, and resulting in a low HMF conversion rate of 40.2% and an FDCA yield of 36.2% when used alone as a catalyst. To address this issue, the hydrophilic property of g-C_3N_4 was utilized to prevent the aggregation of CoPz, thus increasing the availability of active sites.

In the CoPz/g-C3N4 synergistic catalytic system, HMF was efficiently oxidized to FDCA with a remarkable selectivity of over 97% at pH 6.86 and 9.18, and less than 10% selectivity of FDCA was observed at a lower pH of 4.01. The pH value was also found to significantly influence product selectivity, highlighting its importance in the process. This research presents an innovative method for the oxidation of HMF to FDCA through photocatalysis, showing excellent efficiency in gentle environments. Additionally, Yang et al. explored the oxidation of HMF using decatungstate as a photocatalyst under visible light, resulting in the highest FDCA production of merely 5.8% in CH_3CN. The effectiveness of the anion of decatungstic acid as a photocatalyst relies on the synthesis quality, which may be impacted by its counterion and specific additives (Yang et al. 2020). Among the additives examined, NaBr, DMSO, and 1-butyl-3-methyl-imidazolium chloride were found to inhibit the photoinduced polymerization of HMF and enhance the carbon balance yield of the photocatalytic oxidation process. This underscores the importance of optimizing synthesis conditions for improved efficiency in photocatalysis. Water, specifically in the form of concentrated acid solutions, was found to have a positive regulatory role in the photocatalytic oxidation process.

The addition of hydrogen bromide (HBr) was identified as the most effective additive, leading to significant improvements in the overall efficiency of the reaction. With the presence of HBr, the process resulted in a 67.1% yield of DFF, a 5.8% yield of FDCA, and an impressive 87.7% yield of balanced carbon products. These findings highlight the importance of water and specific additives in enhancing the photocatalytic oxidation process for the synthesis of valuable chemicals. By understanding the impact of different components on the reaction outcomes, researchers can optimize conditions to maximize yields and improve overall efficiency. Utilizing HBr as an additive can significantly boost the production of key products, such as DFF and FDCA, leading to more sustainable and cost-effective chemical synthesis processes. The improved outcome observed in this study can be mostly credited to the increased interaction between HBr and decatungstate, where HBr acts as a donor and decatungstate as an acceptor. This interaction allows for the selective oxidation of HBr by photo-excited decatungstate, leading to the generation of Br radicals. This enhanced interaction and selective oxidation process play a crucial role in the overall performance and effectiveness of the system being studied. As a result, Br radicals assist in activating oxygen to produce $\bullet O_2^-$, playing a key role in enhancing the DFF yield. On the contrary, a hybrid material consisting of PPy/ZnO was fabricated through a polymerization method by combining polypyrrole (PPy) with zinc oxide (ZnO) (Gonzalez-Casamachina et al. 2020). Using oxygen as an oxidizing agent and ZnO/PPy as a photocatalyst, visible light was utilized to partially oxidize HMF to FDCA in aqueous solution.

The selectivity of this reaction was found to be 30.7% at 25°C, although it decreased with increasing temperature. In the composite photocatalyst, poly(pyrrole) (PPy) played an essential role by serving as both an electron donor and a photosensitizer. This enhancement boosted the speed of electron transfer within the substance, broadening the scope of visible light response and ultimately enhancing the catalytic capability. In summary, the study demonstrated the effectiveness of utilizing ZnO/PPy as a photocatalyst for the partial oxidation of HMF to FDCA under visible light. The selectivity of the reaction was influenced by temperature, with higher temperatures leading to decreased selectivity. PPy's role in the composite material was highlighted as essential for enhancing the electron transfer rate and broadening the response range of visible light. This research provides valuable insights into the development of efficient and selective photocatalysts for organic transformations.

3.4 BIOCATALYTIC HMF-TO-FDCA OXIDATION

Biocatalytic transformation has the characteristics of mild reaction conditions and high selectivity compared with chemical transformation. In recent years, there has been a growing interest in exploring the biocatalytic conversion of 5-hydroxymethylfurfural (HMF). HMF is a key platform chemical derived from biomass that has potential applications in the production of biofuels and value-added chemicals. Biocatalysis offers a more sustainable and environmentally friendly alternative to traditional chemical methods for converting HMF into useful products. By using enzymes or whole cells as catalysts, biocatalytic processes can achieve high efficiency and selectivity, leading to lower energy consumption and waste generation. One of the key advantages of biocatalysis is the ability to carry out specific transformations under mild conditions, allowing for the production of high-value compounds with minimal by-products. Enzymes can be tailored or engineered to target specific reactions, providing a level of control and precision that is difficult to achieve with chemical catalysts. Furthermore, biocatalytic processes are often more cost-effective and resource-efficient, as enzymes can be recycled and reused multiple times, reducing the need for expensive reagents and improving overall process economics. Overall, the growing interest in the biocatalytic transformation of HMF reflects the increasing recognition of biocatalysis as a powerful tool for sustainable chemical production. By harnessing the power of enzymes and microbial cells, researchers can develop efficient and environmentally friendly processes for converting biomass-derived feedstocks into valuable chemicals. With continued advancements in biocatalyst screening and engineering, the potential for biocatalysis to revolutionize the way we manufacture chemicals from renewable resources is promising. The biotransformation process of HMF was first discovered during fermentation. During the pretreatment of lignocellulose, many compounds were produced that affect the growth and metabolism of microorganisms, thus affecting the yield and efficiency of the fermentation process. It has been found that the use of microbiological biological detoxification and biological elimination can specifically eliminate fermentation inhibitors, without the need to add other chemicals and produce degradable waste (Rosatella et al. 2011; Liu et al. 2004). It can replace physical and chemical detoxification methods, and in this process, many microorganisms have been found that can oxidize HMF.

(1) Microorganism Catalysis

The microbial species that can convert and utilize HMF are different, the HMF oxidative metabolic pathways are similar among these microorganisms. Koopman et al. screened two gene clusters (*hmf*-ABCDE and *hmf*-FGH'H) via transposon mutation, and the functional expression of the two gene clusters in *P. putida* S12 was combined with the 2-furanoformic acid substitution pathway analyzed by Trudgill et al. (Koopman et al. 2010; Trudgill et al. 1969). The metabolic pathway of HMF in *Cupriavidus* HMF14 under aerobic conditions was discovered, and the metabolic pathway was characterized at the gene level. HMF underwent initial oxidation to HMFCA, followed by subsequent oxidation to FDCA facilitated by the FAD-dependent oxidase HMFO.

Zhang et al. took HMF as the only carbon source to study the metabolism of *Amorphotheca resinae* ZN1 under different oxygen levels, which more completely explained the metabolic pathway of Koopman et al. (Ran et al. 2014). Under aerobic conditions, the HMF was first converted to corresponding alcohols and then oxidized to aldehydes under the action of alcohol dehydrogenase or aldehyde reductase, and then continuously oxidized to acid, entering the furfural metabolic pathway, oxidizing to α-ketoglutaric acid and entering the tricarboxylic acid cycle. HMF was degraded into alcohols and could not form carboxylic acids in the absence of oxygen. Therefore, sufficient oxygen is a prerequisite for the complete metabolism of HMF. Koopman et al. functionally expressed the metabolic pathway of HMF in *Cupriavidus* HMF14 in *P. putida* S12, which could theoretically be used in different hosts (Koopman et al. 2010). This provides a theoretical basis for the metabolism of toxic HMF inhibitors by many microorganisms and lays a foundation for further improving the hydrolysis efficiency of lignocellulose. However, the specific properties of some enzymes in this pathway limit the application of this route. For example, the *hmf*-ABC-encoded furanyl-CoA dehydrogenase requires A specific molybdopterin co-factor, which limits the application of the metabolic pathway of furan aldehydes in hosts with different molybdopterin synthesis routes, especially in yeast, which is the most widely used in industry. It cannot synthesize molybdopterin. The oxygen-dependent HMF oxidoreductase can only be used in aerobic processes, while most industrial fermentation is carried out in a microoxygen environment (Wierckx et al. 2011).

(2) Whole-Cell Catalysis

The rise of "green chemistry" has sparked growing interest in the production of large quantities of chemicals through the biocatalytic oxidation of HMF. There are two kinds of biocatalytic oxidation of HMF: whole-cell catalysis and extracellular catalysis by separating enzymes. Wierckx et al. took advantage of the fact that redox enzyme was limited to aerobic conditions, expressed *C.basilensis* redox enzyme in solvent-tolerant strain *P. putilda* 10S12, and used whole cell catalysis to oxidize HMF to obtain 30 g/L FDCA (Koopman et al. 2010). The FDCA yield can reach 97%, and the conversion of HMF is close to 100%, but the reaction time is as long as 144 h.

(3) Enzyme Catalysis

Extracellular enzyme oxidation can act on higher substrate concentration, high product recovery, and low water consumption. Therefore, under the circumstance of clear HMF oxidase gene information, extracellular catalytic oxidation of HMF to prepare downstream fine chemicals via constructing engineering bacteria to express HMF oxidase has increasingly become the focus of research. The extracellular enzyme-catalyzed oxidation of HMF to produce downstream fine furan chemicals also has a good yield, but there are problems such as insufficient substrate concentration, the inability of a single enzyme to complete the complete oxidation of hydroxyl and aldehyde groups, and long conversion time, resulting in the use of extracellular enzyme oxidation process is limited to laboratory research.

In 2012, Krystof et al. proposed the concept of *in situ* oxidation of HMF using peroxic acid. Since peroxidic acid is highly oxidizing and difficult to store and transport, it uses the Baeyer-Villiger reaction principle to provide the acyl group with fat ester and react with hydrogen peroxide under the catalysis of lipase to generate peroxidic acid and oxidize HMF. However, the generated peroxidic acid preferratively oxidizes the aldehyde group rather than the hydroxymethyl group. To completely oxidize the aldehyde group and hydroxyl group of HMF, the researchers used a one-pot, two-step method, first using TEMPO to oxidize HMF to produce DFF intermediates, and then using peroxic acid to oxidize DFF to FDCA. Peroxic acid can completely oxidize DFF (50 mmol/L) to produce FDCA. The FDCA yield can reach 83%. This process achieves the compatibility of solvent systems for chemical and enzyme-catalyzed reactions. When using pure organic reaction, DFF was completely converted to FDCA (48%) and FFCA (52%), which also provides a technology for selective oxidation of FFCA.

Qin et al. improved the Krystof experiment in 2015, which used Galactose oxidase (GO) to replace TEMPO and lipase in tandem oxidation of HMF (Qin et al 2015). GO collaborates with catalase and horseradish peroxidase to oxidize HMF to produce DFF, with the highest yield of 91%. The DFF and remaining HMF were concentrated, extracted by ethyl acetate, and oxidized by lipase to produce FDCA. The FDCA yield of the final product of the one-pot two-step oxidation reaction can reach 88%, which is slightly higher than that of the Krystof reaction (82%), but the reaction time is long. Mckenna et al. achieved better results based on the previous study and adopted the Galactose oxidase M_{3-5} (GOaseM_{3-5}) and Aldehyde oxidase (PaoABC) was the first time to oxidize HMF with two enzymes in one pot to prepare FDCA McKenna et al 2015). GOaseM_{3-5} and PaoABC were used to oxidize HMF with phosphate buffer of pH 7.6 as reaction medium and air oxygen as oxidant at 37 °C for 1 h, and the FDCA yield reached 97%. However, GOaseM_{3-5} has a weak ability to oxygenate HMFCA, so the one-pot method is limited to a lower substrate concentration range. In addition, the authors can completely oxidize HMF by combining the two enzymes in tandem under the same conditions, and the highest substrate concentration can reach 100 mmol/L, which is the highest substrate concentration reported so far, and 50–100 times higher than the previously reported enzymes HMFO (2–4 mmol/L) and AAO (3 mmol/L). It is a green, sustainable, and developable method of chemical catalysis. Compared with AAO, UPO, and HMFO, PaoABC is active on the aldehyde group itself, rather than oxidizing the aldehyde hydrate, thus reducing the reaction time, and can completely transform HMF (50 mmol/L) within 8 h, and the product FDCA yield is greater than 99%.

3.5 CONCLUSION AND PERSPECTIVE

The effective pathway for improving the oxidation of HMF to FDCA and other valuable intermediates has been successfully confirmed. It is essential to thoroughly study and understand the mechanisms behind these processes to optimize production and increase efficiency. Overcoming challenges and barriers in the oxidation reaction of HMF is pivotal for achieving higher yields of FDCA and other valuable products. By implementing improved pathways and strategies, researchers can further advance the synthesis of these important intermediates for various industrial applications. This validation provides a solid foundation for future studies and developments in this area of research. The significance of these compounds extends beyond economic benefits in biomass-based products to include reducing reliance on fossil resources, advancing green chemistry objectives, and promoting environmental conservation. At present, the electrochemical oxidation of HMF provides a hopeful substitute for FDCA synthesis compared to alternative catalytic approaches. Enhancing the yield of FDCA necessitates the use of more robust, stable, and active electrocatalysts. The use of photocatalysis offers a gentle means for valorizing biomass, given its eco-friendly nature and utilization of solar energy. Nonetheless, research findings are limited and warrant further extensive lab investigations. In contrast, biocatalysis entails longer reaction times and harsher reaction conditions. The exploration of large-scale HMF oxidation through electrocatalysis, photocatalysis, and biocatalysis remains a promising area of research. To achieve a harmonious integration of green chemistry and economic feasibility, three main obstacles should be addressed going forward: excessive expenses associated with processing, inadequacies in catalytic mechanisms, and a thorough assessment of these procedures. The production of FDCA heavily relies on the oxidation of HMF. Given that the price of HMF is comparable to, or even surpasses that of FDCA, the raw material itself warrants attention in industrial settings.

Additionally, it is important to further investigate the use of different lignocellulosic biomass sources, including cellulose, glucose, and fructose, for the synthesis of FDCA or other valuable oxidative intermediates in a unified process. Rational design of multifunctional catalysts to facilitate the conversion of lignocellulose resources into target compounds in a single step is imperative. Despite the need for specialized equipment and certain limitations, the economic aspects of green, efficient, and sustainable processes should not be overlooked. There is a need to prioritize large-scale chemical production as many current methods, while technically viable, are not economically feasible. The integration of multidisciplinary approaches encompassing thermo-, photo-, electro-, and biocatalytic mechanisms holds promise for the development of valuable biomass-derived intermediates. By leveraging these diverse approaches and incorporating multifunctional heterogeneous catalysts, progress can be made toward achieving sustainable development goals in the field of biorefinery.

REFERENCES

Barwe, Stefan, Jonas Weidner, Steffen Cychy, Dulce M. Morales, Stefan Dieckhöfer, Dennis Hiltrop, Justus Masa, Martin Muhler, Wolfgang Schuhmann. 2018. "Electrocatalytic oxidation of 5-(Hydroxymethyl)furfural using high-surface-area nickel boride." *Angewandte Chemie International Edition* 57(35). Wiley-VCH Verlag: 11460–11464. doi: 10.1002/anie.201806298.

Cai, Jiaying, Hong Ma, Junjie Zhang, Qi Song, Zhongtian Du, Yizheng Huang, Jie Xu. 2013. "Gold nanoclusters confined in a supercage of Yzeolite for aerobic oxidation of HMF under mild conditions." *Chemistry-A European Journal* 19(42). Wiley-VCH Verlag: 14215–14223. doi: 10.1002/chem.201301735.

Cai, Meng, Sha Ding, Bradley Gibbons, Xiaozhou Yang, Matthew C. Kessinger, Amanda J. Morris. 2020. "Nickel(ii)-modified covalent-organic framework film for electrocatalytic oxidation of 5-hydroxymethylfurfural (HMF)." *Chemical Communications* 56(92). Royal Society of Chemistry: 14361–14364. doi:10.1039/d0cc02206c.

Casanova, Onofre, Sara Iborra, Avelino Corma. 2009. "Biomass into chemicals: Aerobic oxidation of 5-hydroxymethyl-2-furfural into 2,5-furandicarboxylic acid with gold nanoparticle catalysts." *ChemSusChem* 2(12). Wiley-VCH Verlag: 1138–1144. doi: 10.1002/cssc.200900137.

Chadderdon, David J., Le Xin, Ji Qi, Yang Qiu, Phani Krishna, Karren L. Moreb, Wenzhen Li. 2014. "Electrocatalytic oxidation of 5-hydroxymethylfurfural to 2,5-furandicarboxylic acid on supported Au and Pd bimetallic nanoparticles." *Green Chemistry* 16(8). Royal Society of Chemistry: 3778–3786. doi: 10.1039/c4gc00401a.

Chen, Hao, Jiatuan Wang, Yuan Yao, Zihao Zhang, Zhenzhen Yang, Jing Li, Kequan Chen, Xiuyang Lu, Pingkai Ouyang, Jie Fu. 2019. "Cu–Ni bimetallic hydroxide catalyst for efficient electrochemical conversion of 5-hydroxymethylfurfural to 2,5-furandicarboxylic acid." *ChemElectroChem* 6(23). Wiley-VCH Verlag: 5797–5801. doi: 10.1002/celc.201901366.

Deng, Xiaohui, Geyang Xu, Yuejiao Zhang, Lei Wang, Jiujun Zhang, Jianfeng Li, Xianzhu Fu, Jingli Luo. 2021. "Understanding the roles of electrogenerated Co^{3+} and Co^{4+} in selectivity-tuned 5-hydroxymethylfurfural oxidation." *Angewandte Chemie International Edition* 60(37). Wiley-VCH Verlag: 20698–20705. doi: 10.1002/anie.202108955.

Donoeva, Baira, Nazila Masoud, Petra E. de Jongh. 2017. "Carbon support surface effects in the gold-catalyzed oxidation of 5-hydroxymethylfurfural." *ACS Catalysis* 7(7). American Chemical Society: 4581–4591. doi: 10.1021/acscatal.7b00829.

Gonzalez-Casamachin, Diego Alexander, Javier Rivera De la Rosa, Carlos J. Lucio–Ortiz, Ladislao Sandoval-Rangel, Carlos D. García. 2020. "Partial oxidation of 5-hydroxymethylfurfural to 2,5-furandicarboxylic acid using O_2 and a photocatalyst of a composite of ZnO/PPy under visible light: Electrochemical characterization and kinetic analysis." *Chemical Engineering Journal* 393. Elsevier B.V.: 124699. doi: 10.1016/j.cej.2020.124699.

Gorbanev, Yury Y., Søren Kegnæs, Anders Riisager. 2011. "Selective aerobic oxidation of 5-hydroxymethylfurfural in water over solid ruthenium hydroxide catalysts with magnesium-based supports." *Catalysis Letters* 141(12). Springer: 1752–1760. doi: 10.1007/s10562-011-0707-y.

Gupta, Navneet Kumar, Shun Nishimura, Atsushi Takagaki, Kohki Ebitani. 2011. "Hydrotalcite-supported gold-nanoparticle-catalyzed highly efficient base-free aqueous oxidation of 5-hydroxymethylfurfural into 2,5-furandicarboxylic acid under atmospheric oxygen pressure." *Green Chemistry* 13(4). Royal Society of Chemistry: 824–827. doi: 10.1039/c0gc00911c.

Hansen, Thomas S., Irantzu Sádaba, Eduardo J. García-Suárez, Anders Riisager. 2013. "Cu catalyzed oxidation of 5-hydroxymethylfurfural to 2,5-diformylfuran and 2,5-furandicarboxylic acid under benign reaction conditions." *Applied Catalysis A-General* 456. Elsevier B.V.: 44–50. doi: 10.1016/j.apcata.2013.01.042.

Heidary, Nina, Nikolay Kornienko. 2020. "Electrochemical biomass valorization on gold-metal oxide nanoscale heterojunctions enables investigation of both catalyst and reaction dynamics with operando surface-enhanced Raman spectroscopy." *Chemical Science* 11(7). Royal Society of Chemistry: 1798–1806. doi: 10.1039/d0sc00136h.

Holzhäuser, Fabian Joschka, Tobias Janke, Fatma Öztas, Cornelia Broicher, Regina Palkovits. 2020. "Electrocatalytic oxidation of 5-hydroxymethylfurfural into the Monomer 2,5-furandicarboxylic acid using mesostructured nickel oxide." *Advanced Sustainable Systems* 4(10). Wiley-VCH Verlag: 1900151. doi: 10.1002/adsu.201900151.

Hu, Kang, Man Zhang, Biying Liu, Zhiyu Yang, Ruiqi Li, Kai Yan. 2021. "Efficient electrochemical oxidation of 5-hydroxymethylfurfural to 2,5-furandicarboxylic acid using the facilely synthesized 3D porous WO_3/Ni electrode." *Molecular Catalysis* 504. Elsevier B.V.: 111459. doi: 10.1016/j.mcat.2021.111459.

Huang, Xin, Jinliang Song, Manli Hua, Zhenbing Xie, Shuaishuai Liu, Tianbin Wu, Guanying Yang, Buxing Han. 2020. "Enhancing the electrocatalytic activity of CoO for the oxidation of 5-hydroxymethylfurfural by introducing oxygen vacancies." *Green Chemistry* 22(3). Royal Society of Chemistry: 843–849. doi: 10.1039/c9gc03698a.

Jain, Archana, Subash C. Jonnalagadda, Kandalam V. Ramanujachary, Amos Mugweru. 2015. "Selective oxidation of 5-hydroxymethyl-2-furfural to furan-2,5-dicarboxylic acid over spinel mixed metal oxide catalyst." *Catalysis Communications* 58. Elsevier B.V.: 179–182. doi: 10.1016/j.catcom.2014.09.017.

Kandambeth, Sharath, Kaushik Dey, Rahul Banerjee. 2018. "Covalent organic frameworks: Chemistry beyond the structure." *Journal of the American Chemical Society* 141(5). American Chemical Society: 1807–1822. doi: 10.1021/jacs.8b10334.

Kang, Myung Jong, Heesun Park, Jonggeon Jegal, Sung Yeon Hwang, Young Soo Kang, Hyun Gil Cha. 2019. "Electrocatalysis of 5-hydroxymethylfurfural at cobalt based spinel catalysts with filamentous nanoarchitecture in alkaline media." *Applied Catalysis B: Environmental* 242. Elsevier B.V.: 85–91. doi: 10.1016/j.apcatb.2018.09.087.

Kang, Myung Jong, Hye Jin Yu, Hyun Sung Kim, Hyun Gil Cha. 2020. "Deep eutectic solvent stabilised Co–P films for electrocatalytic oxidation of 5-hydroxymethylfurfural into 2,5-furandicarboxylic acid." *New Journal of Chemistry* 44(33). Royal Society of Chemistry: 14239–14245. doi: 10.1039/d0nj01426e.

Koopman, Frank, Nick Wierckx, Johannes H. de Winde, Harald J. Ruijssenaars. "Identification and characterization of the furfural and 5-(hydroxymethyl)furfural degradation pathways of Cupriavidus basilensis HMF14." *PNAS* 107(11). American Chemical Society: 4919–4924. doi: 10.1073/pnas.0913039107.

Koopman, Frank, Nick Wierckx, Johannes H. de Winde, Harald J. Ruijssenaars. 2010. "Efficient whole-cell biotransformation of 5-(hydroxymethyl)furfural into FDCA, 2,5-furandicarboxylic acid." *Bioresource Technology* 101(16). Elsevier B.V.: 6291–6296. doi: 10.1016/j.biortech.2010.03.050.

Kröger, Martin, Ulf Prüße, Klaus-Dieter Vorlop. 2000. "A new approach for the production of 2,5–furandicarboxylic acid by in situ oxidation of 5–hydroxymethylfurfural starting from fructose." *Topics in Catalysis* 13. Springer: 237–242. doi: 10.1023/A:1009017929727

Kunnikuruvan, Sooraj, Nisanth N. Nair. 2019. "Mechanistic insights into the Brønsted acid-catalyzed dehydration of *β*-d-glucose to 5-hydroxymethylfurfural under ambient and subcritical conditions." *ACS Catalysis* 9(9). American Chemical Society: 7250–7263. doi: 10.1021/acscatal.9b03298.

Liu, Biying, Shaojun Xu, Man Zhang, Xin Li, Donato Decarolis, Yuqian Liu, Yuchen Wang, Emma K. Gibson, C. Richard A. Catlow, Kai Yan. 2021. "Electrochemical upgrading of biomass-derived 5-hydroxymethylfurfural and furfural over oxygen vacancy-rich NiCoMn-layered double hydroxides nanosheets." *Green Chemistry* 23(11). Royal Society of Chemistry: 4034–4043. doi: 10.1039/d1gc00901j.

Liu, Xinyu, Min Zhang, Zhaohui Li. 2020. "CoOx-MC (MC = Mesoporous Carbon) for highly efficient oxidation of 5-hydroxymethylfurfural (5-HMF) to 2,5-furandicarboxylic acid (FDCA)." *ACS Sustainable Chemistry & Engineering* 8(12). American Chemical Society: 4801–4808. doi: 10.1021/acssuFigureng.9b07443.

Liu, Z. L., P. J. Slininger, B. S. Dien, M. A. Berhow, C. P. Kurtzman, S. W. Gorsich. 2004. "Adaptive response of yeasts to furfural and 5-hydroxymethylfurfural and new chemical evidence for HMF conversion to 2,5-bis-hydroxymethylfuran." *Microbiology Biotechnology* 31(8). Oxford University Press: 345–352. doi:10.1007/s10295-004-0148-3.

Lu, Yuxuan, Chungli Dong,Yucheng Huang, Yuqin Zou, Zhijuan Liu, Yanbo Liu, Yingying Li, Nihan He, Jianqiao Shi, Shuangyin Wang. 2020. "Identifying the geometric site dependence of spinel oxides for the electrooxidation of 5-hydroxymethylfurfural." *Angewandte Chemie International Edition* 59(43). Wiley-VCH Verlag: 19215–19221. doi: 10.1002/anie.202007767.

Lu, Yuxuan, Chungli Dong, Yucheng Huang, Yuqinzou, Yanboliu, Yingying Li, Nana Zhang, Wei Chen, Ling Zhou, Hongzhen Lin, Shuangyin Wang. 2020. "Hierarchically nanostructured NiO-Co_3O_4 with rich interface defects for the electro-oxidation of 5-hydroxymethylfurfural." *Science China Chemistry* 63(7). Science Press: 980–986. doi: 10.1007/s11426-020-9749-8.

Luo, Jian, Qiuhua Zhang, Javier Garcia-Martinez, Steven L. Suib. 2008. "Adsorptive and acidic properties, reversible lattice oxygen evolution, and catalytic mechanism of cryptomelane-type manganese oxides as oxidation catalysts." *Journal of the American Chemical Society* 130(10). American Chemical Society: 3198–3207. doi: 10.1021/ja077706e.

McKenna, Shane M., Silke Leimkühler, Susanne Herter, Nicholas J. Turner, Andrew J. Carnell. 2015. "Enzyme cascade reactions: Synthesis of furandicarboxylic acid (FDCA) and carboxylic acids using oxidases in tandem." *Green Chemistry* 17(6). Royal Society of Chemistry: 3271–3275. doi: 10.1039/c5gc00707k.

Mei, Nan, Bing Liu, Judun Zheng, Kangle Lv, Dingguo Tang, Zehui Zhang. 2015. "A novel magnetic palladium catalyst for the mild aerobic oxidation of 5-hydroxymethylfurfural into 2,5-furandicarboxylic acid in water." *Catalysis Science & Technology* 5(6). Royal Society of Chemistry: 3194–3202. doi: 10.1039/c4cy01407c.

Meng, Ye, Song Yang, Hu Li. 2022. "Electro- and photocatalytic oxidative upgrading of biobased 5-hydroxymethylfurfural." *ChemSusChem* 15(13). Wiley-VCH Verlag: e202102581. doi: 10.1002/cssc.202102581.

Nam, Do-Hwan, Brandon J. Taitt, Kyoung-Shin Choi. 2018. "Copper-based catalytic anodes to produce 2,5-furandicarboxylic acid, a biomass-derived alternative to terephthalic acid." *ACS Catalysis* 8(2). American Chemical Society: 1197–1206. doi: 10.1021/acscatal.7b03152.

Nishimura, Shun, Yusuke Yakita, Madoka Katayama, Koichi Higashimine, Kohki Ebitani. 2013. "The role of negatively charged Au states in aerobic oxidation of alcohols over hydrotalcite supported AuPd nanoclusters." *Catalysis Science & Technology* 3(2). Royal Society of Chemistry: 351–359. doi: 10.1039/c2cy20244a.

Pasini, Thomas, Marco Piccinini, Magda Blosi, Rosa Bonelli, Stefania Albonetti, Nikolaos Dimitratos, Jose A. Lopez-Sanchez, Meenakshisundaram Sankar, Qian He, Christopher J. Kiely, Graham J. Hutchings, Fabrizio Cavani. 2011. "Selective oxidation of 5-hydroxymethyl-2-furfural using supported gold-copper nanoparticles." *Green Chemistry* 13(8). Royal Society of Chemistry: 2091–2099. doi: 10.1039/c1gc15355b.

Pham, Hoang Minh, Myung Jong Kang, Kyung-An Kim, Chang Gyu Im, Sung Yeon Hwang, Hyun Gil Cha. 2020. "Which electrode is better for biomass valorization: $Cu(OH)_2$ or CuO nanowire?" *Korean Journal of Chemistry Engineering* 37(3). Korean Institute Chemical Engineers: 556–562. doi: 10.1007/s11814-020-0474-9.

Qin,Yezhi Yanmei Li, Minhua Zong, Hong Wu, Ning Li. 2015. "Enzyme-catalyzed selective oxidation of 5-hydroxymethylfurfural (HMF) and separation of HMF and 2,5-diformylfuran using deep eutectic solvents." *Green Chemistry* 17(7). Royal Society of Chemistry: 3718–3722. doi: 10.1039/c5gc00788g.

Ran, Hong, Jian Zhang, Qiuqiang Gao, Zhanglin Lin, Jie Bao. 2014. "Analysis of biodegradation performance of furfural and 5-hydroxymethylfurfural by Amorphotheca resinaeZN1." *Biotechnology for Biofuels*, 7. BioMed Central: 51. doi: 10.1186/1754-6834-7-51.

Rass, Hicham Ait, Nadine Essayem, Michèle Besson Rass. 2013. "Selective aqueous phase oxidation of 5-hydroxymethylfurfural to 2,5-furandicarboxylic acid over Pt/C catalysts: influence of the base and effect of bismuth promotion." *Green Chemistry* 15(8). Royal Society of Chemistry: 2240–2251. doi: 10.1039/c3gc40727f.

Ribeiro, Marcelo L., Ulf Schuchardt. 2003. "Cooperative effect of cobalt acetylacetonate and silica in the catalytic cyclization and oxidation of fructose to 2,5-furandicarboxylic acid." *Catalysis Communications* 4(2). Elsevier B.V.: 83–86. doi: 10.1016/S1566-7367(02)00261-3.

Rosatella, Andreia A., Svilen P. Simeonov, Raquel F. M. Frade, Carlos A. M. Afonso. 2011. "5-Hydroxymethylfurfural (HMF) as a building block platform: Biological properties, synthesis and synthetic applications." *Green Chemistry* 13(4). Royal Society of Chemistry: 754–793. doi: 10.1039/c0gc00401d.

Saha, Basudeb, Dinesh Gupta, Mahdi M. Abu-Omar, Arindam Modak, Asim Bhaumik. 2013. "Porphyrin-based porous organic polymer-supported iron (III) catalyst for efficient aerobic oxidation of 5-hydroxymethyl-furfural into 2,5-furandicarboxylic acid." *Journal of Catalysis* 299. Academic Press Inc Elsevier Science: 316–320. doi: 10.1016/j.jcat.2012.12.024.

Sahu, Ramakanta, Paresh L. Dhepe. 2014. "Synthesis of 2,5-furandicarboxylic acid by the aerobic oxidation of 5-hydroxymethyl furfural over supported metal catalysts."

Reaction Kinetics Mechanisms and Catalysis 112. Springer: 173–187. doi: 10.1007/s11144-014-0689-z.

Siankevich, Sviatlana, Georgios Savoglidis, Zhaofu Fei, Gabor Laurenczy, Duncan T. L. Alexander, Ning Yan, Paul J. Dyson. 2014. "A novel platinum nanocatalyst for the oxidation of 5-hydroxymethylfurfural into 2,5-furandicarboxylic acid under mild conditions." *Journal of Catalysis* 315. Academic Press Inc Elsevier Science: 67–74. doi: 10.1016/j.jcat.2014.04.011.

Siyo, Baraa, Matthias Schneider, Marga-Martina Pohl, Peter Langer, Norbert Steinfeldt. 2014. "Synthesis, characterization, and application of PVP-Pd NP in the aerobic oxidation of 5-hydroxymethylfurfural (HMF)." *Catalysis Letters* 144(3). Springer: 498–506. doi: 10.1007/s10562-013-1186-0.

Taitt, Brandon J., Do-Hwan Nam, Kyoungshin Choi. 2019. "A comparative study of nickel, cobalt, and iron oxyhydroxide anodes for the electrochemical oxidation of 5-hydroxymethylfurfural to 2,5-furandicarboxylic acid." *ACS Catalysis* 9(1). American Chemical Society: 660–670. doi: 10.1021/acscatal.8b04003.

Trudgill P. W. 1969. "The metabolism of 2-furoic acid by pseudomonas F2." *Biochemistry* 113(4). American Chemical Society: 577–587. doi: 10.1042/bj1130577.

Vuyyuru, Koteswara Rao, Peter Strasser. 2012. "Oxidation of biomass derived 5-hydroxymethylfurfural using heterogeneous and electrochemical catalysis." *Catalysis Today* 195(1). Elsevier B.V.: 144–154. doi: 10.1016/j.cattod.2012.05.008.

Wang, Changlong, Hans-Josef Bongard, Mingquan Yu, Ferdi Schüth. 2021. "Highly ordered mesoporous Co_3O_4 electrocatalyst for efficient, selective, and stable oxidation of 5-hydroxymethylfurfural to 2,5-furandicarboxylic acid." *ChemSusChem* 14(23). Wiley-VCH Verlag: 5199–5206. doi: 10.1002/cssc.202002762.

Wang, Jialu, Xian Zhang, Guozhong Wang, Yunxia Zhang, Haimin Zhang. 2020. "Sustainable 2,5-furandicarboxylic synthesis by a direct 5-hydroxymethylfurfural fuel cell based on a bifunctional PtNiSx catalyst." *Chemical Communications* 56(88). Royal Society of Chemistry: 13611–13614. doi: 10.1039/d0cc06087a.

Wierckx, Nick, Frank Koopman, Harald J. Ruijssenaars, Johannes H. de Winde. 2011. "Microbial degradation of furanic compounds: biochemistry, genetics, and impact". *Applied Microbiology Biotechnology* 92(6). Springer: 1095–1105. doi: 10.1007/s00253-011-3632-5.

Xu, Shuai, Peng Zhou, Zehui Zhang, Changjun Yang, Bingguang Zhang, Kejian Deng, Steven Bottle, Huaiyong Zhu. 2017. "Selective oxidation of 5-hydroxymethylfurfural to 2,5-furandicarboxylic acid using O_2 and a photocatalyst of Co-thioporphyrazine bonded to g-C_3N_4." *Journal of the American Chemical Society* 139(41). American Chemical Society: 14775–14782. doi: 10.1021/jacs.7b08861.

Yang, Bo, Wenwei Hu, Feifei Wan, Chao Zhang, Zaihui Fu, Anqun Su, Mengke Chen, Yachun Liu. 2020. "Adjusting effect of additives on decatungstate-photocatalyzed HMF oxidation with molecular oxygen under visible light illumination." *Chemical Engineering Journal* 396. Elsevier B.V.: 125345. doi: 10.1016/j.cej.2020.125345.

Zhang, Yingchuan, Guangri Jia, Wenchao Wang, Liqun Jiang, Zhengxiao Guo. 2024. "Photocatalytic upgrading of 5-hydroxymethylfurfural–aerobic or anaerobic?." *Green Chemistry* 26(6). Royal Society of Chemistry: 2949–2966. doi: 10.1039/D3GC04814D.

Zhang, Zehui, Judun Zhen, Bing Liu, Kangle Lva, Kejian Deng. 2015. "Selective aerobic oxidation of the biomass-derived precursor 5-hydroxymethylfurfural to 2,5-furandicarboxylic acid under mild conditions over a magnetic palladium nanocatalyst." *Green Chemistry* 17(2). Royal Society of Chemistry: 1308–1317. doi: 10.1039/c4gc01833h.

Zhou, Bo, Chungli Dong, Yucheng Huang, Nana Zhang, Yandong Wu, Yuxuan Lu, Xu Yue, Zhaohui Xiao, Yuqin Zou, Shuangyin Wang. 2021. "Activity origin and alkalinity effect of electrocatalytic biomass oxidation on nickel nitride." *Journal of Energy Chemistry* 61. Elsevier B.V.: 179–185. doi: 10.1016/j.jechem.2021.02.026.

Zhou, Hua, Xu, Huanghui, Liu Yun. 2019. "Aerobic oxidation of 5-hydroxymethylfurfural to 2,5-furandicarboxylic acid over Co/Mn-lignin coordination complexes-derived catalysts." *Applied Catalysis B-Environmental* 244. Elsevier B.V.: 965–973. doi: 10.1016/j.apcatb.2018.12.046.

Zhu, Biao, Yang Qin, Jiawei Du, Fazhi Zhang, Xiaodong Lei. 2021. "Ammonia etching to generate oxygen vacancies on $CuMn_2O_4$ for highly efficient electrocatalytic oxidation of 5-hydroxymethylfurfural." *ACS Sustainable Sustainable Chemistry & Engineering* 9(35). American Chemical Society: 11790–11797. doi: 10.1021/acssuFigureng.1c03256.

4 (Photo)Catalytic Oxidation of 5-Hydroxymethylfurfural to 2,5-Diformylfuran

Neeraj Sharma and
Shunmugavel Saravanamurugan

4.1 INTRODUCTION

The contemporary progression of civilization has led to a significant rise in the utilization of conventional fossil fuels, resulting in a pronounced energy crisis and the emergence of environmental pollution concerns. Recently, there has been a notable surge in interest in the pursuit of renewable energy derived from the plentiful source of lignocellulosic biomass (LCB) (Aggarwal et al. 2021; Saini et al. 2022). Considerable attention has been devoted to the compound 5-hydroxymethylfurfural (HMF), which can be obtained by the dehydration of, for example, glucose – the cellulosic component of LCB. HMF is a platform chemical with immense potential, referred to as a "sleeping giant," ever since its discovery in the late 1800s (Zakrzewska et al. 2011). Upon transformation of this platform molecule, either by reductive or oxidative processes, a wide range of value-added chemicals can be produced (Figure 4.1). For example, hydrogenation of HMF can give various chemicals, such as 2,5-dimethylfuran (DMF), 2,5-dihydroxymethylfuran (DHMF) and 2,5-dimethyltetrahydrofuran (DMTHF), with various applications in industrial polymerization or esterification processes including artificial fiber, polyurethane foam and resin (Sun et al. 2023). On the other hand, oxidation of HMF can produce 2,5-furandicarboxylic acid (FDCA), which has been listed as one of the top-12 value-added chemicals from biomass by the US Department of Energy due to its huge application in replacing terephthalic acid in polyethylene terephthalate (PET) plastic to produce polyethylene furanoate (PEF) plastic (Bozell et al. 2010). The transformation of HMF into FDCA occurs via two potential intermediates: 2,5-diformylfuran (DFF) or 5-hydroxymethyl-2-furancarboxylic acid (HMFCA). However, the transformation of HMF to the above-mentioned furanic products typically requires catalytic engineering and harsh reaction conditions, including elevated temperature (≥100 °C), high-pressure oxygen and strong oxidants (e.g. TEMPO). In contrast, photocatalysis exhibits an alternative and environmental-friendly catalytic approach that

 DOI: 10.1201/9781003464839-4

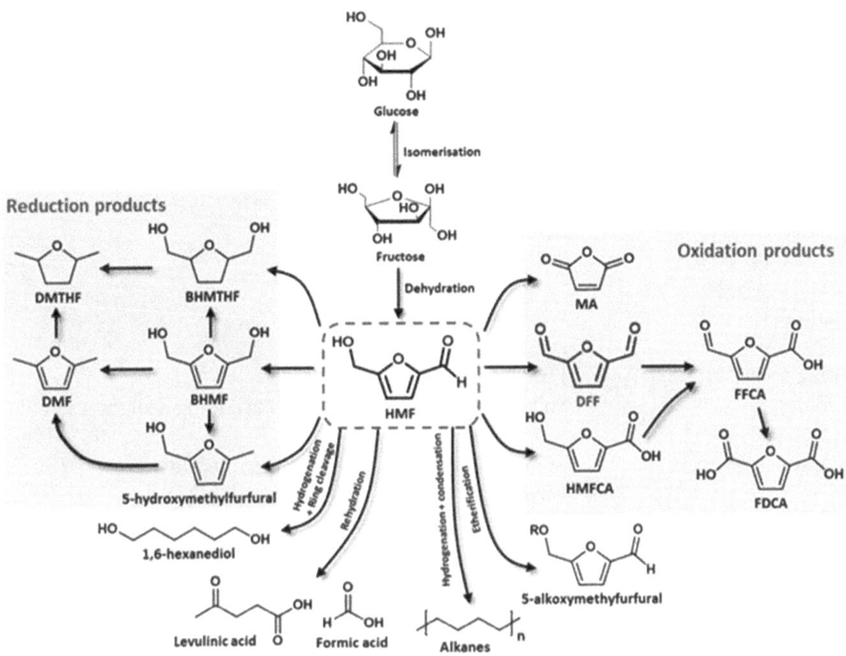

FIGURE 4.1 Transformation of HMF to various value-added products.

enables cost-effective and net-zero transformations of chemicals using solar energy, which is mainly discussed in this chapter (Zhang et al. 2024).

4.2 INDUSTRIAL RELEVANCE

This chapter exclusively focuses on the production of DFF, a valuable intermediate to FDCA formation, via the photocatalytic oxidation of HMF. As mentioned earlier, HMF is a crucial platform chemical with a market value of US$ 62 million in 2022 and is projected to reach US$ 69 million by 2029 at a compound annual growth rate (CAGR) of 1.5%. However, nowadays, the main focus of the transformation of HMF is towards FDCA because of its substantial global market volume and essential precursor for producing PEF plastic (www.24chemicalresearch.com/reports/201094/hydroxymethylfurfural-market-2023-2030-893). At the same time, considerable attention has also been paid to DFF production due to its versatility as both a monomer and a starting material in the production of many substances, such as foams, adhesives, furan urea resins, macrocyclic ligands, pharmaceuticals, fluorescent materials, and many polymeric materials (Pal et al. 2019). DFF is a furan derivative that exists as a crystalline solid at room temperature, and its chemical and physical properties are listed in Table 4.1 (Dai 2021).

Although remarkable efforts have been devoted to DFF production from various starting materials such as cellulose, glucose, fructose or HMF, owing to the high

TABLE 4.1
The Chemical and Physical Properties of DFF (Dai 2021)

Item	Content
CAS registry number	823-82-5
Chemical name	2,5-Furandicarbaldehyde or 2,5-Diformylfuran
Molecular formula	$C_6H_4O_3$
Molecular weight	124.094
Description	White or yellow solid
Boiling point	276.8 °C at 760 mmHg
Melting point	110°C
Solubility	Soluble in chloroform and ethyl acetate and slightly soluble in water
Condition to Avoid	Light Sensitive, Air Sensitive, Heat Sensitive
Density	1.298 g cm^{-3}
Flash point	124.1°C
Refractive index	1.585
Enthalpy of Vaporization	51.5 ± 3.0 kJ/mol
Molar Refractivity	32.1 ± 0.3 cm^3
Surface Tension	47.9 ± 3.0 dyne/cm
Applications	adhesives, foams, binders, new furan materials, and drug synthesis
PubChem Substance ID	87568835

price of starting material or the uneconomic overall process, the large-scale production of DFF is not widely reported. Still, a few companies, such as AVA Biochem, Amadis Chemical, Suzhou Gaide and Ningbo Biomass & Biotechnology, produce and sell DFF on a large scale for industrial applications. (www.verifiedmarketreports.com/product/diformylfuran-dff-market/) Using HMF as a starting material, Shaanxi Dideu Medichem Co. Ltd is producing DFF at €1000 per tonne, which is quite uneconomical. Therefore, it is necessary to find sustainable, greener, and cost-effective processes to produce DFF on a large scale (Dawes 2022).

4.3 BACKGROUND ON THE OXIDATION OF HMF TO DFF

Several approaches have been employed in the oxidation of HMF to DFF, such as the use of homogeneous oxidants (e.g. NaOCl or pyridinium chlorochromate), which are often used in stoichiometric amounts and lead to the generation of various by-products, wastes and environmental pollution (Shinde et al. 2019). Consequently, various catalytic approaches have been explored to enhance this upgrading process, considering environmental and economic factors, and all have their advantages and disadvantages, as shown in Figure 4.2. Many investigations have been conducted on converting HMF to DFF using the traditional thermocatalytic oxidation approach, achieving a more than 95% DFF yield (Antonyraj et al. 2013; Lin et al. 2021). However, these thermocatalytic reactions are usually performed at high reaction

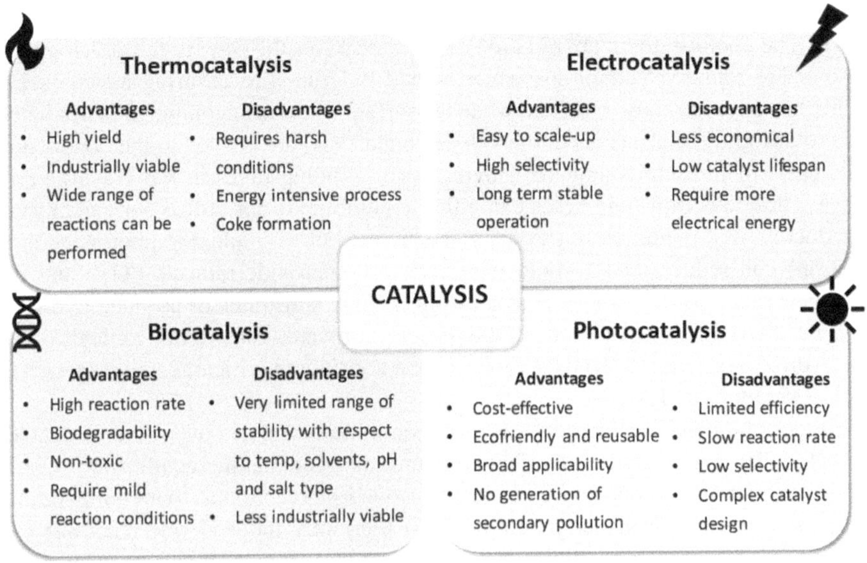

FIGURE 4.2 Advantages and disadvantages of various catalytic methods.

temperatures (80–150°C) and/or pressures (up to 30 bar) using high boiling solvents (e.g. DMSO or DMF), thus causing considerable energy consumption that is not considered environmentally friendly (Li et al. 2021; Pal et al. 2020).

In contrast to the thermocatalytic approach, the biocatalytic oxidation of HMF to DFF exhibits a reduced reaction temperature while maintaining a significantly high degree of product selectivity (Zhang et al. 2021). Nevertheless, several formidable concerns are being confronted, such as the high cost associated with enzyme catalyst construction, the generation of waste, and the stability and recyclability issues (Su et al. 2021). Other than these, another approach for catalytic upgradation is electrocatalytic oxidation, which is an economical and clean strategy using electrode potential as the main driving force under mild conditions. Although using this approach, the reaction rate and selectivity of the target product can be tuned by conveniently controlling the electrode potential, circuit current, and other parameters. However, using costly electrodes, conductive and stable electrolytes, and fabricating suitable reactors for the particular type of reaction are significant roadblocks to this approach (Novaes et al. 2021; Meng et al. 2022). Therefore, based on the approaches mentioned above, it is necessary to explore and understand alternative novel, green, and economic strategies to upgrade bio-based platform chemicals to value-added chemicals under mild conditions.

4.4 SIGNIFICANCE OF PHOTOCATALYSIS AND DFF PRODUCTION

One readily accessible, abundant, inexpensive, safe, and plentiful energy source is sunlight. Due to the fact that only sunlight is required for the completion of the reaction, "photocatalysis" is considered to be a low-cost and eco-friendly alternative option. The photocatalytic upgradation of biomass paves the way for the development of solar bio-refinery technology, which would be crucial to ensuring a sustainable future. In a photocatalytic reaction, when the surface of the semiconductor is irradiated with solar light, the electrons in the valence band (VB) are excited to the conduction band (CB) of the semiconductor, leaving corresponding holes in VB (Zhang et al. 2019). These electrons and holes can either recombine if not utilized or can engage in reduction and oxidation reactions, respectively. For example, the photogenerated electrons can reduce the O_2 molecules to form superoxide radicals ($^{\bullet}O_2^-$) and the photogenerated holes extract electrons from organic substrates or produce hydroxyl radicals ($^{\bullet}OH$) as strong oxidants in the presence of water. Superoxide radicals ($^{\bullet}O_2^-$) and hydroxyl radicals ($^{\bullet}OH$) have always been considered critical reactive species in photo-oxidative catalysis.

Further, the considerable advantage of using photocatalysis over other processes is that under light irradiation, the apparent thermodynamic equilibrium would change and favor the occurrence of the corresponding reaction at room temperature. Therefore, the side products generated from thermally induced side reactions can be easily avoided as the reaction occurs at room temperature. Driven by the benefits mentioned above, photocatalysis has been significantly developed in recent decades and is widely used in many applications, such as pollutant degradation, water splitting, biomass upgrading, and organic synthesis. Hence, photocatalysis emerges as a particularly appealing approach for the selective oxidation and upgradation of HMF to DFF (Granone et al. 2018).

The first synthesis of DFF was carried out in 1985, resulting in a 71% yield, and this process entailed three sequential steps: aldehyde protection, formylation, and deprotection, using furfural as the starting material (Carpenter et al. 1985). However, such synthesis methods have several downsides, such as multistep synthesis, good to moderate yields, and the use of hazardous and polluting reagents. Therefore, greener, more sustainable, renewable, and more economic catalytic approaches are explored for DFF production. As discussed earlier, the photocatalytic approach seems promising for sustainably producing DFF out of all catalytic methods. Therefore, this chapter systematically and comprehensively summarizes the recent advances in the photocatalytic oxidation of HMF to DFF (Scheme 4.1). Also, the catalytic performance of different photocatalysts, including metal oxides, doped metal oxides, heterojunctions, carbon-based materials and metal sulfides photocatalysts, is delineated.

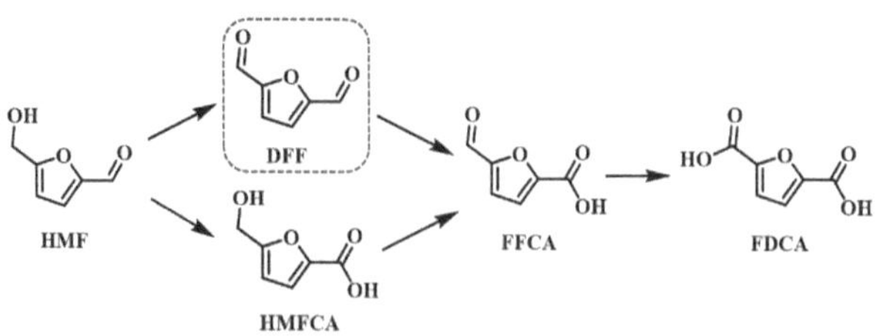

SCHEME 4.1 Oxidation of HMF to DFF and other value-added products.

4.5 PHOTOCATALYSTS FOR THE SELECTIVE OXIDATION OF HMF TO DFF

In the photocatalytic oxidation process of HMF, the considerable reactivity shown by the superoxide radicals ($^{\bullet}O_2^-$) and hydroxyl radicals ($^{\bullet}OH$) species often results in uncontrolled oxidation. Over the course of recent years, to control these radicals and the oxidation process, various photocatalysts have been developed, which can be classified into four major categories: metal oxides, metal sulfides, carbon-based materials and modified photocatalysts. The catalytic performance of these catalysts under various reaction conditions is discussed in detail, in addition to the general reaction mechanisms, as shown in Figure 4.3.

4.5.1 Metal Oxide-Based Photocatalysts for the Oxidation of HMF to DFF

Since the discovery of the Fujishima and Honda effect in 1972, TiO_2 has garnered massive attention for its application in various fields, such as photo-assisted H_2

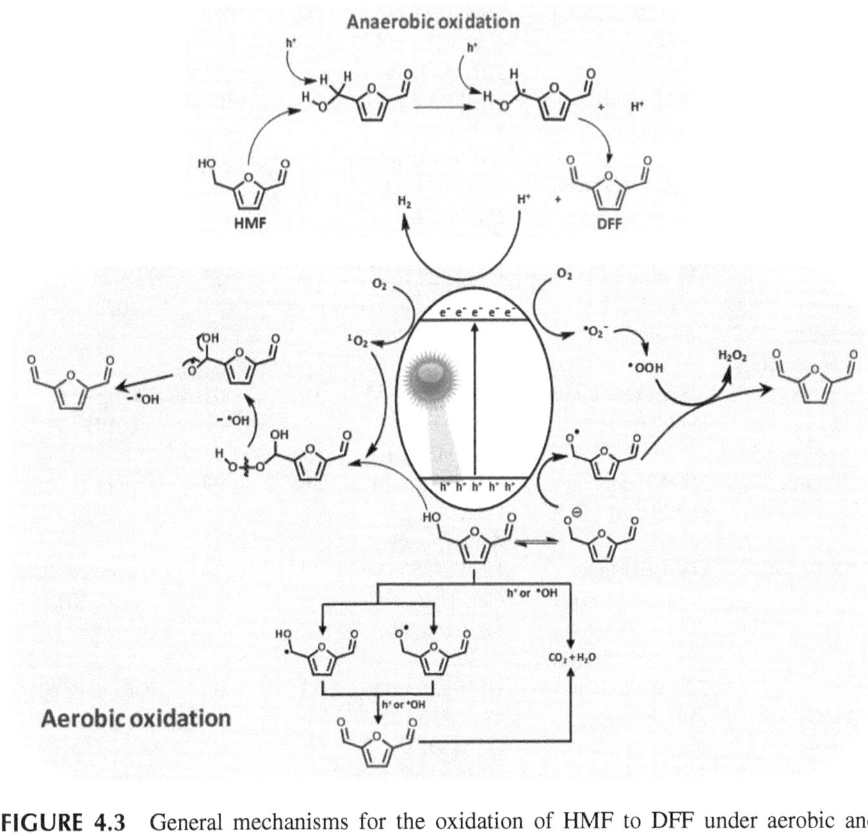

FIGURE 4.3 General mechanisms for the oxidation of HMF to DFF under aerobic and anaerobic conditions.

evolution, pollutant degradation and photocatalytic reactions (Fujishima and Honda 1972). The use of TiO_2 for the photocatalytic oxidation of biomass-derived furanics, such as HMF, was first studied in 2013 (Table 4.2) (Yurdakal et al. 2013). It was observed that the different crystal phases of TiO_2 oxidize HMF to DFF but with low selectivity (~10–25%) due to the uncontrolled oxidation and direct mineralization of HMF to CO_2 due to the formed hydroxyl radicals (•OH). To improve the selectivity of DFF, thermally treated nitrogen-doped TiO_2 was prepared and found that doping of N onto the surface of TiO_2 facilitates the inhibition of the generation of hydroxyl radicals (•OH), thus reducing the decomposition of DFF and enhancing its selectivity (30–40%) to some extent (Krivtsov et al. 2017). To enhance the DFF yield and selectivity further, attempts have been made by altering the synthesis approaches of

TABLE 4.2
Photocatalytic Oxidation of HMF to DFF with Metal Oxide-Based Materials

Catalyst	Lamp	Reaction Conditions	Conv. (%)	Sel. (%)	Ref.
Anatase TiO_2	Fluorescent lamp (λ = 365 nm)	HMF: 0.5 mM Cat: 0.2 g L^{-1} Air, H_2O, 4h	20	26	(Yurdakal et al. 2013)
N-TiO_2	500W lamp (λ = 365 nm)	HMF: 3.2 mM Cat: 1 g L^{-1} Air, H_2O, 4h	NA	30–40	(Krivtsov et al. 2017)
SGH-TiO_2	green LED lamp (λ = 515 nm)	HMF: 1 mM Cat: 1 g L^{-1} O_2, CH_3CN, 4h	59	87	(Khan et al. 2021)
TN-4 (vacancy-induced anatase TiO_2)	6W blue LED lamp (LEDs, λ = 365 nm)	HMF: 25 mM Cat: 0.5 g L^{-1} Air, H_2O, 24h	75	75	(Jang et al. 2023)
Nb_2O_5-800	300W xenon lamp (>400 nm)	HMF: 0.1 mM Cat: 10 g L^{-1} O_2, $PhCF_3$, 10h	19	91	(Zhang et al. 2017)
Ce-NbO-3	300W xenon lamp (>420 nm)	HMF: 20 mM Cat: 10 g L^{-1} Air, H_2O, 6h	36	99	(Zou et al. 2021)
MnO_2	UV LED light (λ=365 nm)	HMF: 0.5 mM Cat: 1 g L^{-1} O_2, ACN, 4h, 39 °C	99	100	(Giannakoudakis et al. 2019)
Fe(III)/Bi_2MoO_6	500 W Xe lamp (>400 nm)	HMF: 20 mM Cat: 10 g L^{-1} O_2, H_2O, 8h	33	95	(Xue et al. 2019)
Rose like Bi_2WO_6	300W xenon lamp (>420 nm)	HMF: 31.7 mM Cat: 4 g L^{-1} O_2, ACN, 10h	58	99	(Kumar et al. 2021)

TiO_2, such as electrochemical (Ulyankina et al. 2020) spray-freeze-drying (Lolli et al. 2018) and microemulsion-spray-freeze-drying (Allegri et al. 2020). However, a lower catalytic activity was observed due to higher degradation of HMF with a low DFF selectivity (10–50%) in water as solvent under UV light irradiation.

Pure TiO_2 has a high energy band gap, thus absorbing UV light and generating various unselective radicals, leading to over-oxidation and/or mineralization to CO or CO_2. This indicates the need to reduce the band-gap energy and broaden the absorbance range to visible light by modifying/doping TiO_2 and consequently fine-tuning its oxidation property. In connection with this, Ag nanoparticles supported TiO_2 for the photocatalytic oxidation of HMF under visible light irradiation was reported and found that the 2.5% Ag/TiO_2 reduced using H_2 at a temperature of 473 K was favorable for HMF oxidation due to the enhanced localized surface plasmon resonance (LSPR) effect induced by strong metal support interaction (SMSI) (Zhang et al. 2022). However, DFF was not the desired product, while highly selective HMFCA production (96.7%) was observed due to the oxidation of the carbonyl group of HMF by superoxide radicals ($^{\bullet}O_2^-$) generated in the system. These results provide an idea of how to design a catalyst to produce selective superoxide radicals ($^{\bullet}O_2^-$) rather than hydroxyl radicals ($^{\bullet}OH$) to achieve the desired product with high selectivity while lowering the mineralization of the substrate and product.

Investigating further to improve the selectivity of DFF, anatase TiO_2 was prepared via the sol-gel method and found that substrate HMF adsorbs on the surface hydroxyl groups of TiO_2, thus forming the ligand to metal charge transfer complex (LMCT) and shifting the absorption to the visible region (Khan et al. 2021). The LMCT complex activates TiO_2, giving a high DFF selectivity of 87% with 59% HMF conversion under visible light irradiation (λ = 515 nm). Another promising approach for HMF oxidation with vacancy-induced core-shell anatase TiO_2 nanoparticles via a simple base treatment after synthesis to generate oxygen vacancy densified shell and stoichiometric core was reported, thus accelerating reaction rate by ~580% compared to untreated TiO_2 (Jang et al. 2023). A high yield of DFF with high selectivity in water as a solvent was achieved under UV irradiation after 6 h, but prolonging the reaction time led to further oxidation of DFF to FDCA (~40% yield). It has been speculated that the hydroxyl radicals generated from the by-product H_2O_2 can react with DFF to give FDCA.

Researchers have also explored other photoactive metal oxides, such as Nb_2O_5, ZnO, and MnO_2 for HMF oxidation. Because of the spectral response solely in the UV region and the wide band gap, Nb_2O_5 (Eg>3.2 eV) has been rarely employed as a photocatalyst. However, it was found that HMF could also interact with the surface of Nb_2O_5 to form alcoholate species similar to the TiO_2 system, resulting in the red shift of the absorption edge to 410 nm (Zhang et al. 2017). A high selectivity of DFF (90.6%) with 19.2% of HMF conversion was obtained with benzotrifluoride as the solvent under visible light, better than under UV light. Interestingly, it was observed that as polar solvents such as water can compete with reactants for the adsorption at catalytic sites, less polar solvents such as benzotrifluoride with good oxygen dissolving capacity can enhance photocatalytic activity. Similarly, Ce was doped into the lattice of Nb_2O_5 (Ce-NbO_3) to improve the photocatalytic activity and induce oxygen vacancy (Zou et al. 2021). For the selective oxidation of HMF to DFF, Pt as a co-catalyst with Ce-NbO_3 was employed under visible light irradiation, achieving a nearly quantitative selectivity of DFF (99%) with a 36% yield. The induced oxygen vacancies contributed to the adsorption and activation of HMF and oxygen molecules. Further, these oxygen vacancy sites facilitated the charge transfer process between the reactants and the catalyst under light irradiation, thus forming the LMCT complex.

Apart from TiO_2 and Nb_2O_5, encouragingly, MnO_2 nanorods prepared via an ultra-slow controlled precipitation approach for HMF oxidation was also reported, achieving a 99% yield of DFF without employing any external oxidant or base (Giannakoudakis et al. 2019). MnO_2 is well documented in traditional thermal catalytic oxidation reactions. But, in this case, possessing a large surface area and interstitial spaces on MnO_2 nanorods afforded a higher catalytic activity by facile activation of HMF and oxygen molecules on the catalyst surface. Also, due to the variable oxidation states of manganese, nanorods can undergo redox cycles between Mn^{4+} and Mn^{3+} states, enabling easy electron transfer, the activation of molecules, and higher activity. Further, other metal oxide-based catalysts, such as modified ZnO (ZnO/polypyrrole composite) (Gonzalez-Casamachin et al. 2020) and hydrogen-treated Au-doped ZnO (Zhu et al. 2022), were also reported as photocatalysts for the production of FDCA from HMF. However, these studies showed the transformation of HMF to FDCA proceeding through DFF as an intermediate, indicating the over-oxidation to form FDCA.

Other than the metal oxides mentioned above, Bi-based ternary metal oxides have also garnered attention in recent years for photocatalytic applications. In line with this, Fe(III) grafted Bi_2MoO_6 nanosheets (Fe(III)/BMO), in which the Fe(III) clusters were in close contact with the support, leading to better photogenerated charge separation, was reported (Xue et al. 2019). A high selectivity of DFF (95%) with a 33% HMF conversion in an aqueous medium was achieved with O_2 as an oxidant. Scavenger studies confirm the generation of superoxide radicals ($^{\bullet}O_2^-$) from O_2 responsible for the selective oxidation of HMF. Further, to compare with other Bi-based ternary metal oxides, three Bi-based oxides, Bi_2WO_6, $BiVO_4$, and Bi_2MoO_6, prepared via the hydrothermal approach for the oxidation of HMF, were reported (Kumar et al. 2021). The rose-like Bi_2WO_6 showed higher activity and selectivity than $BiVO_4$ and Bi_2MoO_6 for photocatalytic oxidation of HMF to DFF because of its appropriate band gap (2.7–2.8 eV), high surface acidity and larger specific surface area. Under visible light, a DFF yield of 58% with a selectivity of >99% in acetonitrile was obtained with rose-like Bi_2WO_6. The above studies disclosed that the photo-generated charge carriers and superoxide radicals ($^{\bullet}O_2^-$) played a key role in the selective photocatalytic oxidation of HMF to DFF.

To sum up this section, although TiO_2 is mainly used as a photocatalyst, other metal oxides, such as MnO_2 and Nb_2O_5, are also found to be effective for the selective oxidation of HMF to DFF. Though metal oxide-based photocatalysts have many advantages, such as recyclability and stability, they usually have light absorption in the UV region and low quantum efficiency. Such drawbacks can be overcome to a considerable extent by introducing/doping auxiliary metals (e.g. Ag, Pt and Au) and nonmetals (e.g. N and S) on the surface or in the lattice of the semiconductors (the corresponding metal oxide) to generate selective superoxide radical ($^{\bullet}O_2^-$) and thereby increasing the selectivity of DFF while suppressing mineralisation.

4.5.2 Carbon-Based Photocatalysts for the Oxidation of HMF to DFF

Carbon-based materials have been reported for several applications, such as catalysts for H_2 production, adsorbents in purification systems and electronic devices (Cao et al. 2016). Since the discovery of its application in photocatalyst for the reduction of water to produce hydrogen in 2009, graphitic carbon nitride (g-C_3N_4) has gained much attention and is well explored in various fields, such as water splitting, CO_2 reduction, organic pollutants degradation and organic substrate oxidation (Sabzehmeidani et al. 2021). Being a carbon-based material, graphitic carbon nitride has multiple advantages compared to other photocatalysts, such as excellent chemical stability, visible light absorption owing to adequate band gap (2.7 eV) and ease of modification. The most significant advantage of carbon nitride is that it has a higher ability to activate O_2 to produce superoxide radicals as it has more negative conduction potential (–1.3 eV), leading to much easier formation of the superoxide radical ($^{\bullet}O_2^-$). In addition, the position of the g-C_3N_4 valence band (1.4 eV) makes it challenging to generate hydroxyl radical ($^{\bullet}OH$), a nonselective strong oxidant in the reaction mixture (Zhang et al. 2020). The carbon-based photocatalysts for the oxidation of HMF to DFF are listed in Table 4.3.

The potential of utilizing various synthesized g-C_3N_4 materials derived from melamine for the photocatalytic oxidation of HMF in an aqueous medium was investigated, yielding a low DFF selectivity (30%) (Krivtsov et al. 2017). Further, thermally exfoliating the photocatalyst resolves the issue of low surface area and eliminates the NH_2 sites, which are not conducive for HMF oxidation to DFF, thus improving selectivity to 45% under artificial solar irradiation. Also, it has been observed that the DFF selectivity was enhanced to 50% with 40% HMF conversion when performing the reaction in natural sunlight. The same group extended their study in 2018 and revealed that blocking NH_2 sites of thermally exfoliated g-C_3N_4 by forming an adduct with H_2O_2 through hydrogen bonding can hinder the interaction of HMF and NH_2 sites on the catalyst surface and thus suppress mineralization of HMF and DFF.

This resulted in enhancing the DFF selectivity to 88% under natural sunlight irradiation (Ilkaeva et al. 2018). Unlike these reports, g-C_3N_4, prepared via calcination of melamine, dispersed in water and subsequent calcination, was reported, thus enhancing the surface area and other properties of the photocatalyst, such as light-harvest ability and suppressing the recombination of the photoexcited electron-hole pair.

Employing fewer polar solvents, such as acetonitrile and trifluorotoluene, significantly improved the selectivity of DFF to 86% with a 31% conversion (Wu et al. 2017). In continuation, the Z-scheme heterojunction material was developed by doping g-C_3N_4 with WO_3 (4.7%). Due to heterojunction promoting the charge separation, a high DFF selectivity of 87% with 27% HMF conversion was accomplished under the irradiation of visible light (>400 nm) (Zhang et al. 2019). Further, as thermal exfoliation is an energy-intensive process due to high-temperature treatment, various

TABLE 4.3
Photocatalytic Oxidation of HMF to DFF with Carbon-Based Materials

Catalyst	Lamp	Reaction Conditions	Conv. (%)	Sel. (%)	Ref.
g-C_3N_4	Fluorescent lamps (340–420 nm)	HMF: 0.5 mM Cat: 0.33 g L^{-1} Air, H_2O, 4h	69	43	(Krivtsov et al. 2017)
PCN-H_2O_2 (TEO)	Natural sunlight	HMF: 0.5 mM Cat: 0.67 g L^{-1} Air, H_2O, 4h	37	83	(Ilkaeva et al. 2018)
metal-free g-C_3N_4	300W xenon lamp (>400 nm)	HMF: 20 mM Cat: 10 g L^{-1} O_2, ACN:$PhCF_3$, 6h	31	86	(Wu et al. 2017)
WO_3/ g-C_3N_4	300W xenon lamp (>400 nm)	HMF: 20 mM Cat: 10 g L^{-1} O_2, ACN:$PhCF_3$, 6h	27	87	(Zhang et al. 2019)
TE-520	fluorescent lamps (340–420 nm)	HMF: 0.5 mM Cat: 0.33 g L^{-1} O_2, H_2O, 4h	58	38	(Akhundi et al. 2017)
Pt/SGCN	LED solar simulator (>400 nm)	HMF: 10 mM Cat: 1 g L^{-1} N_2, H_2O, 48h	38	>99	(Battula et al. 2019)
MXene/ g-C_3N_4	>400 nm	HMF: 5 mM Cat: 10 g L^{-1} O_2, $PhCF_3$, 10h	93	97	(Wang et al. 2020)
ZnPp-C_3N_4-TE	Natural sunlight	HMF: 0.5 mM Cat: 0.66 g L^{-1} Air, H_2O, 4h	73	36	(García-López et al. 2021)
UCNT *With TEOA	300W xenon lamp (≥420 nm)	HMF: 10 mM Cat: 1 g L^{-1} H_2O, 5h	48	95	(Bao et al. 2022)
1% PEPE-C_3N_4	300W Xenon lamp	HMF: 100 mM Cat: 2 g L^{-1} vacuum, H_2O, 3h	1952 μmolg^{-1}h^{-1}		(Jing et al. 2023)
PHI	AM 1.5G illumination (>420 nm)	HMF: 10 mM Cat: 1 g L^{-1} vacuum, H_2O	49 μmolg^{-1} h^{-1}		(Chen et al. 2023)
Au-Ru/rGO	300 W xenon lamp	HMF: 50 mM Cat: 5 g L^{-1} 5 bar O_2, Toluene, 80°C, 8h	96	95	(Ma et al. 2018)
Cs_2SnBr_6/ rGO	300W xenon lamp (>400 nm)	HMF: 5 mM Cat: 1.5 g L^{-1} O_2, ACN, 6h	>99	88	(Yin et al. 2023)
metal-free polyimide (PI)	LED light	HMF: 6.3 mM Cat: 6.67 g L^{-1} Air, ACN, 40 min	61	91	(Chu et al. 2023)

approaches, such as chemical ultrasonic irradiation, hydrothermal, ball-milling and thermal exfoliation for modifying g-C_3N_4 to improve its activity were explored (Akhundi et al. 2017). However, a low selectivity to DFF (38%) with 58% HMF conversion was obtained with a thermally exfoliated (520 °C) catalyst. However, it was still better than other methods; chemical ultrasonic irradiation (27% HMF conversion and 32% DFF selectivity), hydrothermally treated g-C_3N_4 showed (26% HMF conversion with 21% DFF selectivity) and ball-milled g-C_3N_4 (18% HMF conversion and 37% DFF selectivity). Further, coupling with H_2 production, 38% DFF with >99% selectivity was produced using Pt-doped mesoporous carbon nitride as a photocatalyst. Tetraethylorthosilicate (TEOS) was used as a hard template for producing carbon nitride, causing enhanced surface area (250 m^2/g) and, together with Pt as co-catalyst, a high DFF selectivity along with H_2 production in aqueous medium was achieved (Battula et al. 2019). Like WO_3 doped g-C_3N_4, a noble metal-free photocatalyst of 2D MXene/g-C_3N_4 (MX/CN) was prepared for the selective conversion of HMF to DFF. Due to the appropriate band structure, high charge separation and transfer efficiency of the MXene component, which can effectively suppress the recombination of photogenerated charges and the generation of active species (superoxide radical), high yield (90%) and selectivity (97%) of DFF was achieved under visible light irradiation (Wang et al. 2020).

In 2017, Xu et al. discovered the interaction between CoPz and g-C_3N_4, facilitating the singlet oxygen production and reducing hydroxyl radical formation. At an acidic pH (4.01), photocatalytic oxidation of HMF was improved, yielding 80% DFF. On the other hand, at a neutral or basic condition (pH = 6.86 or 9.18), the primary product was a further oxidized product, FDCA, resembling that of a noble metal catalyst under thermal conditions. Further, it has also been demonstrated that unmodified g-C_3N_4 could convert O_2 into active hydroxyl radicals ($^{\bullet}OH$) and oxidize HMF into CO_2 and H_2O under light. However, combining g-C_3N_4 with CoPz under light irradiation excites CoPz, which could convert O_2 adsorbed on the surface into singlet oxygen radical ($^1O_2$), resulting in a highly selective target product, in line with the previous reports related to the oxidation reaction via singlet oxygen radicals (Xu et al. 2017; Wang et al. 2016). Later, in 2020, thermally etched g-C_3N_4 and impregnated with Zn(II) based substituted porphyrins and tested for HMF oxidation, affording 36% selectivity with 73% HMF conversion at a pH of 6 under the sun as a source of light. The rationale behind the low catalytic activity could be due to the production of non-selective radicals rather than selective singlet oxygen or superoxide radicals. (García-López et al. 2021)

Improving the synthesis strategies of g-C_3N_4 as a photocatalyst, flexible ultrathin g-C_3N_4 nanosheets based on thermal treatment and loaded with 3 wt% Pt as a co-catalyst was reported (Bao et al. 2022). It was observed that the OH group of HMF can be vertically adsorbed on the site of sp2-hybridized nitrogen, causing a twist in the hydroxymethyl group structure and reducing the dehydrogenation energy barrier. This interaction leads to easy hydrogen production (92.0 μmol g^{-1} h^{-1}) from water, coupled with selective conversion of HMF (48%) with a DFF selectivity of 95% under visible light irradiation. In line with this, a three-dimensional carbon nitride, prepared via a co-modification approach of doping oxygen and generating nitrogen vacancies based on hydrothermal and thermal polymerization reaction of dicyandiamide and polyethylene glycol – polypropylene glycol – polyethylene glycol triblock copolymer (PEPE-C_3N_4), was reported. The 1% PEPE-C_3N_4 photocatalyst showed the best photocatalytic performance 5.3 and 2.4 times (1177 μmol g^{-1} h^{-1} for H_2 and 358 μmol g^{-1} h^{-1} for DFF generation) higher compared to unmodified g-C_3N_4, respectively. Due to the suitable band-edge position of the catalyst, HMF did not undergo mineralization to final oxidized product CO_2, paving the way to achieve 100% DFF selectivity (Jing et al. 2023). Similarly, a carbon vacancy-modified melon-PHI allotropic carbon nitride composite was reported, demonstrating an exceptional photocatalytic performance for simultaneous H_2 evolution and HMF oxidation (Chen et al. 2023). The intrinsic defect of the material facilitated a facile photo-charge separation and generated additional active sites in the carbon nitride to boost the performances of photocatalytic redox reactions, affording 41 and 49 μmol g^{-1} h^{-1} for H_2 evolution from water splitting coupled with selective oxidation of HMF to DFF, respectively.

Other than carbon nitride, carbon-based materials as photocatalysts have also been developed and investigated for HMF oxidation. In line with this, reduced graphene oxide (rGO) as a support for bimetallic Au-Ru (Au-Ru/rGO) for HMF oxidation was reported, exhibiting an excellent photocatalytic performance and affording 91% DFF yield. As a result of the synergistic interaction between Au and Ru, the Au-Ru/rGO composite exhibited a greater capacity for adsorbing O_2 and HMF compared to unmodified rGO. Further, the electron transfer process, likely occurring from Ru to Au under the light illumination, was the reason behind enhancing the production of superoxide radicals ($^{\bullet}O_2^-$) and diverting the reaction pathway towards DFF instead of HMFCA. Moreover, this catalyst exhibited to produce DFF directly from fructose in a one-pot approach, giving a maximum of 86% (Ma et al. 2018).

Similarly, the lead-free perovskite Cs_2SnBr_6/rGO composite was prepared via photoreduction, that is, by dispersing both components in saturated HBr and H_3PO_2 solution under UV light irradiation (Yin et al. 2023). This resulted in not only enhancing the charge separation and charge transfer but also activating O_2 on the surface of the catalyst, leading to generating singlet oxygen ($^1O_2$) and superoxide radicals ($^{\bullet}O_2^-$), confirmed by EPR studies, which synergistically oxidizes HMF, giving an 88% DFF selectivity with 99.5% conversion. Another interesting material based on metal-free polyimide (PI) was synthesized by Chu et al. via thermal polycondensation of melem and pyromellitic dianhydride (PMDA) at 325 °C for 4 h. Fine-tuning the band structure and the valence band position of PI while maintaining an appropriate band gap with sufficient visible-light response could be achieved by adjusting the molar ratio of amine and anhydride. Interestingly, it was found that PI with moderate photo-oxidation capability showed the best performance of giving 61% HMF conversion with 91% DFF selectivity and a high apparent quantum yield (AQY) of 1.13%, nearly 7 times higher than that of g-C_3N_4 (Chu et al. 2023). This study opens a new window to explore other photo-active polymers and copolymers for the oxidation of HMF and biomass-derived substrates, as these can be easily tuned to get desirable properties, such as adequate band gap, lower valence band position and better charge transfer ability.

To sum up the carbon-based photocatalysts, g-C_3N_4 is mainly used as a photocatalyst due to numerous advantages such as low cost and its excellent properties like 2D layered morphology, resistance to photocorrosion and stability towards the external environment such as temperature, solvent and pH (Mazzanti et al. 2020). Further, g-C_3N_4 can be easily fine-tuned via surface modification that can prevent the generation of hydroxyl radicals (•OH), favoring the mineralization of HMF to CO_2 and H_2O. However, similar to metal oxides, it can be observed that the solvents with low polarity and high oxygen solubilization capacity are favorable for the photocatalytic oxidation of HMF.

4.5.3 Metal Sulfide-Based Photocatalysts for the Oxidation of HMF to DFF

Metal sulfides have recently gained considerable attention due to their inherent properties. Compared to typical metal oxide semiconductors, metal sulfides have a narrow band gap, excellent optical and electrical properties, and can be synthesized in various crystal forms. In addition, suitable sideband positions, along with the properties mentioned above, make them an enticing alternative as photocatalysts. However, the major limitation of these sulfide-based materials is poor stability and photocorrosion problems, somewhat hindering their application in photocatalytic oxidation (Zhang et al. 2021). Therefore, researchers have explored many methods to reduce photocorrosion, thus improving its photocatalytic stability, such as constructing heterogeneous structures, introducing co-catalysts and nonmetallic anionic groups and doping metals (Li et al. 2020). The photocatalytic oxidation of HMF to DFF with sulfide-based materials are shown in Table 4.4. In line with this, it was demonstrated that ultra-thin Ni/CdS nanosheets was an efficient catalyst for the oxidation of HMF, along with a simultaneous generation of H_2 in water (Han et al. 2017). Though a 100% DFF selectivity was achieved with Ni/CdS, due to its inhibitory effect on the aldehyde group, the oxidation of the alcohol group in HMF was impeded, resulting in a poor conversion (20%) even after a longer reaction time (22 h).

S vacancy-rich metal-sulfide material, P-doped $Zn_xCd_{1-x}S$, was reported using a hydrothermal method and hydrazine hydrate as a reducing agent (Ye et al. 2018). The bimetallic $Zn_{0.5}Cd_{0.5}S$-P catalyst showed a high rate of 419 μmol h^{-1} g^{-1} for H_2 evolution via water splitting reaction, which is 72 and 7.5 times higher than that of ZnS-P and CdS. Furthermore, the rate of H_2 evolution increased to 786 μmol h^{-1} g^{-1} when replacing water oxidation with HMF oxidation. Although the selectivity of 65% towards DFF with 40% HMF conversion was not that promising, this study demonstrated that introducing biomass-derived substrate (HMF) in a splitting reaction can enhance the hydrogen production rate. Further, in 2019, to improve the catalytic activity of CdS, a nitrate additive ($LiNO_3$) in acetonitrile solution as a redox medium was employed to promote the selective oxidation of HMF to DFF at the rate of 2.6 mmol h^{-1} (Dimeglio et al. 2019). Employing nitrate additives can generate a nitrate radical ($NO_3^•$) during the reaction, confirmed by *in situ* EPR experiment, acting as a mediator and thus promoting alcohol oxidation. In the absence of a nitrate additive, no HMF oxidation was observed, emphasizing the promoting role of nitrate species. However, using external additives could enhance the oxidation activity and rate of the reaction, but, on the other hand, the process encounters cumbersome handling and recyclability issues.

TABLE 4.4
Photocatalytic Oxidation of HMF to DFF with Sulfide-Based Materials

Catalyst	Lamp	Reaction Conditions	Conv. (%)	Sel. (%)	Ref.
Ni/CdS	Blue LED (440–460 nm)	HMF: 10 mM Cat: 1 g L^{-1} O_2, H_2O, 22h	20	100	(Han et al. 2017)
$Zn_{0.5}Cd_{0.5}S$-P	White LED light	HMF: 15.8 mM Cat: 0.2 g L^{-1} Ar, H_2O, 8h	40	65	(Ye et al. 2018)
Nitrate mediated CdS	Blue LED (450 nm)	HMF: 250 mM Cat: 20 g L^{-1} O_2, H_2O	NA	>99	(Dimeglio et al. 2019)
CC/$ZnIn_2S_4$	Xenon lamp (400–780 nm)	HMF: 20 mM Cat: 1 g L^{-1} Air, ACN	~2980 μmolg^{-1}		(Qiu et al. 2022)
$Cd_xIn_yS_{(x+1.5y)}$	300W xenon lamp (>420 nm)	HMF: 31.7 mM Cat: 4 g L^{-1} Air, H_2O, 6h	69	63	(Zhang et al. 2022)
$Ru_{3.7}$-CdS	300W xenon lamp	HMF: 10 mM Cat: 1 g L^{-1} Ar, DMF, 20h	81	92	(Xia et al. 2022)
Pt/CdS	Blue LED (450 nm)	HMF: 100 mM Cat: 5 g L^{-1} N_2, H_2O: ACN, 12h	95	99	(Wang et al. 2023)
0.25% Au_2S@ CdS NWs	Blue LED (445 nm)	HMF: 10 mM Cat: 1 g L^{-1} Ar, ACN, 4h	100	~95	(Kim et al. 2023)
Co_1/ZIS	300W xenon lamp (>400 nm)	HMF: 5 mM Cat: 1 g L^{-1} Air, ACN, 4h	99	92	(Tan et al. 2023)
SA-PDI/ $ZnIn_2S_4$	300W xenon lamp	HMF: 2 mM Cat: 2 g L^{-1} O_2, ACN, 4h	98	90	(Guo et al. 2024)
cocatalyst-free $ZnIn_2S_4$	300W xenon lamp	HMF: 5 mM Cat: 2 g L^{-1} O_2, ACN, 1h	91	99	(Zhao et al. 2022)

In line with the above studies, controlled growth of $ZnIn_2S_4$ nanosheets over carboxy-cellulose-derived carbon dots, prepared via the hydrothermal treatment of carboxylated cellulose followed by *in situ* growth of $ZnIn_2S_4$, was reported (Qiu et al. 2022). The carboxylic acid group present on the surface of carbon dots have an ionic interaction with cations (Zn^{2+} and In^{3+}) of $ZnIn_2S_4$, leading to a relatively uniform wrapping of $ZnIn_2S_4$ nanosheets on the small carbon dots. In conjunction with the reinforced visible-light capture and good conductivity of carbon dots, the resultant

$CC/ZnIn_2S_4$ showed the photocatalytic activity of DFF evolution up to 2980 μmol g^{-1}, about 3.4 times higher than that of pristine $ZnIn_2S_4$. Further, $Cd_xIn_yS_{(x+1.5y)}$ ternary sulfide with the Cd: In a ratio of greater than 1:2 via a one-step hydrothermal method, comprised of one-dimensional and three-dimensional heterojunction structures of CdS nanorods and cubic $CdIn_2S_4$, respectively, was synthesised (Zhang et al. 2022). One-dimensional CdS nanorods can effectively promote the separation and migration of photogenerated carriers, enhancing the photocatalytic efficiency of $Cd_xIn_yS_{(x+1.5y)}$ materials for HMF oxidation. Among the synthesized samples, $Cd_{1.5}In_2S_{4.5}$ showed the best activity with 69% HMF conversion and 63% DFF selectivity and exhibited recyclability of 4 times. In this study, due to the production of H_2O_2, DFF could undergo further oxidation to form FDCA; therefore, a small amount of FDCA (10% selectivity) was observed.

In an attempt to improve the activity of CdS quantum dots (QDs), a Ru complex was anchored on the surface of QDs through the coordination method, enabling the efficient extraction of the photogenerated holes from CdS QDs to generate hydroxyl radicals (•OH), thus enhancing selective oxidation of HMF to DFF under Ar atmosphere (Xia et al. 2022). Earlier, it was assumed that hydroxyl radicals (•OH) caused the degradation of HMF, but in this case, it was presumed that hydroxyl radicals (•OH) generated by photocatalyst could be responsible for the selective DFF production (>90%) under the Ar atmosphere. On the other hand, under air atmosphere, superoxide radicals ($•O_2^-$) radicals were generated, selectively producing HMFCA from HMF oxidation, manifesting switchable oxidation of HMF depending on the reaction atmosphere. Further, it was found that doping Pt over CdS with a photoreduction method enhanced the dehydrogenation ability of the photocatalyst (Wang et al. 2023). For example, 1 wt% Pt over CdS gave a complete conversion of HMF with >99% selectivity of DFF, along with stoichiometric amount hydrogen generation. It was observed that the Pt nanoparticles interacting with the furan ring of HMF facilitated the fast transfer of protons from the hydroxymethyl group of HMF to Pt NPs in the close vicinity for further reduction to generate hydrogen.

Similarly, the decoration of CdS nanowires with Au_2S using a simple ion exchange method was reported, and it was found that 0.25% Au_2S/CdS efficiently oxidized HMF by directly utilizing photogenerated holes under anaerobic conditions. Under aerobic conditions, singlet oxygen radical ($^1O_2$) is produced from oxygen, acting as an active species for HMF conversion to DFF, as the mechanism proposed previously (Xu et al. 2017). Under both conditions, a high yield (~ 95%) was obtained, but a higher reaction rate was observed under anaerobic conditions, speculating that the higher oxidation rate using photogenerated holes than involving oxygen radicals produced under aerobic conditions (Kim et al. 2023). Other attempts have been made to replace doping precious metals with nonprecious metals due to their cost-effectiveness and abundance. In line with that, an ultralow-amount (0.08 wt%) Co single atoms on $ZnIn_2S_4$ nanosheets (Co_1/ZIS) was prepared, exhibiting an outstanding catalytic activity under visible light irradiation and air atmosphere, affording 92% DFF selectivity with an HMF conversion of 99%.

Compared to other nonprecious metals, Co showed the best charge transfer ability for photogenerated charge carriers and also provided the active site to promote the adsorption/activation of O_2 to generate reactive oxygen species ($\bullet O_2^-$ and $^1O_2$), thereby leading to its superior photocatalytic performance compared to the pristine $ZnIn_2S_4$ counterpart (Tan et al. 2023). Rather than using metals (e.g. Pt and Ag) as co-catalysts, self-assembled perylene diimide (SA-PDI) supramolecule as an oxygen reduction co-catalyst on $ZnIn_2S_4$ was employed (Guo et al. 2024). The functionalized perylene imide supramolecule (1%) was *in situ* loaded onto the preformed $ZnIn_2S_4$ and displayed as an active site similar to the metal (e.g. Pt and Ag) active site and, thus, accelerating the generation of superoxide radical by absorbing oxygen while concentrating the photogenerated electrons of $ZnIn_2S_4$. Notably, the HMF transformation frequency with SA-PDI/$ZnIn_2S_4$ was 1952 μmol g^{-1} h^{-1}, 2–50 times higher than previously reported techniques for the photocatalytic aerobic oxidation of HMF (Guo et al. 2024 and references therein) Using no auxiliary metal or co-catalyst is generally considered to be advantageous, and in line with that, cocatalyst-free $ZnIn_2S_4$ was found to be an excellent photocatalyst, yielding 91.1% HMF conversion with a DFF selectivity of 99.4% (Zhao et al. 2022). The rationale behind this is that the spatial accumulation of photogenerated electrons and holes on the (110) and (102) crystal facets trigger a two-electron oxygen reduction reaction ($2e^-$ ORR) for H_2O_2 generation and HMF oxidation to DFF, respectively. Simultaneously, *in situ* generated hydroxyl radicals ($^{\bullet}OH$) formed by the decomposition of H_2O_2 caused an irreversible photocorrosion of Zn-S by forming Zn-O. The spontaneous substitution of oxygen with sulphur using Na_2S has been proven to improve the photostability of $ZnIn_2S_4$.

To sum up, metal sulfide-based photocatalysts, due to their excellent electrical and optical properties and ability to be used under various conditions such as aerobic and anaerobic conditions, make them appealing and intriguing photocatalysts to be explored further for future studies. However, the stability of those materials remains elusive; though various attempts have been made to produce robust sulfide-based materials and regenerate the catalyst, more stable and efficient catalysts are required for photocatalytic oxidation reactions.

4.5.4 Modified Photocatalysts by Constructing Heterojunction

As discussed earlier, conventional photocatalysts, such as TiO_2, CdS, ZnO and g-C_3N_4, have many advantages; at the same time, these materials possess some disadvantages, such as fast recombination of photogenerated charges, low efficiency for light absorption, inadequate active sites, inability for charge transfer or the fatal disadvantage of photocorrosion. Thus, the construction of heterojunction is one of the approaches to overcome these disadvantages of the photocatalysts mentioned above to a considerable extent. Conventionally, the heterojunction systems can be divided into three types: (i) type I (straddling gap) heterojunctions, where the band gaps of the two semiconductors are overlapped; (ii) type II (staggered gap) heterojunctions where band gaps are staggered; and (iii) type III (broken gap) heterojunctions where band gaps are uncrossed. Other than these are the Z-scheme, S-scheme and *p-n* heterojunction, the band gap is also staggered, but the transfer route of electrons and holes is different, as shown in Figure 4.4.

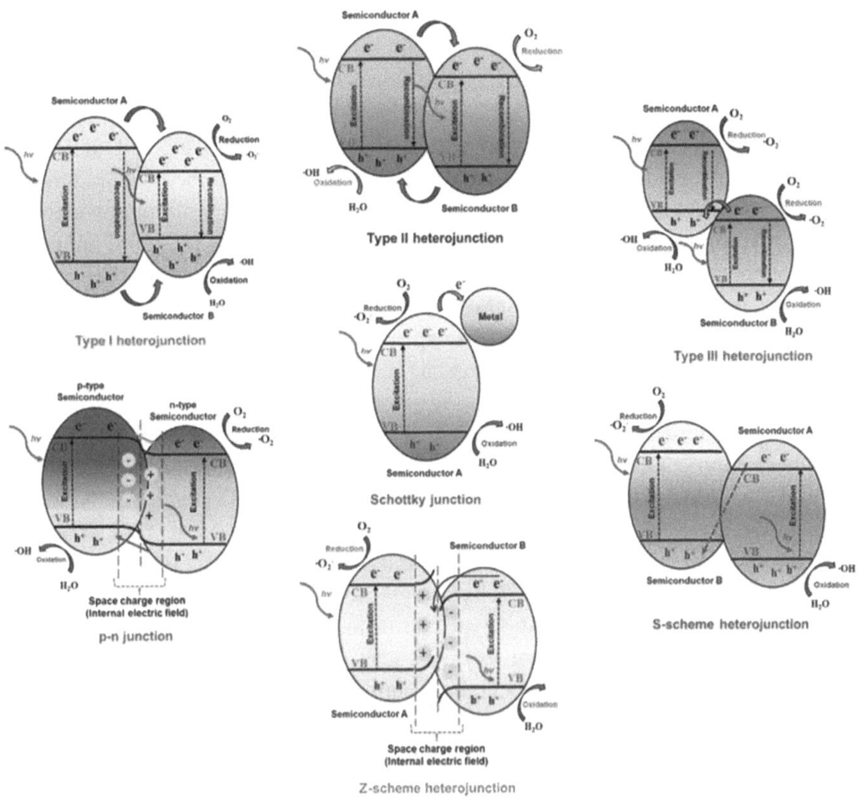

FIGURE 4.4 Schematic representation of different types of heterojunctions. (Gnanaguru et al. 2024.)

As shown earlier, a Z-scheme heterojunction by doping WO_3 (4.7%) on g-C_3N_4 was synthesized for selective oxidation of HMF (Zhang et al. 2019). Considering a similar approach, novel 2D/2D relative *p-n* heterojunction CdS/g-C_3N_4 with Z-scheme (or S-scheme) by *in situ* growing ultrasmall 2D CdS nanoflakes on the surface of 2D g-C_3N_4 nanosheets was prepared (Wu et al. 2020). Undoped g-C_3N_4 exhibited low conversion (30%) with very good selectivity to DFF (77%) for HMF oxidation, while CdS alone showed high conversion (87%) with low selectivity (39%) towards DFF. Interestingly, CdS/g-C_3N_4 composites with a ratio of 1:1 displayed excellent photocatalytic activity, giving 89% HMF conversion with 91% DFF selectivity along with higher H_2 evolution under visible light irradiation, indicating a concerted role of g-C_3N_4 and CdS. Similarly, a Nb_2O_5/g-C_3N_4 composite with type II heterojunction was reported and found to be affording 72% HMF conversion with 84% DFF selectivity in acetonitrile under sunlight irradiation, 1.5 and 3.3 times higher than that of pure Nb_2O_5 and g-C_3N_4, respectively (Sharma et al. 2022). The construction of heterojunction changes the surface properties of the photocatalyst, thus reducing the recombination of photogenerated carriers, promoting charge separation, and enhancing visible light utilization by the electrons transferring from g-C_3N_4 to Nb_2O_5 with holes immigrating from Nb_2O_5 to g-C_3N_4

In a similar fashion, Bi_2WO_6 over mesoporous graphitic C_3N_4 (mpg-C_3N_4) was also investigated to obtain Z-scheme heterojunction-based material (Cheng et al. 2021). It was observed that the photogenerated electron from the CB of Bi_2WO_6 moved towards the VB of mpg–C_3N_4 and thus recombined with the hole, and eventually, the composite material generated superoxide radicals ($^{\bullet}O_2^-$) mainly by electrons excited on mpg–C_3N_4. Therefore, the concerted effect of Bi_2WO_6 and mpg-C_3N_4 on photogenerated charge separation through the charge transfer between the two semiconductors contributed to the photocatalytic activity, giving 59% HMF conversion with 84% DFF selectivity. The less explored V_2O_5 as a photocatalyst by making a composite with g-C_3N_4 using the ultrasonic method to get a Z-scheme-based charge transfer material was also investigated (Wang et al. 2023). It was found that the 5% $V_2O_5{\cdot}nH_2O$ (hydrated vanadium pentoxide (HVO) over g-C_3N_4 (HVO/ g-C_3N_4) gave the highest photocatalytic performance with a 36% yield and an 84% selectivity of DFF under visible light irradiation. The interaction between HVO and g-C_3N_4 led to a notable abundance of oxygen vacancies in the composite, facilitating oxygen adsorption, promoting the harvest of visible light, and hastening the generation of hydroxyl radicals. As it was observed that both hydroxyl radicals ($^{\bullet}OH$) and superoxide radicals ($^{\bullet}O_2^-$), substantiated by ESR studies, are the primary active species in the selective oxidation of HMF, the presence of oxygen vacancies emerges as a significant factor contributing to the enhanced photocatalytic activity of HVO/g-C_3N_4.

Because of its attractive properties and photocatalytic performance, like carbon nitride, metal sulfides can also be used to construct various heterojunctions. Following that, Z-scheme-based $Nb_2O_5/ZnIn_2S_4$ nanocomposite was reported by growing Nb_2O_5 nanospheres on the petals of $ZnIn_2S_4$ microspheres via a facile hydrothermal approach (Wang et al. 2020). The materials possessing a direct Z-scheme heterojunction can generally prevent the recombination of photogenerated electrons and holes in the composite. The 8% $Nb_2O_5/ZnIn_2S_4$ composite displayed excellent photocatalytic performance, giving 86% HMF conversion with 88% DFF selectivity along with a higher rate of H_2 generation compared to pristine Nb_2O_5 and $ZnIn_2S_4$. Similarly, a Z-scheme photocatalyst system, $Zn_{0.5}Cd_{0.5}S/xMnO_2$, was developed, which has the optimum band structure suitable for efficient visible-light-assisted photocatalytic H_2 generation via water splitting integrated with selective oxidation of HMF to DFF (Dhingra et al. 2020). Photocatalytic investigations revealed good catalytic activity with $Zn_{0.5}Cd_{0.5}S/1\%MnO_2$, affording a DFF yield of 46% and a simultaneous H_2 generation rate of 1322 μmol g^{-1} after 24 h, 9 and 4 times higher than those of parent $Zn_{0.5}Cd_{0.5}S$, respectively.

Likewise, 12.5% $MoS_2/CdIn_2S_4$ flower-like type Z heterojunction photocatalyst was reported (Zhu et al. 2023), and it was found to have the highest activity for photocatalytic oxidation of HMF to DFF production among various loading of MoS_2 (7.5% to 17.5%). $MoS_2/CdIn_2S_4$ heterojunction reduced the recombination of photogenerated carriers, promoted charge separation, and enhanced visible light utilization by electrons migrating from the CB of MoS_2 to the VB of $CdIn_2S_4$, thus giving a good DFF yield (52%). The Schottky junction by *in situ* fabricating $CdIn_2S_4$

over $Ti_3C_2F_x$ MXene ultra-thin nanosheets was also prepared, which provides extra stability along with better charge separation, and tested for HMF oxidation, giving 60.3% DFF yield under visible light irradiation (Zhang et al. 2023). The same group synthesized another 2D-3D heterojunction structure of black phosphorus (BP)/$CdIn_2S_4$ for the selective oxidation of HMF to DFF. This S-type heterojunction showed high photogenerated carrier separation and migration efficiency, giving a fair DFF yield (42%) with 1.5% BP/$CdIn_2S_4$ (Zhang et al. 2023). The studies referred to above show that carbon nitride is one of the most common carbon-based photocatalysts, and $CdIn_2S_4$ is also the most utilized bimetallic metallic sulfide; intriguingly, the combination of these two would enhance its catalytic activity.

With regard to the above, the $CdIn_2S_4$/g-C_3N_4 heterojunctions composite for photocatalytic oxidation of HMF to DFF was reported, achieving 65% HMF conversion with 92% DFF selectivity (Cui et al. 2023). The VB potential of the constructed heterojunctions did not reach the oxidation potential of $^{\bullet}OH/OH^-$, thus avoiding the massive generation of strongly oxidizing active species hydroxyl radicals ($^{\bullet}OH$) and selectively producing superoxide radicals ($^{\bullet}O_2^-$) radical, which was responsible for a high selectivity of DFF. Hence, the construction of various heterojunctions improves the photocatalytic performance, mainly due to the heterostructure facilitating the rapid separation and transfer of photogenerated electrons. The construction of heterojunctions can be a promising approach, among others (such as precious metal doping, doping of organic moieties, and surface modification via various methods as mentioned in Sections 4.5.1–4.5.3), for developing photocatalysts for various oxidation reactions, including selective oxidation of HMF to DFF.

4.5.5 Photocatalytic Oxidation of HMF to DFF with Miscellaneous Materials

In addition to the photocatalyst systems described above, other photocatalyst systems have also been considered for oxidizing HMF to DFF (Table 4.5). In 2020, Zhang et al. discovered that HMF can be converted to DFF using lead halide perovskite ($MAPbBr_3$) as a photocatalyst under aerobic conditions with visible light exposure. The combination of superoxide radicals ($^{\bullet}O_2^-$), singlet oxygen radicals ($^1O_2$), and photogenerated electron and hole resulted in 100% HMF conversion with 90% DFF selectivity (Zhang et al. 2020). Further, in 2020, another interesting material, that is, thiophene-containing covalent triazine framework on mesoporous silica (SBA-15) for generating active oxygen radical species, was synthesized (Ayed et al. 2020). It was observed that this photocatalyst gave around 57% HMF conversion with 100% DFF selectivity in water under visible light irradiation. EPR studies confirmed that singlet oxygen radicals ($^1O_2$) and superoxide radicals ($^{\bullet}O_2^-$) act as active species for the selective oxidation of HMF, paving the way for exploring other potential covalent organic and metal-organic frameworks towards selective photocatalytic oxidation reactions. Similarly, an electron transport channel was constructed to generate selective oxidation radicals by combining g-C_3N_4, WO_3 and MnO_2 (Qian et al. 2022). CN-WO_3@MnO_2 material was found to have efficient electron transfer pathways and ample oxygen adsorption sites. Moreover, the adequate transportation

TABLE 4.5
Photocatalytic Oxidation of HMF to DFF with Modified Materials Possessing Heterojunction

Catalyst	Lamp	Reaction Conditions	Conv. (%)	Sel. (%)	Ref.
50% $CdS/g\text{-}C_3N_4$	300W xenon lamp (>420 nm)	HMF: 5 mM Cat: 10 g L^{-1} O_2, $PhCF_3$, 2h	90	91	(Wu et al. 2020)
$Nb_2O_5/g\text{-}C_3N_4$	Natural sunlight	HMF: 1 mM Cat: 4.5 g L^{-1} Air, ACN, 5h	72	84	(Sharma et al. 2022)
12%Bi_2WO_6/ mpg–C_3N_4	300W xenon lamp	HMF: 0.1 mM Cat: 10 g L^{-1} O_2, ACN:$PhCF_3$, 8h	59	84	(Cheng et al. 2021)
$V_2O_5 \cdot nH_2O$/ g-C_3N_4	>400 nm	HMF: 0.1 mmol O_2, ACN, 4h	43	80	(Wang et al. 2023)
$ZnIn_2S_4$/ Nb_2O_5	300W xenon lamp	HMF: 10 mM Cat: 10 g L^{-1} O_2, $PhCF_3$, 3h	86	88	(Wang et al. 2020)
1% MnO_2 /$Zn_{0.5}Cd_{0.5}S$	white LED bulb (≥400 nm)	HMF: 15.8 mM Cat: 1 g L^{-1} N_2, H_2O, 24h	46	100	(Dhingra et al. 2020)
MoS_2/ $CdIn_2S_4$	LED lamp (>420 nm).	HMF: 0.5 mM Cat: 1 g L^{-1} H_2O, 10h	62	81	(Zhu et al. 2023)
$Ti_3C_2F_x$ MXene/ $CdIn_2S_4$	xenon lamp (>400 nm)	HMF: 10 mM Cat: 4 g L^{-1} Air, H_2O, 12h	79	77	(Zhang et al. 2023)
black phosphorus/ $CdIn_2S_4$	xenon lamp (>420 nm)	HMF: 31.7 mM Cat: 4 g L^{-1} Air, H_2O, 6h	50	83	(Zhang et al. 2023)
$CdIn_2S_4$/ g-C_3N_4	300W xenon lamp (>420 nm)	HMF: 5 mM Cat: 10 g L^{-1} Air, H_2O, 12h	65	92	(Cui et al. 2023)

of photogenerated electrons between CN-WO_3 and MnO_2 effectively suppressed the nonselective hydroxyl radicals (•OH) production. Simultaneously, the oxygen adsorption sites on the surface of MnO_2 of CN-WO_3@MnO_2 increased the efficiency of generating superoxide radicals ($^{\bullet}O_2^-$) and singlet oxygen radicals ($^1O_2$), affording 78% HMF conversion with 62% DFF yield.

Further, Yang et al. found that decatungstate has a strong donor-acceptor interaction with HBr, enhancing visible light-triggered oxidation of HMF, giving a 67% yield of DFF in acetonitrile under an O_2 atmosphere with ambient conditions (Yang et al. 2020). Gong et al. prepared interfacial Pt-O bonded Pt-Ov-BiOBr catalysts, which selectively produce superoxide radicals ($^{\bullet}O_2^-$) for the selective oxidation of

HMF. Due to the environment of unsaturated coordination on Ov-BiOBr, Pt can form a Pt-O bond as an electron transport channel to transfer photogenerated electrons from Ov-BiOBr to Pt for efficient generation of superoxide radicals ($^{\bullet}O_2^-$). Meanwhile, photogenerated h^+ could gather at the interface between Pt and Ov-BiOBr contact to oxidize HMF to carbon-centred radicals (Gong et al. 2023). From all these materials, it could be concluded that new materials that can selectively produce superoxide radicals ($^{\bullet}O_2^-$) and have interesting properties such as oxygen vacancies, better electron transport channels and electron transfer ability can be explored for the selective oxidation of HMF.

4.6 CONCLUSION AND PERSPECTIVE

In conclusion, from this chapter, one can understand that photocatalyst oxidation ability can be varied through the engineering of band gap, band-edge positions and surface modifications to generate selective oxidation radicals, leading to enhanced selectivity to DFF and other desired oxidation products. Recent photocatalytic approaches identify that employing co-catalysts or additives improved catalytic performance and also enhanced the photostability of innate catalysts. Using solar energy as an example as a source of light for the oxidation of the hydroxymethyl group of HMF and other bio-based substrates paves the way for employing natural light on a large scale and possibly decreasing the cost of the process. *In situ* techniques, such as EPR and DRIFT spectroscopy, and control experiments using scavengers are advantageous in identifying intermediates, active sites, and reactive species responsible for photocatalytic oxidation and elucidating mechanisms and to improve the photocatalytic activity further. However, the further development of photocatalytic materials with more stability is required to enhance their efficiency for the selective oxidation of HMF to DFF.

Despite substantial research efforts, sunlight-driven photocatalysis is still regarded as an emerging field that is not yet developed enough for effective commercialization as the availability of high technology readiness level prototypes and pilot setups are scarce. Although recent success in photocatalytic hydrogen evolution and CO_2 reduction has gained the interest of researchers, the demand for stable photocatalysts with high quantum efficiency remains a significant roadblock in the development of the photocatalysis process. The following efforts should be considered to increase the efficiency and scope of photocatalytic oxidation processes:

(i) The development of inventive heterogeneous photocatalysts with high activity, high quantum efficiency/yield, broad range light absorption region, selectivity towards a desired product with high yield; notably, the stability of the material is vital for industrial applications.

(ii) The mechanisms of photocatalytic alcohol oxidation, the hydroxymethyl group of HMF, are not well-understood, particularly the activation of C-H bonds and mineralization mechanism to CO_2 and water by hydroxyl radicals ($^{\bullet}OH$).

(iii) The critical role of co-catalysts/additives in photocatalytic oxidation should be explored further, especially since noble metals are the most employed co-catalysts in HMF oxidation, necessitating the development of low-cost co-catalysts or co-catalyst-free photocatalysts

(iv) Conventional photocatalysts have a low quantum efficiency and absorb light mainly in the UV region, thus emphasising the development of photocatalysts containing heterojunctions or surface modification for using a wide spectral range.

(v) Also, the focus is highly required on designing robust, simple and suitable photocatalytic reactors; as of today, most photocatalytic approaches encounter the challenges of scaling up the process due to a lack of adaptability and agility to perform numerous processes without requiring major disassembly.

ACKNOWLEDGEMENTS

We sincerely thank the Department of Biotechnology (DBT), New Delhi, India, for the financial support and infrastructural facilities provided through CIAB. NS gratefully thanks CIAB for financial support through the senior research fellowship and the Indian Institute of Science Education and Research (IISER) Mohali for PhD registration.

REFERENCES

Aggarwal, Nidhi, Priyanka Pal, Neeraj Sharma, and Shunmugavel Saravanamurugan. 2021. "Consecutive Organosolv and Alkaline Pretreatment: An Efficient Approach Toward the Production of Cellulose from Rice Straw." *ACS Omega* 6 (41). American Chemical Society: 27247–58. doi:10.1021/acsomega.1c04030.

Akhundi, Anise, Elisa I. García-López, Giuseppe Marcì, Aziz Habibi-Yangjeh, and Leonardo Palmisano. 2017. "Comparison between Preparative Methodologies of Nanostructured Carbon Nitride and Their Use as Selective Photocatalysts in Water Suspension." *Research on Chemical Intermediates* 43 (9). Springer Netherlands: 5153–68. doi:10.1007/s11164-017-3046-9.

Allegri, Alessandro, Valeriia Maslova, Magda Blosi, Anna Luisa Costa, Simona Ortelli, Francesco Basile, and Stefania Albonetti. 2020. "Photocatalytic Oxidation of HMF under Solar Irradiation: Coupling of Microemulsion and Lyophilization to Obtain Innovative TiO_2-Based Materials." *Molecules* 25 (22). MDPI AG: 5225. doi:10.3390/molecules25225225.

Antonyraj, Churchil A., Jaewon Jeong, Bora Kim, Seunghan Shin, Sangyong Kim, Kwan Young Lee, and Jin Ku Cho. 2013. "Selective Oxidation of HMF to DFF Using Ru/γ-Alumina Catalyst in Moderate Boiling Solvents toward Industrial Production." *Journal of Industrial and Engineering Chemistry* 19 (3): 1056–59. doi:10.1016/j.jiec.2012.12.002.

Ayed, Cyrine, Wei Huang, Gönül Kizilsavas, Katharina Landfester, and Kai A. I. Zhang. 2020. "Photocatalytic Partial Oxidation of 5-Hydroxymethylfurfural (HMF) to 2,5-Diformylfuran (DFF) Over a Covalent Triazine Framework in Water." *ChemPhotoChem* 4 (8). Wiley-VCH Verlag: 571–76. doi:10.1002/cptc.202000070.

Bao, Xiaolei, Mu Liu, Zeyan Wang, Dujuan Dai, Peng Wang, Hefeng Cheng, Yuanyuan Liu, Zhaoke Zheng, Ying Dai, and Baibiao Huang. 2022. "Photocatalytic Selective Oxidation of HMF Coupled with H_2 Evolution on Flexible Ultrathin g-C_3N_4 Nanosheets with Enhanced N-H Interaction." *ACS Catalysis* 12 (3). American Chemical Society: 1919–29. doi:10.1021/acscatal.1c05357.

Battula, Venugopala Rao, Arpna Jaryal, and Kamalakannan Kailasam. 2019. "Visible Light-Driven Simultaneous H_2 Production by Water Splitting Coupled with Selective Oxidation of HMF to DFF Catalyzed by Porous Carbon Nitride." *Journal of Materials Chemistry A* 7 (10). Royal Society of Chemistry: 5643–49. doi:10.1039/c8ta10926e.

Bozell, Joseph J., and Gene R. Petersen. 2010. "Technology Development for the Production of Biobased Products from Biorefinery Carbohydrates—the US Department of Energy's 'Top 10' Revisited." *Green Chemistry* 12 (4): 539–55. doi:10.1039/b922014c.

Cao, Shaowen, and Jiaguo Yu. 2016. "Carbon-Based H_2-Production Photocatalytic Materials." *Journal of Photochemistry and Photobiology C: Photochemistry Reviews* 27. Elsevier B.V: 72–99. doi:10.1016/j.jphotochemrev.2016.04.002.

Carpenter, Andrew J., and Derek J Chadwick. 1985. "Chemoselective Protection of Heteroaromatic Aldehydes as Imidazolidine Derivatives. Preparation of +Substituted Furan-and Thiophene-2-Carboxaldehydes via Metallo-Imidazolidine Intermediates." *Tetrahedron* 41 (18). Elsevier: 3803–12. doi: 10.1016/S0040-4020(01)91401-2

Chen, Chih Chia, Dai Ling Tsai, Hsing Ting Liu, and Jih Jen Wu. 2023. "Carbon Vacancy-Modified Carbon Nitride Allotropic Composite for Solar Hydrogen Generation Coupled with Selective Oxidation of 5-Hydroxymethylfurfural." *ACS Sustainable Chemistry and Engineering* 11 (16). American Chemical Society: 6435–44. doi:10.1021/acssuschemeng.3c00363.

Cheng, Lili, Danyao Huang, Yun Zhang, and Ying Wu. 2021. "Photocatalytic Selective Oxidation of HMF to DFF over Bi_2WO_6/mpg–C_3N_4 Composite under Visible Light." *Applied Organometallic Chemistry* 35 (11). John Wiley and Sons Ltd: e6404. doi:10.1002/aoc.6404.

Chu, Sheng, Jingjing Shao, Hongyu Qu, Xintie Wang, Rui Xiao, and Huiyan Zhang. 2023. "Band Structure Engineering of Polyimide Photocatalyst for Efficient and Selective Oxidation of Biomass-Derived 5-Hydroxymethylfurfural." *ChemSusChem* 16 (19). John Wiley and Sons Inc. doi:10.1002/cssc.202300886.

Cui, Jifang, Zhihao Yu, Linhao Sun, Ming Zhang, Mengyan Guo, Jian Xiong, Yina Qiao, Rui Zhang, and Xuebin Lu. 2023. "Photocatalytic Selective Oxidation of Biomass-Derived 5-Hydroxymethylfurfural to 2,5-Diformylfuran under Ambient Conditions over $CdIn_2S_4$/g-C_3N_4 Heterojunctions." *Catalysis Science and Technology* 13 (20). Royal Society of Chemistry: 5892–901. doi:10.1039/d3cy00794d.

Dai, Jinhang. 2021. "Synthesis of 2,5-Diformylfuran from Renewable Carbohydrates and Its Applications: A Review." *Green Energy and Environment* 6 (1). KeAi Publishing Communications Ltd: 22–32. doi:10.1016/j.gee.2020.06.013.

Dawes, Jacob. 2022. Techno-economic evaluation for FDPA, a novel bobased monomer. https://fse.studenttheses.ub.rug.nl/27936/1/Integrationproject_JacobDawes.pdf.

Dhingra, Suman, Tripti Chhabra, Venkata Krishnan, and C. M. Nagaraja. 2020. "Visible-Light-Driven Selective Oxidation of Biomass-Derived HMF to DFF Coupled with H2Generation by Noble Metal-Free Zn0.5$Cd_{0.5}S/MnO_2$ Heterostructures." *ACS Applied Energy Materials* 3 (7). American Chemical Society: 7138–48. doi:10.1021/acsaem.0c01189.

Dimeglio, John L., Andrew G. Breuhaus-Alvarez, Siqi Li, and Bart M. Bartlett. 2019. "Nitrate-Mediated Alcohol Oxidation on Cadmium Sulfide Photocatalysts." *ACS Catalysis* 9 (6). American Chemical Society: 5732–41. doi:10.1021/acscatal.9b01051.

Fujishima, Akira and Honda, Kenichi. 1972. "Electrochemical Photolysis of Water at a Semiconductor Electrode." *Nature*, 238, 37–38.

García-López, Elisa I., Francesca Rita Pomilla, Ermelinda Bloise, Xiang fei Lü, Giuseppe Mele, Leonardo Palmisano, and Giuseppe Marcì. 2021. "C_3N_4 Impregnated with Porphyrins as Heterogeneous Photocatalysts for the Selective Oxidation of 5-Hydroxymethyl-2-Furfural

Under Solar Irradiation." *Topics in Catalysis* 64 (13–16). Springer: 758–71. doi:10.1007/s11244-020-01293-0.

Giannakoudakis, Dimitrios A., Vaishakh Nair, Ayesha Khan, Eleni A. Deliyanni, Juan Carlos Colmenares, and Konstantinos S. Triantafyllidis. 2019. "Additive-Free Photo-Assisted Selective Partial Oxidation at Ambient Conditions of 5-Hydroxymethylfurfural by Manganese (IV) Oxide Nanorods." *Applied Catalysis B: Environmental* 256 (November). Elsevier. doi:10.1016/j.apcatb.2019.117803.

Gnanaguru, Mario Vino Lincy, Vishal Kumar Parida, Makarand M. Ghangrekar, Ashok Kumar Gupta, and Shamik Chowdhury. 2024. "Insights into the Performance of Binary Heterojunction Photocatalysts for Degradation of Refractory Pollutants." *Environmental Science and Pollution Research*, 31 (8) January: 11349–70. doi:10.1007/s11356-023-31592-4.

Gong, Ming, Hui Zhao, Chengsi Pan, Yuming Dong, Yingxin Guo, Haixia Li, Jiawei Zhang, Guangli Wang, and Yongfa Zhu. 2023. "Highly Selective Photocatalytic Oxidation of 5-Hydroxymethylfurfural by Interfacial Pt-O Bonding Pt-Ov-BiOBr." *New Journal of Chemistry* 47 (15). Royal Society of Chemistry: 7118–26. doi:10.1039/d3nj00498h.

Gonzalez-Casamachin, Diego Alexander, Javier Rivera De la Rosa, Carlos J. Lucio-Ortiz, Ladislao Sandoval-Rangel, and Carlos D. García. 2020. "Partial Oxidation of 5-hydroxymethylfurfural to 2,5-Furandicarboxylic Acid Using O_2 and a Photocatalyst of a Composite of ZnO/PPy under Visible-Light: Electrochemical Characterization and Kinetic Analysis." *Chemical Engineering Journal* 393 (August). Elsevier B.V. doi:10.1016/j.cej.2020.124699.

Granone, L. I., F. Sieland, N. Zheng, R. Dillert, and D. W. Bahnemann. 2018. "Photocatalytic Conversion of Biomass into Valuable Products: A Meaningful Approach?" *Green Chemistry* 20 (6). Royal Society of Chemistry: 1169–92. doi:10.1039/c7gc03522e.

Guo, Yingxin, Bing Liu, Jiawei Zhang, Guangli Wang, Chengsi Pan, Hui Zhao, Chuntao Wang, Feiyang Yu, Yuming Dong, and Yongfa Zhu. 2024. "Perylene Imide Supermolecule Promote Oxygen to Superoxide Radical for Ultrafast Photo-Oxidation of 5-Hydroxymethylfurfural." *Applied Catalysis B: Environmental* 340 (January). Elsevier B.V. doi:10.1016/j.apcatb.2023.123217.

Han, Guanqun, Yan Huan Jin, R. Alan Burgess, Nicholas E. Dickenson, Xiao Ming Cao, and Yujie Sun. 2017. "Visible-Light-Driven Valorization of Biomass Intermediates Integrated with H_2 Production Catalyzed by Ultrathin Ni/CdS Nanosheets." *Journal of the American Chemical Society* 139 (44). American Chemical Society: 15584–87. doi:10.1021/jacs.7b08657.

Ilkaeva, M., I. Krivtsov, E. I. García-López, G. Marcì, O. Khainakova, J. R. García, L. Palmisano, E. Díaz, and S. Ordóñez. 2018 "Selective Photocatalytic Oxidation of 5-Hydroxymethylfurfural to 2,5-Furandicarboxaldehyde by Polymeric Carbon Nitride-Hydrogen Peroxide Adduct." *Journal of Catalysis* 359 (March). Elsevier: 212–22. doi:10.1016/j.jcat.2018.01.012

Jang, Woo Sung, Vy Ngoc Pham, Sang Hyeok Yang, Jaeyoon Baik, Hangil Lee, and Young Min Kim. 2023. "Enhanced Selective Photocatalytic Oxidation of a Bio-Derived Platform Chemical with Vacancy-Induced Core-Shell Anatase TiO2 Nanoparticles." *Applied Catalysis B: Environmental* 322 (March). Elsevier doi:10.1016/j.apcatb.2022.122140.

Jing, Liquan, Meng Xie, Yuanguo Xu, Chun Tong, Yanhua Song, Xia Du, Heng Zhao, et al. 2023. "O-Doped and Nitrogen Vacancies 3D C3N4 Activation of Peroxydisulfate for Pollutants Degradation and Transfer Hydrogenation of Nitrophenols with Water." *Separation and Purification Technology* 314 (June). Elsevier doi:10.1016/j.seppur.2023.123540.

Khan, Ayesha, Michael Goepel, Adam Kubas, Dariusz Łomot, Wojciech Lisowski, Dmytro Lisovytskiy, Ariadna Nowicka, Juan Carlos Colmenares, and Roger Gläser. 2021.

"Selective Oxidation of 5-Hydroxymethylfurfural to 2,5-Diformylfuran by Visible Light-Driven Photocatalysis over In Situ Substrate-Sensitized Titania." *ChemSusChem* 14 (5). John Wiley and Sons Inc: 1351–62. doi:10.1002/cssc.202002687.

Kim, Jeongeun, Jiwon Choi, Nurwarrohman Andre Sasongko, Myeongkee Park, Myung Jong Kang, and Hyun Sung Kim. 2023. "Highly Selective Photocatalytic Conversion of 5-Hydroxymethyl-2-Furfural to 2,5-Diformylfuran over Gold Ion Exchanged Cadmium Sulfide Nanowire." *Advanced Sustainable Systems* 7 (8). John Wiley and Sons Inc. doi:10.1002/adsu.202300102.

Krivtsov, Igor, Elisa I. García-López, Giuseppe Marcì, Leonardo Palmisano, Zakariae Amghouz, José R. García, Salvador Ordóñez, and Eva Díaz. 2017. "Selective Photocatalytic Oxidation of 5-Hydroxymethyl-2-Furfural to 2,5-Furandicarboxyaldehyde in Aqueous Suspension of g-C_3N_4." *Applied Catalysis B: Environmental* 204 (May). Elsevier: 430–39. doi:10.1016/j.apcatb.2016.11.049.

Krivtsov, Igor, Marina Ilkaeva, Eduardo Salas-Colera, Zakariae Amghouz, José R. García, Eva Díaz, Salvador Ordóñez, and Silvia Villar-Rodil. 2017. "Consequences of Nitrogen Doping and Oxygen Enrichment on Titanium Local Order and Photocatalytic Performance of TiO_2 Anatase." *Journal of Physical Chemistry C* 121 (12). American Chemical Society: 6770–80. doi:10.1021/acs.jpcc.7b00354.

Kumar, Abhinav, and Rajendra Srivastava. 2021. "Rose-like Bi2WO6Nanostructure for Visible-Light-Assisted Oxidation of Lignocellulose-Derived 5-Hydroxymethylfurfural and Vanillyl Alcohol." *ACS Applied Nano Materials* 4 (9). American Chemical Society: 9080–93. doi:10.1021/acsanm.1c01679.

Li, Chen, Jiang Li, Ling Qin, Piaoping Yang, and Dionisios G. Vlachos. 2021. "Recent Advances in the Photocatalytic Conversion of Biomass-Derived Furanic Compounds." *ACS Catalysis* 11 (18). American Chemical Society: 11336–59. doi:10.1021/acscatal.1c02551.

Li, Lei, Changfa Guo, Junling Shen, Jiqiang Ning, Yijun Zhong, and Yong Hu. 2020. "Construction of Sugar-Gourd-Shaped CdS/Co1-XS Hollow Hetero-Nanostructure as an Efficient Z-Scheme Photocatalyst for Hydrogen Generation." *Chemical Engineering Journal* 400 (November). Elsevier. doi:10.1016/j.cej.2020.125925.

Lin, Kun, Yi Andrew, Wen Da Oh, Meng Wei Zheng, Eilhann Kwon, Jechan Lee, Jia Yin Lin, Xiaoguang Duan, and Farshid Ghanbari. 2021. "Aerobic Oxidation of 5-Hydroxymethylfurfural into 2,5-Diformylfuran Using Manganese Dioxide with Different Crystal Structures: A Comparative Study." *Journal of Colloid and Interface Science* 592 (June). Academic Press Inc.. 416–29. doi:10.1016/j.jcis.2021.02.030.

Lolli, Alice, Valeriia Maslova, Danilo Bonincontro, Francesco Basile, Simona Ortelli, and Stefania Albonetti. 2018. "Selective Oxidation of HMF via Catalytic and Photocatalytic Processes Using Metal-Supported Catalysts." *Molecules* 23 (11). MDPI: 2792. doi:10.3390/molecules23112792.

Ma, Ben, Yingyong Wang, Xiaoning Guo, Xili Tong, Chang Liu, Yunwei Wang, and Xiangyun Guo. 2018. "Photocatalytic Synthesis of 2,5-Diformylfuran from 5-Hydroxymethyfurfural or Fructose over Bimetallic Au-Ru Nanoparticles Supported on Reduced Graphene Oxides." *Applied Catalysis A: General* 552 (February). Elsevier: 70–6. doi:10.1016/j.apcata.2018.01.002.

Mazzanti, Stefano, and Aleksandr Savateev. 2020. "Emerging Concepts in Carbon Nitride Organic Photocatalysis." *ChemPlusChem* 85 (11). Wiley-VCH Verlag: 2499–517. doi:10.1002/cplu.202000606.

Meng, Ye, Song Yang, and Hu Li. 2022. "Electro- and Photocatalytic Oxidative Upgrading of Bio-Based 5-Hydroxymethylfurfural." *ChemSusChem* 15 (13). John Wiley and Sons Inc. doi:10.1002/cssc.202102581.

Novaes, Luiz F. T., Jinjian Liu, Yifan Shen, Lingxiang Lu, Jonathan M. Meinhardt, and Song Lin. 2021. "Electrocatalysis as an Enabling Technology for Organic Synthesis." *Chemical Society Reviews* 50 (14). The Royal Society of Chemistry: 7941–8002. doi:10.1039/D1CS00223F.

Pal, Priyanka, and Shunmugavel Saravanamurugan. 2019. "Recent Advances in the Development of 5-Hydroxymethylfurfural Oxidation with Base (Nonprecious)-Metal-Containing Catalysts." *ChemSusChem* 12 (1). Wiley-VCH Verlag: 145–63. doi:10.1002/cssc.201801744.

Pal, Priyanka, and Shunmugavel Saravanamurugan. 2020. "Heterostructured Manganese Catalysts for the Selective Oxidation of 5-Hydroxymethylfurfural to 2,5-Diformylfuran." *ChemCatChem* 12 (8). Wiley Blackwell: 2324–32. doi:10.1002/cctc.202000086.

Qian, Hengli, Qidong Hou, Weijie Zhang, Yifan Nie, Ruite Lai, Huiru Ren, Guanjie Yu, Xinyu Bai, Haozhi Wang, and Meiting Ju. 2022. "Construction of Electron Transport Channels and Oxygen Adsorption Sites to Modulate Reactive Oxygen Species for Photocatalytic Selective Oxidation of 5-Hydroxymethylfurfural to 2,5-Diformylfuran." *Applied Catalysis B: Environmental* 319 (December). Elsevier doi:10.1016/j.apcatb.2022.121907.

Qiu, Jianhao, Lu Zhang, Dingliang Dai, Guanglu Xia, and Jianfeng Yao. 2022. "Cellulose-Derived Carbon Dot-Guided Growth of $ZnIn_2S_4$ Nanosheets for Photocatalytic Oxidation of 5-Hydroxymethylfurfural into 2,5-Diformylfuran." *ChemSusChem* 15 (10). John Wiley and Sons Inc. doi:10.1002/cssc.202200399.

Sabzehmeidani, Mohammad Mehdi, Sahar Mahnaee, Mehrorang Ghaedi, Hadi Heidari, and Vellaisamy A. L. Roy. 2021. "Carbon Based Materials: A Review of Adsorbents for Inorganic and Organic Compounds." *Materials Advances* 2 (2). Royal Society of Chemistry: 598–627. doi:10.1039/d0ma00087f.

Saini, Ms Kanika, Sahil Kumar, Hu Li, Srinivasarao Arulananda Babu, and Shunmugavel Saravanamurugan. 2022. "Advances in the Catalytic Reductive Amination of Furfural to Furfural Amine: The Momentous Role of Active Metal Sites." *ChemSusChem* 15 (7). John Wiley and Sons Inc. doi:10.1002/cssc.202200107.

Sharma, Shelja, Sandeep Kumar, Senthil Murugan Arumugam, Muthukumaran Palanisami, Vijayakumar Shanmugam, and Sasikumar Elumalai. 2022. "Nb2O5/g-C3N4 Heterojunction Facilitates 2,5-Diformylfuran Production via Photocatalytic Oxidation of 5-Hydroxymethylfurfural under Direct Sunlight Irradiation." *ChemPhotoChem* 6 (2). John Wiley and Sons Inc. doi:10.1002/cptc.202100199.

Shinde, Suhas Hanmant, and Chandrashekhar Vasant Rode. 2019. "2,5-Diformylfuran-an Oxidation Product of 5-Hydroxymethylfurfural." In Shunmugavel Saravanamurugan, Ashok Pandey, Hu Li and Anders Riisager (Eds.), *Biomass, Biofuels, Biochemicals: Recent Advances in Development of Platform Chemicals*, 95–133. Elsevier Inc. doi:10.1016/B978-0-444-64307-0.00004-4.

Su, Ting, Deyang Zhao, Yantao Wang, L. Hongying, Rajender S. Varma, and Christophe Len. 2021. "Innovative Protocols in the Catalytic Oxidation of 5-Hydroxymethylfurfural." *ChemSusChem*. Wiley-VCH Verlag. doi:10.1002/cssc.202002232.

Sun, Huimin, Ruohan Xu, Xiang Jia, Zhongyi Liu, Haijun Chen, and Tianliang Lu. 2023. "Recent Advances in the Photocatalytic Oxidation of 5-Hydroxymethylfurfural to 2,5-Diformylfuran." *Biomass Conversion and Biorefinery*. Springer Science and Business Media Deutschland GmbH: 1–18. doi:10.1007/s13399-022-03727-z.

Tan, Xinying, Shenghe Si, Difei Xiao, Xiaolei Bao, Kepeng Song, Zeyan Wang, Yuanyuan Liu, et al. 2023. "Single Cobalt Atoms Induced Molecular O_2 Activation for Enhanced Photocatalytic Biomass Upgrading on $ZnIn_2S_4$ Nanosheets." *ACS Catalysis* 13 (21). American Chemical Society: 14395–403. doi:10.1021/acscatal.3c03332.

Ulyankina, Anna, Sergey Mitchenko, and Nina Smirnova. 2020. "Selective Photocatalytic Oxidation of 5-HMF in Water over Electrochemically Synthesized TiO_2 Nanoparticles." *Processes* 8 (647). MDPI AG: 647. doi:10.3390/pr8060647.

Wang, Hao, Danyao Huang, Qixuan Zhuge, and Ying Wu. 2023. "Oxygen Vacancies Mediated Enhanced Visible-Light-Driven Photocatalytic Oxidation of Biomass Derived 5-Hydroxymethylfurfural over V_2O_5·nH2O/g-C_3N_4 Composite." *Diamond and Related Materials* 139 (November). Elsevier Ltd. doi:10.1016/j.diamond.2023.110413.

Wang, Hui, Shenlong Jiang, Shichuan Chen, Dandan Li, Xiaodong Zhang, Wei Shao, Xianshun Sun, et al. 2016. "Enhanced Singlet Oxygen Generation in Oxidized Graphitic Carbon Nitride for Organic Synthesis." *Advanced Materials* 28 (32). Wiley-VCH Verlag: 6940–45. doi:10.1002/adma.201601413.

Wang, Jiaqi, Yun Yuan, Kaiyuan Ren, Bingqing Wang, and Zhaohui Li. 2023. "Efficient Acceptorless Dehydrogenation of 5-Hydroxymethylfurfural (HMF) to 2,5-Diformylfuran (DFF) over Pt/CdS under Visible Light." *Journal of Catalysis* 417 (January). Academic Press Inc.: 178–84. doi:10.1016/j.jcat.2022.12.008.

Wang, Xiang Xiang, Sugang Meng, Sujuan Zhang, Xiuzhen Zheng, and Shifu Chen. 2020. "2D/2D MXene/g-C3N4 for Photocatalytic Selective Oxidation of 5-Hydroxymethylfurfural into 2,5-Formylfuran." *Catalysis Communications* 147 (December). Elsevier: doi:10.1016/j.catcom.2020.106152.

Wang, Yuwei, Xianggui Kong, Meihong Jiang, Fazhi Zhang, and Xiaodong Lei. 2020. "A: Z -Scheme $ZnIn_2S_4/Nb_2O_5$ Nanocomposite: Constructed and Used as an Efficient Bifunctional Photocatalyst for H_2 Evolution and Oxidation of 5-Hydroxymethylfurfural." *Inorganic Chemistry Frontiers* 7 (2). Royal Society of Chemistry: 437–46. doi:10.1039/c9qi01196j.

Wu, Huihui, Sugang Meng, Jinfeng Zhang, Xiuzhen Zheng, Ya Xiao Wang, Shifu Chen, Guoqing Qi, and Xianliang Fu. 2020. "Construction of Two-Dimensionally Relative p-n Heterojunction for Efficient Photocatalytic Redox Reactions under Visible Light." *Applied Surface Science* 505 (March). Elsevier doi:10.1016/j.apsusc.2019.144638.

Wu, Qi, Yiming He, Huili Zhang, Zhiyuan Feng, Ying Wu, and Tinghua Wu. 2017. "Photocatalytic Selective Oxidation of Biomass-Derived 5-Hydroxymethylfurfural to 2,5-Diformylfuran on Metal-Free g-C_3N_4 under Visible Light Irradiation." *Molecular Catalysis* 436. Elsevier: 10–8. doi:10.1016/j.mcat.2017.04.012.

Xia, Tong, Wanbing Gong, Yihong Chen, Meilin Duan, Jun Ma, Xiaofeng Cui, Yitao Dai, Chao Gao, and Yujie Xiong. 2022. "Sunlight-Driven Highly Selective Catalytic Oxidation of 5-Hydroxymethylfurfural Towards Tunable Products." *Angewandte Chemie – International Edition* 61 (29). John Wiley and Sons Inc. doi:10.1002/anie.202204225.

Xu, Shuai, Peng Zhou, Zehui Zhang, Changjun Yang, Bingguang Zhang, Kejian Deng, Steven Bottle, and Huaiyong Zhu. 2017. "Selective Oxidation of 5-Hydroxymethylfurfural to 2,5-Furandicarboxylic Acid Using O_2 and a Photocatalyst of Co-Thioporphyrazine Bonded to g-C_3N_4." *Journal of the American Chemical Society* 139 (41). American Chemical Society: 14775–82. doi:10.1021/jacs.7b08861.

Xue, Jinjuan, Chengjuan Huang, Yuqing Zong, Jiandong Gu, Mingxin Wang, and Shuaishuai Ma. 2019. "Fe (III)-Grafted Bi_2MoO_6 Nanoplates for Enhanced Photocatalytic Activities on Tetracycline Degradation and HMF Oxidation." *Applied Organometallic Chemistry* 33 (11). John Wiley and Sons Ltd. doi:10.1002/aoc.5187.

Yang, Bo, Wenwei Hu, Feifei Wan, Chao Zhang, Zaihui Fu, Anqun Su, Mengke Chen, and Yachun Liu. 2020. "Adjusting Effect of Additives on Decatungstate-Photocatalyzed HMF Oxidation with Molecular Oxygen under Visible Light Illumination." *Chemical Engineering Journal* 396 (September). Elsevier doi:10.1016/j.cej.2020.125345.

Ye, Hui Fang, Rui Shi, Xiao Yang, Wen Fu Fu, and Yong Chen. 2018. "P-Doped $Zn_xCd_{1-x}S$ Solid Solutions as Photocatalysts for Hydrogen Evolution from Water Splitting Coupled with Photocatalytic Oxidation of 5-Hydroxymethylfurfural." *Applied Catalysis B: Environmental* 233 (October). Elsevier: 70–9. doi:10.1016/j.apcatb.2018.03.060.

Yin, Liwen, Yaqiang Wu, Xiaolei Bao, Xiaolei Liu, Dujuan Dai, Minghui Zhang, Zeyan Wang, et al. 2023. "Lead-Free Perovskite Cs_2SnBr_6/rGO Composite for Photocatalytic Selective Oxidation of 5-Hydroxymethylfurfural to 2,5-Diformylfuran." *Chemistry – A European Journal* 29 (39). John Wiley and Sons Inc. doi:10.1002/chem.202300999.

Yurdakal, Sedat, Bilge Sina Tek, Oğuzhan Alagöz, Vincenzo Augugliaro, Vittorio Loddo, Giovanni Palmisano, and Leonardo Palmisano. 2013. "Photocatalytic Selective Oxidation of 5-(Hydroxymethyl)-2-Furaldehyde to 2,5-Furandicarbaldehyde in Water by Using Anatase, Rutile, and Brookite TiO_2 Nanoparticles." *ACS Sustainable Chemistry and Engineering* 1 (5): 456–61. doi:10.1021/sc300142a.

Zakrzewska, Małgorzata E., Ewa Bogel-Łukasik, and Rafał Bogel-Łukasik. 2011. "Ionic Liquid-Mediated Formation of 5-Hydroxymethylfurfural-A Promising Biomass-Derived Building Block." *Chemical Reviews* 111 (2). American Chemical Society: 397–417. doi:10.1021/cr100171a.

Zhang, Fubao, Xianmin Wang, Haonan Liu, Chunli Liu, Yong Wan, Yunze Long, and Zhongyu Cai. 2019. "Recent Advances and Applications of Semiconductor Photocatalytic Technology." *Applied Sciences (Switzerland)* 9 (12). MDPI AG: 2489. doi:10.3390/app9122489.

Zhang, Huili, Zhiyuan Feng, Yekun Zhu, Ying Wu, and Tinghua Wu. 2019. "Photocatalytic Selective Oxidation of Biomass-Derived 5-Hydroxymethylfurfural to 2,5-Diformylfuran on WO3/g-C3N4 Composite under Irradiation of Visible Light." *Journal of Photochemistry and Photobiology A: Chemistry* 371 (February). Elsevier B.V.: 1–9. doi:10.1016/j.jphotochem.2018.10.044.

Zhang, Huili, Qi Wu, Cong Guo, Ying Wu, and Tinghua Wu. 2017. "Photocatalytic Selective Oxidation of 5-Hydroxymethylfurfural to 2,5-Diformylfuran over Nb_2O_5 under Visible Light." *ACS Sustainable Chemistry and Engineering* 5 (4). American Chemical Society: 3517–23. doi:10.1021/acssuschemeng.7b00231.

Zhang, Ming, Zhihao Yu, Jian Xiong, Rui Zhang, Xinzhong Liu, and Xuebin Lu. 2022. "One-Step Hydrothermal Synthesis of CdxInyS(X+1.5y) for Photocatalytic Oxidation of Biomass-Derived 5-Hydroxymethylfurfural to 2, 5-Diformylfuran under Ambient Conditions." *Applied Catalysis B: Environmental* 300 (January). Elsevier B.V. doi:10.1016/j.apcatb.2021.120738.

Zhang, Ming, Zhihao Yu, Yuxin Zhang, Linhao Sun, Jifang Cui, Jian Xiong, You Han, and Xuebin Lu. 2023. "Photocatalytic Conversion of 5-Hydroxymethylfurfural to 2,5-Diformylfuran by S-Scheme Black Phosphorus/$CdIn_2S_4$ Heterojunction." *Catalysis Science and Technology* 13 (23). Royal Society of Chemistry: 6640–52. doi:10.1039/d3cy01075a.

Zhang, Ming, Yuxin Zhang, Lei Ye, Zhihao Yu, Runyu Liu, Yina Qiao, Linhao Sun, Jifang Cui, and Xuebin Lu. 2023. "In Situ Fabrication $Ti_3C_2F_x$ MXene/$CdIn_2S_4$ Schottky Junction for Photocatalytic Oxidation of HMF to DFF under Visible Light." *Applied Catalysis B: Environmental* 330 (August). Elsevier doi:10.1016/j.apcatb.2023.122635.

Zhang, Mo, Zheng Li, Xing Xin, Junhao Zhang, Yeqin Feng, and Hongjin Lv. 2020. "Selective Valorization of 5-Hydroxymethylfurfural to 2,5-Diformylfuran Using Atmospheric O_2 and $MAPbBr_3$ Perovskite under Visible Light." *ACS Catalysis* 10 (24). American Chemical Society: 14793–800. doi:10.1021/acscatal.0c04330.

Zhang, Qiaozhi, Zhonghao Wan, Iris K. M. Yu, and Daniel C. W. Tsang. 2021. "Sustainable Production of High-Value Gluconic Acid and Glucaric Acid through Oxidation of

Biomass-Derived Glucose: A Critical Review." *Journal of Cleaner Production* 312. Elsevier Ltd. doi:10.1016/j.jclepro.2021.127745.

Zhang, Weizi, Xinxin Li, Shaoru Liu, Jianhao Qiu, Jiahuan An, Jianfeng Yao, Songlin Zuo, Bo Zhang, Haian Xia, and Changzhi Li. 2022. "Photocatalytic Oxidation of 5-Hydroxymethylfurfural Over Interfacial-Enhanced Ag/TiO_2 Under Visible Light Irradiation." *ChemSusChem* 15 (13). John Wiley and Sons Inc. doi:10.1002/cssc.202102158.

Zhang, Xin, Xingzhong Yuan, Longbo Jiang, Jin Zhang, Hanbo Yu, Hou Wang, and Guangming Zeng. 2020. "Powerful Combination of 2D g-C_3N_4 and 2D Nanomaterials for Photocatalysis: Recent Advances." *Chemical Engineering Journal* 390. Elsevier doi:10.1016/j.cej.2020.124475.

Zhang, Yingchuan, Guangri Jia, Wenchao Wang, Liqun Jiang, and Zhengxiao Guo. 2024. "Photocatalytic Upgrading of 5-Hydroxymethylfurfural – Aerobic or Anaerobic?" *Green Chemistry* 26 (6). Royal Society of Chemistry: 2949–66. doi:10.1039/d3gc04814d.

Zhang, Zhe, Min Wang, Hongru Zhou, and Feng Wang. 2021. "Surface Sulfate Ion on CdS Catalyst Enhances Syngas Generation from Biopolyols." *Journal of the American Chemical Society* 143 (17). American Chemical Society: 6533–41. doi:10.1021/jacs.1c00830.

Zhao, Heng, Dhwanil Trivedi, Morteza Roostaeinia, Xue Yong, Jun Chen, Pawan Kumar, Jing Liu, et al. 2022. "Spatial Charge Separation on the (110)/(102) Facets of Cocatalyst-Free ZnIn2S4 for the Selective Conversion of 5-Hydroxymethylfurfural to 2,5-Diformylfuran." *Green Chemistry* 25 (2). Royal Society of Chemistry: 692–99. doi:10.1039/d2gc04362a.

Zhu, Peng, Weizi Zhang, Qinfang Li, and Haian Xia. 2022. "Visible-Light-Driven Photocatalytic Oxidation of 5-Hydroxymethylfurfural to 2,5-Furandicarboxylic Acid over Plasmonic Au/ZnO Catalyst." *ACS Sustainable Chemistry and Engineering* 10 (27). American Chemical Society: 8778–87. doi:10.1021/acssuschemeng.2c01143.

Zhu, Qian, Yan Zhuang, Hongqing Zhao, Peng Zhan, Cong Ren, Changsheng Su, Wenqiang Ren, Jiawen Zhang, Di Cai, and Peiyong Qin. 2023. "2,5-Diformylfuran Production by Photocatalytic Selective Oxidation of 5-Hydroxymethylfurfural in Water Using MoS_2/$CdIn_2S_4$ Flower-like Heterojunctions." *Chinese Journal of Chemical Engineering* 54 (February). Materials China: 180–91. doi:10.1016/j.cjche.2022.04.018.

Zou, Yingtong, Yezi Hu, Andrej Uhrich, Zewen Shen, Baoxiang Peng, Zhuoyu Ji, Martin Muhler, Guixia Zhao, Xiangke Wang, and Xiaoxiang Xu. 2021. "Steering Accessible Oxygen Vacancies for Alcohol Oxidation over Defective Nb_2O_5 under Visible Light Illumination." *Applied Catalysis B: Environmental* 298 (December). Elsevier B.V. doi:10.1016/j.apcatb.2021.120584.

5 Catalytic Hydrogenation of Furfural

Mei Wu, Jian He, and Hu Li

5.1 INTRODUCTION

Traditionally fossil-based resources have been integral to developing desired fuels and chemicals, driving economic growth (Chheda and Dumesic 2007, Ma et al. 2019). However, these resources are finite in supply, and CO_2 is released into the environment by combustion, bringing about a typical global climate warming (Farmer and Prather 2020, Wang et al. 2019). Before 2016, approximately 85.5% of the world's energy consumption originated from fossil resources. For these reasons, biomass is recognized as a perfect candidate for traditional fossil reserves because of its sustainability on Earth, which is expected to be upgraded into useful fuels and fine chemicals (Elliott et al. 2007). Biomass, particularly lignocellulose, has garnered vast research interest as a competitive candidate in recent decades (Chheda et al. 2007). Lignocellulosic biomass is plentiful, sustainable, and economically viable to produce organic chemicals and liquid biofuels (Wang et al. 2020, Jung et al. 2024). An abundance of lignocellulosic biomass makes it an attractive option for reducing our dependence on finite fossil resources, while its cost-efficiency adds to its appeal. Furthermore, the production of lignocellulosic biomass involves the growth of plants, which naturally consume CO_2 during their life cycle (Fang et al. 2020). This carbon sequestration process makes the utilization of biomass more environmentally friendly compared to the exploitation of fossil resources (Huber et al. 2006). Utilizing lignocellulosic biomass contributes to mitigating CO_2 emissions and makes the future more sustainable and greener.

Lignocellulose is a complicated organic material composed of cellulose, hemicellulose, and lignin (Amarasekara et al. 2023, Rai et al. 2019), and identified as one of the most viable raw materials for biorefinery processes because of the richness and potential for converting into desired products (Yoo and Kim 2020, Morales et al. 2020, Li et al. 2017, Tan et al. 2022). Upgrading lignocellulose into biofuels and platform molecules involves various processes, including pyrolysis, liquefaction, gasification, and hydrolysis, while pyrolysis, liquefaction, and gasification are effective, they often require harsh conditions (Li et al. 2016, Zhou et al. 2021, Wang et al. 2022). An attractive way involves the hydrolysis of cellulose and hemicellulose, which provides access for directionally converting these sustainable resources into next-generation biofuels and platform molecules (Garedew et al. 2020, Brodin et al.

 DOI: 10.1201/9781003464839-5

2017, Kobayashi et al. 2012). Via hydrolysis, the hemicellulose can be valorized into monosaccharides, polyols, furans, and acids (Li et al. 2017, Li et al. 2016, Bozell and Petersen 2010, Wang et al. 2024, Meng et al. 2023). These compounds are often regarded as valuable building blocks for yielding biofuels and platform chemicals. Furfural is particularly noteworthy as it can act as a "bridge" in relation to the biomass and chemical industry, and has garnered significant attention due to its potential as a versatile intermediate for producing broad value-added products (Shivhare et al. 2021, Lucarelli et al. 2019, Luo et al. 2022). In addition, furfural is rated as a key precursor for producing various important compounds in the chemical industry, which include furfural, furanone, furan, furfuryl alcohol, furfuryl acetate, 2-furoyl chloride, furoylglycine and so on (Lv et al. 2022, Hashim et al. 2020, Arumugam et al. 2020, Shrestha et al. 2017, Kumar et al. 2020, Li et al. 2017, Li et al. 2019). Despite its importance, the current global production of furfural is still low (~50%) (Li et al. 2000). In light of the increasing demand for renewable alternatives to fossil-based chemicals, there is growing interest in synthesizing furfural from biomass-derived feedstocks. Along this field, researchers are exploring a variety of effective research approaches and technologies, aiming for maximizing furfural yield and pave the way for a greener and more sustainable future.

Table 5.1 has displayed a summary of the physicochemical properties of furfural, highlighting the characteristics and potential applications in the biofuel and chemicals industry (Gowda et al. 2012, Yan et al. 2013). Chemically, furfural is a typical aromatic molecule, which contains an aldehyde group (–CHO) and a furan-ring. Valorization of furfural into valuable chemicals often involves reactions that exploit these functional groups. Various typical reactions can be performed to upgrade the aldehyde group of furfurals, such as hydrogenation reduction, decarbonylation, reductive amination, oxidation, and acetalization (Xu et al. 2019, Fang et al. 2021, Dong et al. 2020, Ren et al. 2023, Song et al. 2021). On the other things, its furan

TABLE 5.1
Physical Characteristics of Furfural

Characteristics	Specific Value
Molecular formula	$C_5H_4O_2$
CAS number	98-01-01
Molecular weight (g mol^{-1})	96.08
Boiling point (°C)	161.7
Melting point (°C)	–37
Density (g/mL, 25 °C)	1.16
Critical pressure (MPa)	5.502
Solubility in water (g/mL)	8.3
Vapor pressure (mmHg, 55 °C)	13.5
Flash point (°F)	137
Vapor density (vs air)	3.31
LogP	0.41
Dielectric constant (20 °C)	41.9
PH (50g/L, 25 °C)	>=3.0

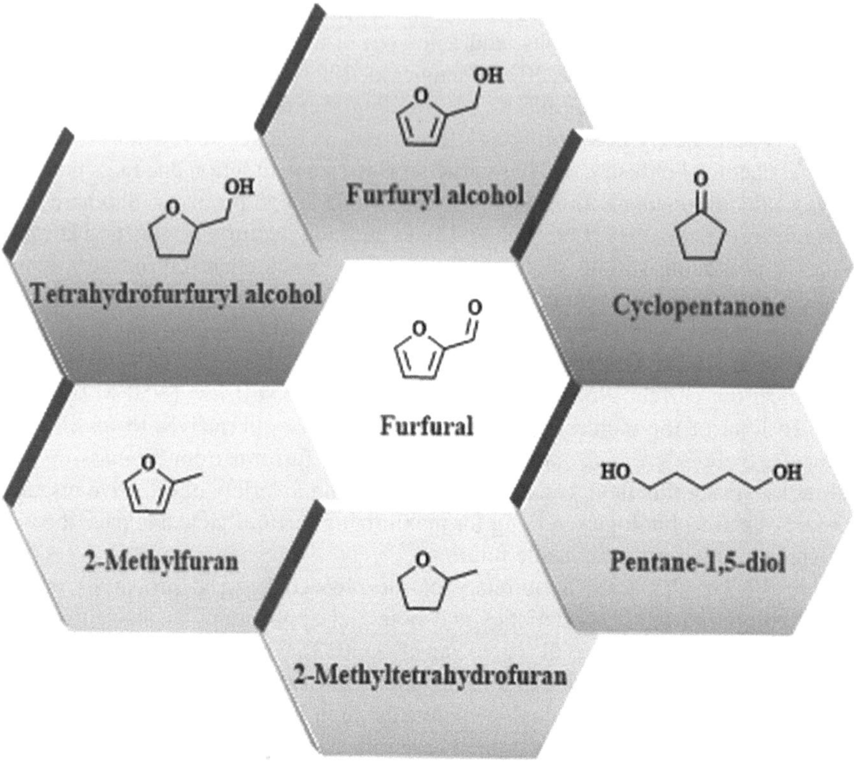

FIGURE 5.1 Several value-added products from furfural hydrogenation.

ring can also be transformed via hydrogenation, ring-opening, alkylation, oxidation, halogenation, and nitration reactions. Figure 5.1 illustrates that several hydrogenation products produced from furfural (e.g., furfuryl alcohol, tetrahydrofurfuryl alcohol, 2-methylfuran, 2-methyltetrahydrofuran, and 1,5-pentanediol) (Tukacs et al. 2017, Yan et al. 2014, Hutchings et al. 2017, Liu et al. 2020, Huang et al. 2017, Li et al. 2023). Except for the hydrogenation and hydrogenolysis reactions, may require rearranging the molecular structure for converting furfural, which involves the shifts of functional groups or carbon atoms within the molecule (Dutta et al. 2021). Currently, the selective hydrogenation catalysts are made up of noble metal catalysts (palladium, platinum, ruthenium, rhodium, iridium, etc.) and a base metal (e.g., Fe or Ni) (Chen et al. 2023, Bu et al. 2012, Shivhare et al. 2021, Sudarsanam et al. 2020). Fe-based catalysts, although inexpensive, are less active and produce iron sludge that is difficult to handle and prone to contamination. Ni-based catalysts usually have low selectivity, poor stability and harsh reaction conditions, while Pd-based catalysts are very favorable for catalyzing hydrogenation reactions (Ma et al. 2019, Climent et al. 2014, Wang et al. 2020), which is due to the more suitable adsorption strength of the empty d orbitals in Pd metal crystals for both hydrogen and most organic functional groups (Grabovskii et al. 2021), which makes Pd-based catalysts active in hydrogenation reactions with most unsaturated functional groups. Zero-valent Pd (Pd^0) is a good nucleophilic reagent that can be readily oxidized to monovalent or divalent Pd (Pd^{1+}, Pd^{2+}). Pd^0 has been extensively studied in chemical synthesis, Pd^{1+} and Pd^{2+} are good electrophilic reagents. Pd^{1+} and Pd^{2+} compounds can react with electron-rich compounds such as olefins and alkynes to form unstable intermediates, i.e., they exhibit vigoroso electrophilicity for olefin or alkyne, which is then attacked and released by nucleophiles for catalytic reactions. However, Pd-based catalysts often have the disadvantage of poor selectivity along with excellent catalytic activity in the selective hydrogenation process.

Thus, to achieving selective transformations in the reductive conversion of furfural requires careful tuning of both the catalyst and experimental conditions, Pd-based catalyst formulations with a focus on noble metals have been screened for this purpose. This chapter provides a literature review and analyses the catalytic activities of Pd-based catalysts, incorporating essential parameters, which include the key evaluating parameters (temperature, pressure, solvent), catalyst formulation (metal type, support material), catalyst preparation methods, and surface properties.

5.2 THE CATALYTIC HYDROGENATION OF FURFURAL

Pd is a famous catalyst due to its high activity with respect to hydrogenation. With Pd-based catalysts, furfural is upgraded to furfuryl alcohol by hydrogenation (Salnikova et al. 2019), and furfural can also undergo further transformations. For instance, tetrahydrofurfuryl alcohol, 2-methyltetrahydrofuran, tetrahydrofuran, cyclopentanone, furfuryl methyl ether, furanic esters, and liquid fuels can be formed through subsequent hydrogenation, hydrogenolysis, rearrangement, etherification, and esterification reactions, either individually or in combination (Zhou et al. 2022, Li et al. 2017, Kosydar et al. 2019, Šivec et al. 2021, Fang et al. 2022, Wang et al. 2020). In this chapter, the catalytic system of furfural hydrogenation based on Pd species was specially studied, and the factors affecting the catalytic performance were emphasized. These factors include the utilization of additives, the screening of catalyst supports, preparation processes, and the exploration of mechanics, as well as catalytic reactivity.

5.2.1 Production of Furfuryl Alcohol

Furfuryl alcohol is identified as one of the most momentous furfural-derived chemical compounds (Figure 5.2), accounting for approximately 65% of the overall production, which is obtained by the –CHO group hydrogenation in furfural (Tarazanov et al. 2020). The primary application of furfuryl alcohol is to produce foundry resins

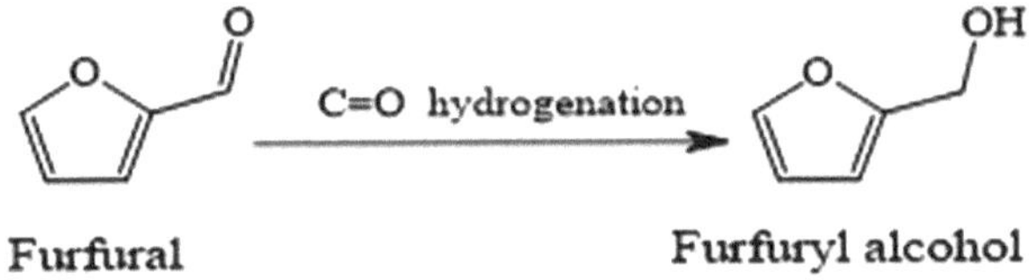

FIGURE 5.2 Catalytic hydrogenation of furfural to furfuryl alcohol.

as raw material. In addition, it also serves as a reactive solvent (Mariscal et al. 2016, Jaswal et al. 2022, Gupta et al. 2018). Furfuryl alcohol can be converted in either the liquid or vapor phase system (Chen et al. 2018, Sharma et al. 2013). The former is typically carried out in a batch reactor and requires pressurized conditions. As a result, this method is not viable in the economy for large-scale applications so far. On the contrary, the latter is more conducive to the large-scale manufacture of furfuryl alcohol.

As previously mentioned, the most steady adsorption conformation of furfural over Pd (111) is typically the furan-ring lying flat and closely combined with the Pd surface, resulting in a low selectivity of furfuryl alcohol on monometallic Pd-based catalysts. Other side reactions like the hydrogenolysis and over-hydrogenation of the ring are difficult to avoid. Consequently, only limited cases are paid attention to furfural hydrogenation into furfuryl alcohol using Pd-based catalysts. From another perspective, by optimizing the loading, Pd-based catalysts emerge to yield favorable reactivity for furfuryl alcohol production, under lower reaction temperatures (≤ 220 °C) and moderate hydrogen pressures (≤5 MPa). According to the research from Zhao and co-workers, 0.105 g 5 wt.% Pd/SiO_2 (Pd supported onto mesoporous SiO_2) was employed as a catalyst, delivering a furfuryl alcohol yield of 70.9% at a conversion rate of 75% (Zhao et al. 2014). In a separate study, a furfuryl alcohol yield of 40.9% at a conversion rate of 61% was obtained using 0.01 g Pd-Ir/SiO_2 consisting of 1 wt.% Pd and 1 wt.% Ir supported on SiO_2. Notably, this reaction was performed at 2 °C but under 8 MPa (Nakagawa et al. 2014). It was speculated that the inclusion of Ir in Pd–Ir/SiO_2 formulation was found to be promoted to enhancing the adsorption of furfural.

The Pd-based materials modified by Cu have demonstrated outstanding catalytic hydrogenation activities for upgrading furfural into furfuryl alcohol. Salnikova et al. utilized 0.1 g 3 wt.% Pd-3 wt.% Cu catalyst, achieving a 95.2% furfuryl alcohol yield and 99.9% conversion under 120 °C within 6 MPa (Salnikova et al. 2020). The high yield of furfuryl alcohol was attributed to Cu having a selective hydrogenation effect on the C=O bond. An analogous result was received in a subsequent study (Biradar et al. 2014), where a 0.5 g catalyst composed of 3 wt.% Pd and 20 wt.% Cu supported on MFI-type silica molecular sieves for upgrading furfural into furfuryl alcohol, and 91% furfuryl alcohol yield was obtained at complete conversion. The high performance of Pd-based materials modified by Cu in furfural hydrogenation is often associated with relatively high Cu loadings. Recently, the catalytic performances for furfural hydrogenation on Pd-Cu catalysts have been explored, and their loading number was screened (Fulajtárova et al. 2015). In this work, using 0.1 g 5 wt.% Pd-5 wt.% Cu/MgO catalyst, achieving an impressive 98.6% yield for furfuryl alcohol at complete conversion.146 It was indicated that the utilization of H_2 promoted the upgrade of furfural, whereas the selective reduction for furfuryl alcohol by HCHO or $NaBH_4$ was enhanced. In addition, MgO as a support is more conducive to the upgrading of furfural in comparison with activated carbon or hydrotalcite due to the MgO bears basicity and the interaction between Pd species and MgO affects the reduction capacity of Cu. It should be noted that Pd-Cu alloy is not formed in the PdCu/MgO. It is speculated that the Pd^0-Cu_2O was believed to play a crucial role in the selectivity of C=O bond hydrogenation. Furthermore, the choice of solvent also influenced compatibility for catalysts. Research has explored the catalytic performance of Pd-based catalysts for furfural hydrogenation under nonconventional environments. The capacity of furfural hydrogenation by utilizing 0.1 g 1 wt.% Pd/meso-AC as a catalyst was excavated, achieving a furfuryl alcohol yield of 56.1% at 79% conversion (Liu et al. 2015). The as-prepared catalyst emerges with a higher furfural conversion rate thanks to the lower mass-transfer limitations in supercritical CO_2 and the hydrogen bears a higher solubility.

The catalytic hydrogenation of furfural to furfuryl alcohol offers potential candidates to overcome the demand for pressurized hydrogen and reduce production costs. In this process, a solvent, typically a secondary alcohol serves as the H-donor. For instance, 2-propanol has been testified to be an ideal H-donor, which makes it particularly favorable for the hydrogenation of furfural (Rahimi et al. 2021, Bertero et al. 2011, Alonso et al. 2010, Alonso et al. 2008). However, some disadvantages associated with the hydrogenation reaction should be considered. For instance, the production of side-products produced by dehydrogenation of the H-donor, together with the unwanted reaction (e.g., acetalization) between the –CHO group in furfural and the solvent can lead to the formation of undesired acetal compounds, thereby impacting the activities of furfuryl alcohol. Scholz et al. utilized in situ reduced 0.5 g 10 wt.% Pd/Fe_2O_3 catalyst for furfural hydrogenation, achieving a furfuryl alcohol yield of 50% at 87% conversion (Scholz et al. 2014). At continuous conditions, employing 2.5 g 2 wt.% Pd/Fe_2O_3 catalyst with a furfural concentration of 0.4 M, a 70% furfuryl alcohol selectivity was received at 81% furfural conversion rate under 180 °C and 2.5 MPa.

5.2.2 Production of Tetrahydrofurfuryl Alcohol

Tetrahydrofurfuryl alcohol as a terminal product after complete hydrogenation of furfural, which has wide applications such as directly acting as a biofuel and solvent, along with serving as a precursor to producing valuable chemicals (Al-Yusufi et al. 2022, Chen et al. 2007). The main industrial way to yield tetrahydrofurfuryl alcohol involves the hydrogenation of furfuryl alcohol from supported Ni-based materials (Lv et al. 2022), whereas the process directly from furfural is more competitive by virtue of its lower energy consumption (Nakagawa et al. 2012, Xu et al. 2015). The production of tetrahydrofurfuryl alcohol from furfural involves two sequential hydrogenation steps (Figure 5.3). Firstly, the C=O bond in furfural is converted to a C–OH bond in furfuryl alcohol. Subsequently, the C=C bond in furfuryl alcohol is hydrogenated to form a C–C bond in tetrahydrofurfuryl alcohol. It is important to note that a minor side reaction can occur, resulting in the sequence of C=C bond hydrogenation before the complete C=O bond is fully hydrogenated.

The C=C bond hydrogenation in furfuryl alcohol is a well-studied reaction, which can be catalyzed by noble metals. To achieve high tetrahydrofurfuryl alcohol selectivity, careful optimization of reaction conditions or specific catalyst formulations was required (Nakagawa et al. 2014). Monometallic Pd-based catalysts loaded on carbon, SiO_2, and $CaCO_3$ have been found to reveal lower

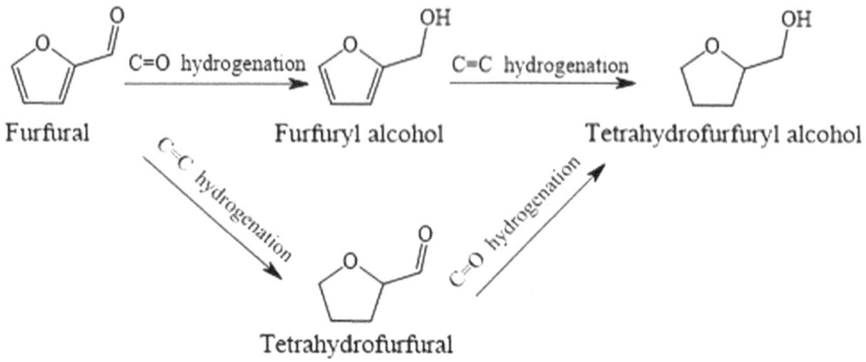

FIGURE 5.3 Catalytic hydrogenation of furfural to tetrahydrofurfuryl alcohol.

selectivity for tetrahydrofurfuryl alcohol at 120–220 °C and 3.5–5 MPa (Biradar et al. 2014, Merat et al. 1990), while the better catalytic activities with Al_2O_3-supported Pd-based materials were observed (BhogeswararaoSrinivas et al. 2015, Lesiak et al. 2014). Biradar et al. reported an impressive catalytic performance on 3 wt.% Pd/MFI catalyst, affording a 95% selectivity for tetrahydrofurfuryl alcohol at a complete conversion (Biradar et al. 2014), and the high activity was ascribed to the utilization of the siliceous MFI-type molecular sieve as a support material. The molecular sieve structure restricts the Pd particles within the cavities and channels, which have a molecular size (5~6 Å), resulting in highly dispersive Pd nanoparticles and a promoted strong interaction.

The introduction of extra metals into Pd-based catalysts can be a reliable strategy to enhance the activities in furfural to tetrahydrofurfuryl alcohol. Taking the Pd-Ir/SiO_2 catalyst as an example, 0.15 g 2 wt.% Pd-2 wt.% Ir/SiO_2 catalyst in water demonstrated outstanding performance, with a tetrahydrofurfuryl alcohol yield of 93% at over 99% conversion, even at demanding reaction parameters (2 °C, 8 MPa) (Nakagawa et al. 2014). In this work, the as-prepared catalyst bearing a higher reactivity was ascribed to the synergistic effect between Pd and Ir. Similarly, Chen et al. utilized 0.2 g Pd-Ni/TiO_2-ZrO_2 to catalytic the furfural hydrogenation (Chen et al. 2015), giving a tetrahydrofurfuryl alcohol yield of 93.4%. Under the circumstances, H_2 was activated by Pd, which boosted the reduction of Ni species to zerovalent sites, leading to an increased Ni active site concentration.

Incorporation of Cu into a Pd-based catalyst has disadvantageous effects for furfural selective converting to tetrahydrofurfuryl alcohol. Typically, the decrease of selectivity for tetrahydrofurfuryl alcohol was observed while adding Cu into a Pd-based catalyst, the reaction tends to the formation of furfuryl alcohol rather than tetrahydrofurfuryl alcohol (Lesiak et al. 2014). Interestingly, with an independent component (Cu or Pd) as a catalyst, a high tetrahydrofurfuryl alcohol yield (96%) could be obtained in the proposed tandem system (Stevens et al. 2010). It is suggested that the separate use of Cu and Pd catalysts in this reaction can better control the selectivity of tetrahydrofurfuryl alcohol.

In another study, the effects of particle size and catalyst preparation conditions on the reactivity of the Pd/TiO_2 catalyst together with the role of colloidal reduction were investigated at lower temperatures (Rogers et al. 2017). Notably, the selectivity of furfuryl alcohol and tetrahydrofurfuryl alcohol is not only decided by the Pd particle size but also by the presence of capping agents adsorbed over Pd. Capping agents can affect the accessibility of reactants to the active sites on the catalyst surface, thereby affecting the selectivity of the hydrogenation reactions. In addition, the highest tetrahydrofurfuryl alcohol yield with the Pd/TiO_2 catalyst in water was gained.

Recently, a 2 g 5 wt.% Pd/C catalyst was applied for upgrading furfural to tetrahydrofurfuryl alcohol, with a 100% selectivity for tetrahydrofurfuryl alcohol, albeit furfural conversions were undesired (<30%) (Merat et al. 1990). In another work, the catalytic activity of the Pd/MnO_x catalyst for furfural hydrogenation was also investigated, a 99.7% selectivity for tetrahydrofurfuryl alcohol was received over a furfural conversion with 45.6% (Zhang et al. 2012). King et al. used 0.5 g 3 wt.% Pd/TiO_2 as a catalyst to upgrade the furfuryl alcohol into tetrahydrofurfuryl alcohol, paying a 69.4% yield at 81.7% furfuryl alcohol conversion under 0.1 MPa. Amongst, the hydrogenation ability of the catalyst was markedly suppressed by adding Sn into Pd/TiO_2 (Pd^{2+} species occupy the dominant position in Pd-Sn/TiO_2) (King et al. 2015). Additionally, a 3% Pd/C catalyst provides the total tetrahydrofurfuryl alcohol yield from furfuryl alcohol but requires higher temperatures and pressures (Biradar et al. 2014). In another work, Gong et al. manufactured a graphitic carbon nitride-supported Pd/CN material and applied it for a catalytic transfer hydrogenation process, paying a 99% selectivity for tetrahydrofurfuryl alcohol at a conversion rate of 64% (Gong et al. 2014).

5.2.3 Production of 2-Methylfuran

2-methylfuran is a vital biofuel from furfural and stands as one of the representative molecules for second-generation fuels, which bears favorable physicochemical properties that make it suitable for blending with gasoline. Compared to ethanol, 2-methylfuran exhibits better combustion performance due to its appropriate content of oxygen (19.5 wt.%) (Jithin et al. 2021, Xu et al. 2016). Apart from the applications mentioned above, 2-methylfuran also acts as a solvent and a valuable feedstock for producing antimalarial drugs (Xu et al. 2016, Hansen et al. 2020, Shahbeik et al. 2023, Corma et al. 2011).

The production of 2-methylfuran involves the hydrogenolysis of the C=O bond (or C–O bond) from furfural (or furfuryl alcohol), and Pd-based catalysts have been employed in this process (Figure 5.4). Dong et al. synthesized a Pd-Cu/SiO_2 catalyst for the hydrogenolysis of furfural, offering a 66.1% 2-methylfuran yield at 88.7% conversion. In comparison, a Cu phyllosilicate and Pd/SiO_2 via physical mixture to synthesize a new catalyst, which possessed a higher conversion in this reaction, giving 65.6% 2-methylfuran selectivity and 99.1% furfural conversion at uniform conditions (Dong et al. 2015). Aldosari et al. tried to use TiO_2 as a support to synthesize 5 wt.% Pd/TiO_2 and applied for furfural hydrogenolysis, which afforded 42.8% selectivity for 2-methylfuran with a 98% furfural conversion under 0.3 MPa (Aldosari

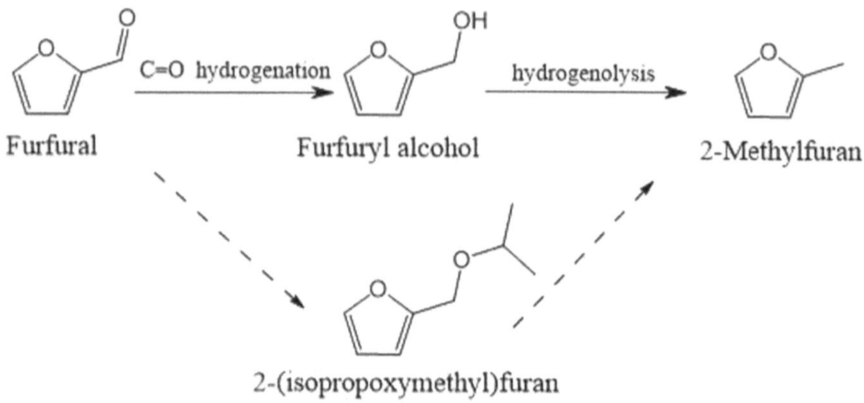

FIGURE 5.4 Catalytic hydrogenation of furfural to 2-methylfuran.

et al. 2016). Notably, the addition of a small quantity (1 wt.%) of Ru to Pd/TiO_2 catalysts shifted the selectivity towards furfuryl alcohol and 2-methylfuran, albeit at lower conversion rates. Moreover, the choice of solvent also demonstrated significant impacts on the catalytic activity and selectivity. The furfural conversion was more than 90% while utilizing 1,2-dichloroethane, toluene, and methanol as solvents, but the undesired byproducts were formed through ring hydrogenation or polymerization reactions. Furthermore, utilizing octane as a solvent facilitated the improvement of selectivity for 2-methylfuran and furfuryl alcohol, albeit at a lower conversion.

In the catalytic transfer hydrogenation, Chang et al. introduced 10 wt.% Cu to the Pd/ZrO_2 catalyst for helping the furfural hydrogenolysis in isopropanol, resulting in a 63.6% 2-methylfuran yield and a 98.5% furfural conversion (Chang et al. 2016). Utilizing TiO_2 as a support on the Pd-based catalyst also showed promising performances (61.2% yield, 100% conversion) for producing 2-methylfuran from furfural, but it required high loadings of Pd. It was speculated that the ZrO_2 in the Pd/ZrO_2 catalyst was believed to facilitate the dissociation and hydrogen transfer from 2-propanol into the C=O bond of furfural via acid or base sites. While the Pd loadings are as high as 5 wt.%, the selectivity for ring hydrogenation of C=C products was remarkably changed. The main hydrogenation/hydrogenolysis pathway for furfural to 2-methylfuran is furfuryl alcohol as an intermediate product, other analogical pathways via intermediate species like 2-(isopropoxymethyl)furan also contributed to form partially 2-methylfuran.

Compared with furfural, it is relatively easier to prepare 2-methylfuran over Pd-based catalysts from furfuryl alcohol. Scholz et al. tried to perform a furfuryl alcohol hydrogenation to 2-methylfuran over a 2 wt.% Pd/Fe_2O_3 catalyst in a continuous reactor, giving a conspicuous improvement in 2-methylfuran yield to 76%, in comparison with 31% in batch conditions. This transformation was attributed to the shorter contact time between furfuryl alcohol and Pd/Fe_2O_3 (Scholz et al. 2014). Pd was also loaded on the TiO_2 by wetness impregnation method to synthesize Pd/TiO_2, 2.58 wt.% Pd/TiO_2 achieved 87.7% 2-methylfuran yield and 93.7% furfuryl alcohol conversion in 1,2-dichloroethane at low pressure with 0.3 MPa (Iqbal et al. 2014). By keeping the H_2 pressure, the conversion of furfuryl alcohol increased by approximately 20%. The high reactivity of Pd/TiO_2 was ascribed to the small size of Pd particles (<2 nm), as well as the primary active species is Pd cations: Pd cations can function as activity sites for –CHO group (C=O bond) in furfural hydrogenation/hydrogenolysis, whereas metallic Pd is more suitable for furan-ring hydrogenation.

In addition, 2-dichloroethane is the most effective solvent, in comparison with acetonitrile, toluene, and octane. Later, the researchers of the same group further offered information on the hydrogenation/hydrogenolysis performance of Pd/TiO_2 (King et al. 2015). Doping Sn into Pd/TiO_2 catalysts was found to have similar activity, whereas ethanol or methanol as solvents would lead to selective form ring-hydrogenolysis products. Recently, Chang et al. utilized 3 wt.% Pd–10 wt.% Cu/ZrO_2 to catalyze the furfuryl alcohol hydrogenation at 220 °C in isopropanol, offering a 98.9% furfuryl alcohol conversion and 64.8% 2-methylfuran yield after 4 h. Meanwhile, 20.1% 2-methyltetrahydrofuran (ring hydrogenation product) was also observed (Chang et al. 2016). In conclusion, achieving a high 2-methylfuran yield with Pd-based catalysts requires careful controlling the state of Pd (oxidized or reduced state), using mild reaction parameters (e.g., low pressure or temperature), and doping additional metal species for the C=O bond (–CHO group) hydrogenation, which contribute to optimizing the reactivity on Pd-based catalysts for furfural/furfuryl alcohol hydrogenation/hydrogenolysis to 2-methylfuran.

5.2.4 Production of 2-Methyltetrahydrofuran

2-methyltetrahydrofuran is a recognized environment-friendly solvent (Al-Shaal et al. 2014, Aycock et al. 2007), which can be easily separated from water due to its immiscibility. It is a desired candidate for tetrahydrofuran, particularly in organometallic chemistry and the conversion of bio-sourced molecules. 2-methyltetrahydrofuran possesses some characteristics (e.g., a higher octane number, good miscibility, low volatility) that make it suitable as a competitive biofuel and gasoline blending additive (Alamillo et al. 2012). 2-methyltetrahydrofuran is primarily produced via two main ways in the current. The main way involves the hydrogenation of levulinic acid to form γ-valerolactone as an intermediate, followed by proceeding with the 2-methylfuran hydrogenation. Another major way involves the direct hydrogenation of 2-methylfuran. In addition to established routes, mentioning the direct formation of 2-methyltetrahydrofuran from other starting materials such as tetrahydrofurfuryl alcohol, furfural, sugars, and lignocellulosic biomass (Figure 5.5) (Date et al. 2017, Mironenko et al. 2022, YangSen et al. 2010).

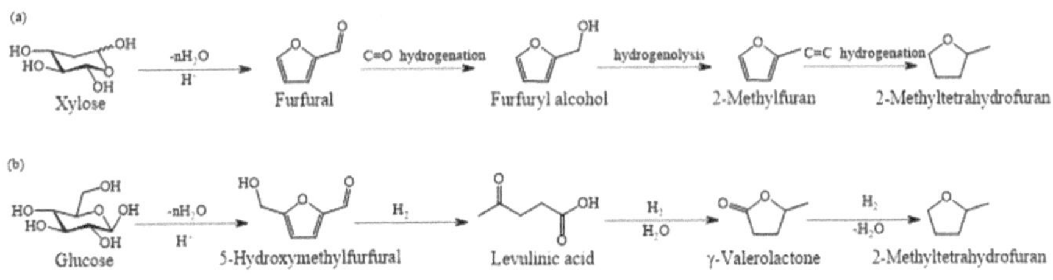

FIGURE 5.5 Catalytic hydrogenation of 2-methyltetrahydrofuran from lignocellulose.

Pd-based catalysts demonstrated high activity and selectivity in the 2-methylfuran hydrogenation to 2-methyltetrahydrofuran. Biswas et al. utilized 0.5 wt.% Pd/C catalyst in the vapor phase for 2-methylfuran hydrogenation, a complete 2-methylfuran conversion, and 100% 2-methyltetrahydrofuran yield were achieved at relatively lower temperatures (≤ 200 °C) and with a H_2/2-methylfuran ratio with 10 in a continuous process. The increase in reaction temperature can transfer the selectivity of the product to the ring-opening product, indicating that the reaction conditions are of great significance for the preferential formation of 2-methyltetrahydrofuran (Biswas et al. 2014).

The catalytic 2-methylfuran hydrogenation to 2-methyltetrahydrofuran using Pd-based catalysts has been investigated (Gong et al. 2014). In this case, a room temperature 2-methylfuran hydrogenation in HCOOH solution applying Pd/CN as the catalyst was analyzed, and an impressive 99% 2-methyltetrahydrofuran yield was obtained with 8 wt.% Pd/CN at 0.1 MPa air pressure. The as-prepared catalyst could be reused 26 times with no obvious activity loss, while the regeneration of the intermediate was achieved by washing with 0.1 M NaOH solution. The high reactivity was ascribed to the Mott-Schottky effect that occurs at the interface between Pd and graphitic carbon nitride, which involves a charge transfer. Other supports without the Mott-Schottky effect (e.g., nitrogen-rich carbon, active carbon, SBA-15) exhibited a weak catalytic performance. In recent work by Chang et al., a Pd/ZrO_2 was applied for the 2-methylfuran hydrogenation reaction with 2-propanol. This reaction proceeded with 220 °C, paying 84% 2-methyltetrahydrofuran yield and 95% 2-methylfuran conversion (Chang et al. 2016).

Forming 2-methyltetrahydrofuran from furfuryl alcohol has indeed been less extensively investigated compared to other reactions and pathways. For instance, Scholz et al. provided some insights into the catalytic furfuryl alcohol hydrogenation into 2-methyltetrahydrofuran. In their work, a 2 wt.% Pd/Fe_2O_3 offered high catalytic performance for the furfuryl alcohol hydrogenation in isopropanol, with a 42% yield 2-methyltetrahydrofuran at a 95% conversion under 0.1 MPa N_2 atmosphere at 165 °C (Scholz et al. 2014). It should be noted that this conversion to 2-methyltetrahydrofuran was achieved using a relatively low Pd loading. Interestingly, higher Pd loadings with 10 wt.% were found to have the benefit of forming tetrahydrofurfuryl alcohol and tetrahydrofuran. In comparison, Pd supported on other materials (e.g., activated carbon, Al_2O_3, MgO) emerged awfully weak activity for furfuryl alcohol hydrogenation to 2-methyltetrahydrofuran at identical reaction parameters.

Converting furfural to 2-methyltetrahydrofuran is indeed an attractive process, which has been investigated in single-reactor/dual-reactor processes. In a study by Chang et al., a 10Cu-5Pd/ZrO_2 was applied for the furfural hydrogenation with isopropanol in a Parr reactor, offering a notable 2-methyltetrahydrofuran yield with 78.8% at complete conversion (Chang et al. 2016). Notably, a lower Pd loading led to insufficient furan-ring hydrogenation, emphasizing the importance of an appropriate catalyst composition. Recently, Date et al. utilized Pd-B/C (Pd/C reduced by $NaBH_4$) as the catalyst for the liquid-phase furfural hydrogenation, achieving 40% 2-methyltetrahydrofuran yield and surpassing 99% furfural conversion under 220 °C at 3.4 MPa (Date et al. 2017). Importantly, the $NaBH_4$ as a reducing agent had a significant impact on the particle size of Pd/C, thereby changing 2-methyltetrahydrofuran selectivity. This as-prepared Pd-B/C exhibited the minimum average particle size of 4.8 nm and the strongest 2-methyltetrahydrofuran selectivity, which was attributed to boron introduced into the Pd lattice. In contrast, Pd/C reduced by HCHO displayed the maximum particle size (22.4 nm) and the weakest 2-methyltetrahydrofuran selectivity.

Two-reactor processes offer a sequential reductive conversion for furfural to 2-methyltetrahydrofuran using separate catalysts, which allow for greater control over the hydrogenolysis/hydrogenation sequence of the C=O bond and the C=C bond. For instance, a tandem reaction with two reactors was described, and the first reactor contained copper chromite as the catalyst and operated at 240 °C, while the later reactor utilized 5 wt.% Pd/C as a catalyst and operated at 300 °C, a 82% 2-methyltetrahydrofuran yield in $scCO_2$ with 15 MPa was achieved (Stevens et al. 2010). Another two-stage-packed reactor system for the reactions mentioned above was also investigated, wherein, the upper reactor employed Cu phyllosilicate as the catalyst, while the bottom reactor utilized Pd/SiO_2, resulting in a 97.1% 2-methyltetrahydrofuran yield and a complete furfural conversion under 180 °C and 0.1 MPa (Dong et al. 2015). This work can accurately control the reaction sequence of the C=O and C=C bonds in the reactor.

5.2.5 Production of 1,5-Pentanediol

1,5-pentanediol as an important chemical intermediate has diverse applications for producing polyesters, polyurethanes, plasticizers, coatings spices, and so on (Zhang et al. 2012, Sun et al. 2016, Khemthong et al. 2021). Currently, the primary pathway of 1,5-pentanediol is derived from unsustainable petroleum resources. More absorbing pathways to produce 1,5-pentanediol from furfural involve the furfural total hydrogenation to tetrahydrofurfuryl alcohol, whereafter, the furan-ring hydrogenolysis or the furfuryl alcohol ring-opening reaction proceeded.

Pd-based catalysts have been confirmed to be effective in furfural total hydrogenation, as mentioned earlier. Furthermore, it has been demonstrated that Pd can act as a promoter to produce 1,5-pentanediol via furfural hydrogenation/hydrogenolysis. For instance, Liu et al. utilized Pd-promoted IrReOx/SiO_2 (0.66 wt.% Pd-Ir-ReOx/SiO_2) as the catalyst for furfural conversion to 1,5-pentanediol with a two-step reaction at 140 °C for 8 h, whereafter, continuing to operate 3 days with 110 °C, finally offering a 71.4% 1,5-pentanediol yield (Liu et al. 2014). In addition, some studies have proved that Pd-based catalysts have the ability to promote the ring-opening reaction of the furan ring, which provides an alternative way of producing pentanediol (Tuteja et al. 2014, Buntara et al. 2012, Sutton et al. 2013).

5.2.6 Production of Cyclopentanone

Cyclopentanone is a multipurpose compound, which utilizes it as an intermediate to produce fungicides, pharmaceuticals, biological products, synthetic rubber, and so on (Hronec and Fulajtarová 2012, Liu et al. 2017, Shao et al. 2022). Currently, the predominantly conversion route of cyclopentanone is the petroleum-chemical route, which uses 1,6-hexanediol, adipic esters, and cyclopentene as starting materials (Akashi et al. 2003, Dubkov et al. 2002). Recently, a new route has been excavated by introducing a ring-rearrangement strategy to produce cyclopentanone from furan derivatives. A 5 wt.% Pd/C catalyst was applied for the conversion of furfural to cyclopentanone, with 67.1% cyclopentanone yield and 97.8% furfural conversion under 160 °C and 3 MPa (Hronec and Fulajtarová 2012). It was observed that the limitation of further reduction for cyclopentanone to cyclopentanol was due to the formation of furfuryl alcohol polymers (in water below 150 °C), and the polymers were strongly adsorbed onto the Pd/C surface, thereby inhibiting consecutive hydrogenations. Consequently, the harsher reaction conditions (175 °C and 8 MPa) resulted in the complete conversion of furfural, but lower cyclopentanone selectivity, which was attributed to furan-ring hydrogenation to form 2-methyltetrahydrofuran and tetrahydrofurfuryl alcohol. Later, this group also excavated the possibility of converting furfuryl alcohol to cyclopentanone, and yet the cyclopentanone yield was only 6.3%, 2-methyltetrahydrofuran and tetrahydrofurfuryl alcohol are the major products (Hronec et al. 2012). Date et al. utilized a silica-supported Pd catalyst Pd/f-SiO_2 to catalytic the furfural one-pot into cyclopentanone, paying 87% yield at 98% furfural conversion (Date et al. 2018).

Pd-Cu catalysts have demonstrated outstanding catalytic performances to produce cyclopentanone. Hronec et al. synthesized 5 wt.% Pd-10 wt.% Cu/C catalyst via electroless plating of Cu in the tartrate carboxylate ligands and applied it for the furfural to cyclopentanone in water, a high cyclopentanone yield of 92.1% was achieved at 98% furfural conversion conditions at 160 °C and 3 MPa (Hronec et al. 2016). In this catalytic system, Pd^0-Cu^+ acted as the active species responsible for the furan-ring rearrangement. According to the experimental results, a mechanism for the conversion of furfural to cyclopentanone in water was presented. Initially, furfural is adsorbed on Cu^+ through the electron pair of the –CHO group, and the hydrogen transfer from adjacent Pd–H sites promotes furfural hydrogenation into furfuryl alcohol. Then, furfuryl alcohol was protonated and rearranged to form 2-cyclopentenone, followed by the 2-cyclopentenone hydrogenation into cyclopentanone (Figure 5.6). It is found that the tiny amount of Co has a cooperative effect with other components of the catalyst and promotes the improvement of the activity. For instance, a catalyst Pd-Co@UiO-66 was prepared via encapsulating bimetallic Pd-Co nanoparticles into the UiO-66 organic framework and applied for manufacturing cyclopentanone from furfural for 120 °C and 3 MPa, offering a 95% cyclopentanone yield at 99% furfural conversion (Wang et al. 2020).

In a word, Pd-based catalysts have shown high activity and selectivity in various hydrogenation/hydrogenolysis or rearrangement reactions for furfural,

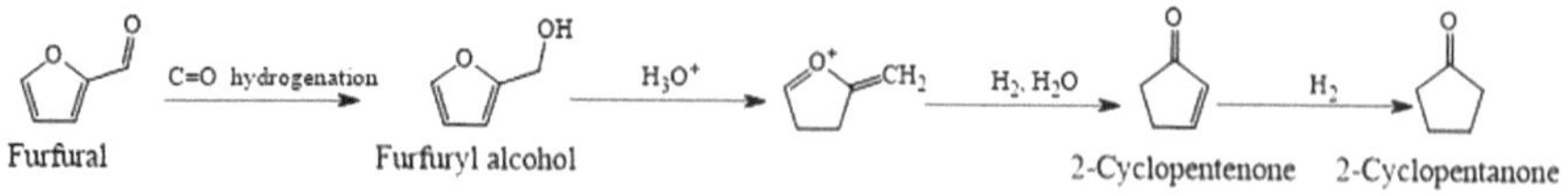

FIGURE 5.6 Catalytic hydrogenative synthesis of cyclopentanone from furfural.

thereby allowing production of desired chemicals and sustainable fuels. For reactions involving the –CHO group, furfuryl alcohol or 2-methylfuran was gained via hydrogenation/hydrogenolysis reaction at mild reaction parameters and low Pd loadings. Introducing Cu as a modifier and incomplete Pd reduction is beneficial to these pathways. On the other hand, producing furan and its derivatives through hydrogenation/hydrogenolysis typically requires higher reaction temperatures, whereas the complete hydrogenation of the above-mentioned compounds needs higher H_2 pressure. Furthermore, Pd-based catalysts also have displayed outstanding catalytic performance for producing liquid fuels from furfural.

5.3 CONCLUSION AND OUTLOOK

This chapter specifically focuses on the research and exploration of experiments conducted on the catalytic hydrogenation for furfural with Pd-based catalysts. The properties of materials applied for catalyzing, the effects of reaction parameters for product selectivity and yield, as well as the possible reaction mechanisms, were discussed. As displayed in Figure 5.7, Pd offers access to a wide range of reaction pathways and is commonly used for upgrading furfural into furfuryl alcohol or furfural hydrogenolysis into 2-methylfuran under mild reaction parameters and low metal loadings. On the other hand, more severe conditions and higher metal loadings are beneficial for producing tetrahydrofurfuryl alcohol, and 2-methyltetrahydrofuran. The ring-hydrogenation and ring-opening reactions may occur during the hydrogenation process due to furfural bears' diverse functional groups. Therefore, it is crucial to identify and develop catalysts to maximize the selectivity for the desired products

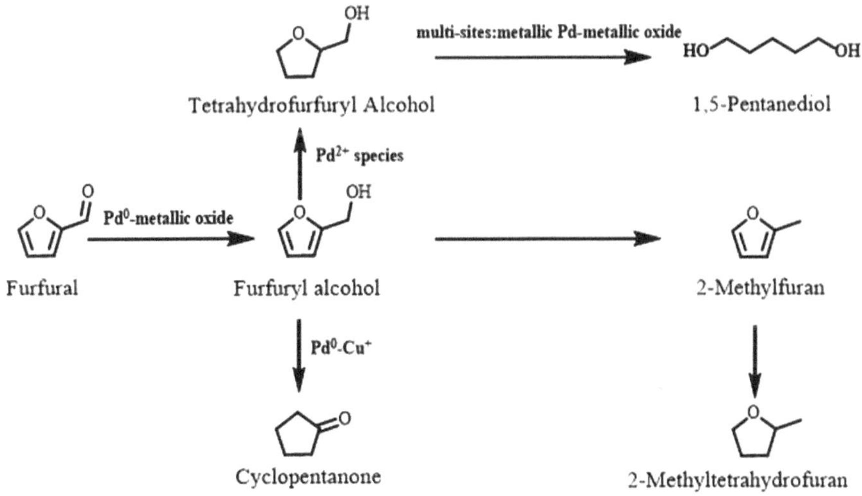

FIGURE 5.7 The catalytic characters or functions of Pd-based catalysts relevant to desired furfural conversions.

and minimize side reactions for unwanted compounds. Utilizing Pd is necessary and is an available catalyst for achieving effective catalytic hydrogenation of biomass-derived compounds. However, the high cost needs to be considered and controlled during the entire reaction, the effective approaches include replacing conventional hydrogen with organic hydrogen donors (e.g., alcohols) to eliminate the desire for high-price and high-pressure reactors, as well as promoting sustainable development by reducing the dependence on carbon-intensive, unsustainable derived hydrogen from natural gas. And further economic analysis and optimization of the overall process are essential to assess the viability and competitiveness of these alternative hydrogen sources.

In most cases, high selectivity is usually accompanied by some loss of activity through surface modifiers or doping with other elements or oxides. Therefore, to achieve high product selectivity without compromising activity, the construction of catalysts with a well-conditioned environment for adjustable coordination is a promising direction of development.

However, the cost and availability of Pd have limited their large-scale application in catalytic processes, leading to a growing interest in exploring non-noble metal-based catalysts for producing furfural to valuable products. By analyzing the change for furfural conversion with the metal content in the catalyst, the breakthrough progress of non-noble metal-based catalysts and the limitations of each catalytic system were revealed. Significant advancements have been achieved with the introduction of Cu for furfural selective upgrading into furfuryl alcohol and 2-methylfuran. However, the activity of these non-noble metal catalysts still needs to improve, particularly under mild reaction conditions. To enhance the catalytic performance, a promising strategy is stabilizing metallic nanoparticles to be safe against harsh reaction conditions commonly encountered in hydrogenation reactions, which can be achieved by carefully selecting the appropriate preparation method and support, thereby improving the stability and catalytic activity.

Maximizing the selectivity of target products and minimizing side reactions for furfural hydrogenation is a complex issue due to the potential for various unwanted reactions. The selectivity of the desired target product is greatly influenced by reaction parameters such as temperature, pressure, contact time, and solvent, as well as the composition of the catalyst. One prospective approach to addressing these challenges is utilizing the composite modes of multi-metallic combinations and non-noble metals, and it can exhibit synergistic effects, which may arise from alloying of the metals. The activity and selectivity were improved in comparison to individual metals. Alternatively, the presence of different metals in a catalyst can facilitate a series of multi-stage tandem reactions occurring on particles containing various metals. Bimetallic catalytic systems, such as Pd-Cu and Pd-Ni, are preferred for total hydrogenation reactions due to their ability to promote the hydrogenation function of Pd while suppressing side reactions like decarbonization or ring-opening reactions. Additionally, the catalyst systems combining Pd with other species such as carbon, carbon nitride, alumina, and magnesia have shown potential for total hydrogenation reactions. These catalysts improved selectivity to desired products and reduced the occurrence of undesired side reactions.

In addition to the typical functions in a chemical process and the limitations (e.g., solubility of H_2 and reactant, and miscibility), solvents can also play a role in liquid-phase processes by either facilitating the reaction or acting as reactants. The ability of reactants, catalysts, and solvents to be hydrophilic or hydrophobic is crucial in determining the adsorption mode of substrate over catalyst. This delicate balance is what governs this process.

The adsorption behavior of the substrate over the catalyst surface is influenced by the hydrophilic/hydrophobic properties of the reactants, catalysts, and solvents. These properties determine the efficiency of the overall process by governing a delicate balance of interactions. In catalytic hydrogenation reactions, alcohols can act as hydrogen donors, the success of the reaction relies on the activation of alcohols by the catalyst's surface sites, which involves the formation of surface alkoxides from oxidic support or oxide-metal interfaces and plays a crucial role in the reaction result. Water is usually the key to the ring-opening mechanism, but it may also lead to unwanted side reactions such as competitive adsorption on surface sites, inducing structural modifications of the catalyst support, as well as causing the leaching of active species. The combination of ionic liquids and conventional solvents or supercritical fluids (e.g., CO_2, alcohols) for designing one-pot multi-step with multiphase reactions provides new opportunities.

Understanding the mechanistic aspects of the adsorption modes of furfural and preferred reaction pathways for specific catalytic systems is crucial for designing more efficient catalysts. The adsorption mode of furfural on the catalyst surface plays a decisive role in the availability and accessibility of the reaction sites, which in turn affects the thermodynamically and kinetically favorable reaction pathways. Different catalysts may exhibit distinct adsorption modes, depending on changes in their surface properties, such as the presence of specific active sites or the carrier material nature. By studying the mechanisms of adsorption, activation, and subsequent reactions during furfural hydrogenation or hydrogenolysis, researchers can confirm the key factors that control selectivity and efficiency. Various experimental techniques, such as spectroscopy, surface science, and kinetics studies, as well as computational methods, are employed to elucidate the mechanistic details of the catalytic processes. By combining experimental observations with theoretical calculations, the energetics and dynamics of the reactions can be deeply understood.

For future research, it is suggested to explore the preparation of Pd-based catalysts using various bimetallic and trimetallic materials, as well as carbon-supported materials, thereby enhancing the reaction activity of furfural hydrogenation. It would be beneficial to synthesize different hybrid materials by combining carbon-based matrices with diverse functional materials. The currently available synthesis techniques include wet chemistry, sol-gel, co-precipitation, and others. In further research, it is necessary to consider the use of carbon-based materials such as carbon dots to synthesize noble metal nanoparticles for various furfural hydrogenation reactions. In addition, the furfural hydrogenation reaction is still in the stage of laboratory theory, and it remains uncertain whether catalytic hydrogenation technology can be applied to achieve furfuryl alcohol and other value-added manufacturing production. Thus, extensive research should be conducted to fully explore the catalytic capabilities and synergistic effects of Pd combined with other available matrices, which provide worthy views and lay the foundation for future advancements and applications.

REFERENCES

Akashi, Tetsu, Sato, Satoshi, Takahashi, Ryoji, Sodesawa, ToshiakiandInui, and Kanichiro. 2003. "Catalytic vapor-phase cyclization of 1,6-hexanediol into cyclopentanone." *Catal. Commun.* 4. Elsevier: 411–416. doi: 10.1016/S1566-7367(03)00095-5.

Alamillo, Ricardo, Tucker, Mark, Chia, Mei, Pagán-Torres, Yomaira and Dumesic, James. 2012. "The selective hydrogenation of biomass-derived 5-hydroxymethylfurfural using heterogeneous catalysts." *Green Chem.* 14. Royal Society of Chemistry: 1413–1419. doi: 10.1039/C2GC35039D.

Aldosari, Obaid F., Iqbal, Sarwat, Miedziak, Peter J., Brett, Gemma L., Jones, Daniel R., Liu, Xi, Edwards, Jennifer K., Morgan, David J., Knight, David K. and Hutchings, Graham J. 2016. "Pd–Ru/TiO2 catalyst – an active and selective catalyst for furfural hydrogenation." *Catal. Sci. Technol.* 6. Royal Society of Chemistry: 234–242. doi: 10.1039/C5CY01650A.

Alonso, Francisco, Riente, Paola, Sirvent, Juan Alberto and Yus, Miguel. 2010. "Nickel nanoparticles in hydrogen-transfer reductions: Characterisation and nature of the catalyst." *Appl. Catal. A-Gen.* 378. Elsevier: 42–51. doi: 10.1016/j.apcata.2010.01.044.

Alonso, Francisco, Riente, Paola and Yus, Miguel. 2008. "Hydrogen-transfer reduction of carbonyl compounds catalysed by nickel nanoparticles." *Tetrahedron Lett.* 49. Elsevier: 1939–1942. doi: 10.1016/j.tetlet.2008.01.097.

Al-Shaal, Mohammad G., Dzierbinski, Adam and Palkovits, Regina. 2014. "Solvent-free γ-valerolactone hydrogenation to 2-methyltetrahydrofuran catalysed by Ru/C: A reaction network analysis." *Green Chem.* 16. Royal Society of Chemistry: 1358–1364. doi: 10.1039/C3GC41803K.

Al-Yusufi, Mohammed, Steinfeldt, Norbert, Eckelt, Reinhard, Atia, Hanan, Lund, Henrik, Bartling, Stephan, Rockstroh, Nils and Köckritz, Angela. 2022. "Efficient base nickel-catalyzed hydrogenolysis of furfural-derived tetrahydrofurfuryl alcohol to 1,5-pentanediol." *ACS Sustain. Chem. Eng.* 10. American Chemical Society: 4954–4968. doi: 10.1021/acssuschemeng.1c08340.

Amarasekara, Ananda S., Nwankwo, Victor C. and Fernando, Harshica. 2023. "Hydroxy sulfonic acid catalyzed hydrolysis of cellulose." *Biofuels.* 15. Taylor & Francis Ltd: 1–7. doi: 10.1080/17597269.2023.2221969.

Aycock, David F. 2007. "Solvent applications of 2-methyltetrahydrofuran in organometallic and biphasic reactions." *Org. Process Res. Dev.* 11. American Chemical Society: 156–159. doi: 10.1021/op060155c.

Bertero, Nicolás M., Trasarti, Andrés F., Apesteguía, Carlos R. and Marchi, Alberto J. 2011. "Solvent effect in the liquid-phase hydrogenation of acetophenone over Ni/SiO2: A comprehensive study of the phenomenon." *Appl. Catal. A-Gen.* 394. Elsevier: 228–238. doi: 10.1016/j.apcata.2011.01.003.

Bhogeswararao, S. and Srinivas, D. 2015. "Catalytic conversion of furfural to industrial chemicals over supported Pt and Pd catalysts." *J Catal.* 327. Academic Press Inc.: 65–77. doi: 10.1016/j.jcat.2015.04.018.

Biradar, Narayan S., Hengne, Amol A., Birajdar, Shobha N., Swami, Rameshwar and Rode, Chandrashekhar V. 2014. "Tailoring the product distribution with batch and continuous process options in catalytic hydrogenation of furfural." *Org. Process Res. Dev.* 18. American Chemical Society: 1434–1442. doi: 10.1021/op500196x.

Biradar, Narayan S., Hengne, Amol M., Birajdar, Shobha N., Niphadkar, Prashant S., Joshi, Praphulla N. and Rode, Chandrashekhar V. 2014. "Single-pot formation of THFAL via catalytic hydrogenation of FFR over Pd/MFI catalyst." *ACS Sustain. Chem. Eng.* 2. American Chemical Society: 272–281. doi: 10.1021/sc400302b.

Biswas, Prakash, Lin, Jiann-Horng, Kang, Jungshik and Guliants, Vadim V. 2014. "Vapor phase hydrogenation of 2-methylfuran over noble and base metal catalysts." *Appl. Catal. A-Gen.* 475. Elsevier: 379–385. doi: 10.1016/j.apcata.2014.01.054.

Bozell, Joseph J and Petersen, Gene R. 2010. "Technology development for the production of biobased products from biorefinery carbohydrates—the US Department of Energy's "Top 10" revisited." *Green Chem.* 12. Royal Society of Chemistry: 539–554. doi: 10.1039/B922014C.

Brodin, Malin, Vallejos, María, Opedal, Mihaela Tanase, Area, María Cristina and Chinga-Carrasco, Gary. 2017. "Lignocellulosics as sustainable resources for production of bioplastics – A review." *J Clean. Prod.* 162. Elsevier: 646–664. doi: 10.1016/j.jclepro.2017.05.209.

Bu, Quan, Lei, Hanwu, Zacher, Alan H., Wang, Lu, Ren, Shoujie, Liang, Jing, Wei, Yi, Liu, Yupeng, Tang, Juming, Zhang, Qin and Ruan, Roger. 2012. "A review of catalytic hydrodeoxygenation of lignin-derived phenols from biomass pyrolysis." *Bioresource Technol.* 124. Elsevier: 470–477. doi: 10.1016/j.biortech.2012.08.089.

Buntara, Teddy, Noel, Sebastien, Phua, Pim Huat, Melián-Cabrera, Ignacio, de Vries, Johannes G. and Heeres, Hero J. 2012. "From 5-hydroxymethylfurfural (HMF) to polymer precursors: Catalyst screening studies on the conversion of 1,2,6-hexanetriol to 1,6-hexanediol." *Top. Catal.* 55. Springer: 612–619. doi: 10.1007/s11244-012-9839-6.

Chang, Xin, Liu, An-Feng, Cai, Bo, Luo, Jin-Yue, Pan, Hui and Huang, Yao-Bing. 2016. "Catalytic transfer hydrogenation of furfural to 2-methylfuran and 2-methyltetrahydrofuran over bimetallic copper–oalladium catalysts." *ChemSusChem.* 9. John Wiley & Sons Ltd: 3330–3337. doi: 10.1002/cssc.201601122.

Chheda, Juben N, and Dumesic, James A. 2007. "An overview of dehydration, aldol-condensation and hydrogenation processes for production of liquid alkanes from biomass-derived carbohydrates." *Catal. Today.* 123. Elsevier: 59–70. doi: 10.1016/j.cattod.2006.12.006.

Chheda, Juben N, Huber, George W and Dumesic, James A. 2007. "Liquid-phase catalytic processing of biomass-derived oxygenated hydrocarbons to fuels and chemicals." *Angew. Chem. Int. Edit.* 46. John Wiley & Sons Ltd: 7164–7183. doi: 10.1002/anie.200604274.

Chen, Bingfeng, Li, Fengbo, Huang, Zhijun and Yuan, Guoqing. 2015. "Tuning catalytic selectivity of liquid-phase hydrogenation of furfural via synergistic effects of supported bimetallic catalysts." *Appl. Catal. A-Gen.* 500. Elsevier: 23–29. doi: 10.1016/j.apcata.2015.05.006.

Chen, Lungang, Liu, Yong, Zhang, Xinghua, Liu, Jianguo, Zhang, Qi and Ma, Longlong. 2023. "Catalytic production of long-chain hydrocarbons suitable for aviation turbine fuel from biomass-derived levulinic acid and furfural." *Fuel.* 334. Elsevier: 126665. doi: 10.1016/j.fuel.2022.126665.

Chen, Xiaochun, Sun, Wei, Xiao, Nan, Yan, Yanjiao and Liu, Shiwei. 2007. "Experimental study for liquid phase selective hydrogenation of furfuryl alcohol to tetrahydrofurfuryl alcohol on supported Ni catalysts." *Chem. Eng. J.* 126. Elsevier: 5–11. doi: 10.1016/j.cej.2006.08.019.

Chen, Shuo, Wojcieszak, Robert, Dumeignil, Franck, Marceau, Eric and Royer, Sébastien. 2018. "How catalysts and experimental conditions determine the selective hydroconversion of furfural and 5-hydroxymethylfurfural." *Chem. Rev.* 118. American Chemical Society: 11023–11117. doi: 10.1021/acs.chemrev.8b00134.

Climent, Maria J., Corma, Avelino and Iborra, Sara. 2014. "Conversion of biomass platform molecules into fuel additives and liquid hydrocarbon fuels." *Green Chem.* 16. Royal Society of Chemistry: 516–547. doi: 10.1039/C3GC41492B.

Corma, Avelino, de la Torre, Olalla, Renz, Michael and Villandier, Nicolas. 2011. "Production of high-quality diesel from biomass waste products." *Angew. Chem. Int. Edit.* 50. John Wiley & Sons Ltd: 2375–2378. doi: 10.1002/anie.201007508.

Date, Nandan S., Kondawar, Sharda E., Chikate, Rajeev C. and Rode, Chandrashekhar V. 2018. "Single-pot reductive rearrangement of furfural to cyclopentanone over silica-supported Pd catalysts." *ACS Omega.* 3. American Chemical Society: 9860–9871. doi: 10.1021/acsomega.8b00980.

Date, Nandan S., Biradar, Narayan S., Chikate, Rajeev C. and Rode, Chandrashekhar V. 2017. "Effect of reduction protocol of Pd catalysts on product distribution in furfural hydrogenation." *ChemistrySelect.* 2. Wiley-VCH Verlag GmbH: 24–32. doi: 10.1002/slct.201601790.

Dong, Chenglong, Wang, Hongtao, Du, Haochen, Peng, Jiebang, Cai, Yang, Guo, Shuai, Zhang, Jianli, Samart, Chanatip and Ding, Mingyue. 2020. "Ru/HZSM-5 as an efficient and recyclable catalyst for reductive amination of furfural to furfurylamine." *Mol. Catal.* 482. Elsevier: 110755. doi: 10.1016/j.mcat.2019.110755.

Dong, Fang, Zhu, Yulei, Ding, Guoqiang, Cui, Jinglei, Li, Xianqing and Li, Yongwang. 2015. "One-step conversion of furfural into 2-methyltetrahydrofuran under mild conditions." *ChemSusChem.* 8. John Wiley & Sons Ltd: 1534–1537. doi: 10.1002/cssc.201500178.

Dubkov, Konstantin A., Panov, Gennady I., Starokon, Evgeny V. and Parmon, Valentin N. 2002. "Non-catalytic liquid phase oxidation of alkenes with nitrous oxide. 2. Oxidation of cyclopentene to cyclopentanone." *React. Kinet. Catal. L.* 77. Springer: 197–205. doi: 10.1023/A:1020372726494.

Dutta, Saikat and Bhat, Navya Subray. 2021. "Catalytic transformation of biomass-derived furfurals to cyclopentanones and their derivatives: A review." *ACS Omega.* 6. American Chemical Society: 35145–35172. doi: 10.1021/acsomega.1c05861.

Elliott, Douglas C. 2007. "Historical developments in hydroprocessing bio-oils." *Energ. Fuel.* 21. American Chemical Society: 1792–1815. doi: 10.1021/ef070044u.

Farmer, Delphine K. and Prather, Kimberly A. 2020. "A chemical perspective on climate: Emerging research into atmospheric chemistry impacts on earth's radiative balance." *Accounts Chem. Res.* 53. American Chemical Society: 2496–2497. doi: 10.1021/acs.accounts.0c00577.

Fang, Ruiqi, Dhakshinamoorthy, Amarajothi, Li, Yingwei and Garcia, Hermenegildo. 2020. "Metal organic frameworks for biomass conversion." *Chem. Soc. Rev.* 49. Royal Society of Chemistry: 3638–3687. doi: 10.1039/D0CS00070A.

Fang, Wenting, Egebo, Jonas, Schill, Leonhard, Chen, Hua and Riisager, Anders. 2022. "Reductive etherification of furfural via hydrogenolysis with Pd-modified aluminum phosphate and formic acid." *Green Chem.* 24. Royal Society of Chemistry: 7346–7349. doi: 10.1039/D2GC02190K.

Fang, Wenting and Riisager, Anders. 2021. "Recent advances in heterogeneous catalytic transfer hydrogenation/hydrogenolysis for valorization of biomass-derived furanic compounds." *Green Chem.* 23. Royal Society of Chemistry: 670–688. doi: 10.1039/D0GC03931D.

Fulajtárova, Katarína, Soták, Tomáš, Hronec, Milan, Vávra, Ivo, Dobročka, Edmund and Omastová, Mária. 2015. "Aqueous phase hydrogenation of furfural to furfuryl alcohol over Pd–Cu catalysts." *Appl. Catal. A-Gen.* 502. Elsevier: 78–85. doi: 10.1016/j.apcata.2015.05.031.

Garedew, Mahlet, Lin, Fang, Song, Bing, DeWinter, Tamara M., Jackson, James E., Saffron, Christopher M., Lam, Chun Hoand Anastas, and Paul T. 2020. "Greener routes to biomass waste valorization: Lignin transformation through electrocatalysis for renewable chemicals and fuels production." *ChemSusChem.* 13. John Wiley & Sons Ltd: 4214–4237. doi: 10.1002/cssc.202000987.

Gong, Ling-Hong, Cai, Yi-Yu, Li, Xin-Hao, Zhang, Ya-Nan, Su, Juan and Chen, Jie-Sheng. 2014. "Room-temperature transfer hydrogenation and fast separation of unsaturated compounds over heterogeneous catalysts in an aqueous solution of formic acid." *Green Chem.* 16. Royal Society of Chemistry: 3746–3751. doi: 10.1039/C4GC00981A.

Grabovskii, A. Stanislav, Akchurin, I. Timur and Dokichev, A. Vladimir. 2021. "Heterogeneous palladium catalysts in the hydrogenation of the carbon-carbon double bond." *Curr. Org. Chem.* 25. Bentham Science Publishers: 315–329. doi: 10.2174/1385272824999201202084812.

Gupta, Kavita, Rai, Rohit K. and Singh, Sanjay K. 2018. "Metal catalysts for the efficient transformation of biomass-derived HMF and furfural to value added chemicals." *ChemCatChem.* 10. Wiley-VCH Verlag GmbH & CO. KGaA: 2326–2349. doi: 10.1002/cctc.201701754.

Hansen, Samuel, Mirkouei, Amin and Diaz, Luis A. 2020. "A comprehensive state-of-technology review for upgrading bio-oil to renewable or blended hydrocarbon fuels." *Renew. Sust. Energ. Rev.* 118. Elsevier: 109548. doi: 10.1016/j.rser.2019.109548.

Hashim, Luqman H., Halilu, Ahmed, Sudarsanam, Putla, Umar, Yahaya Balarabe, Johan, Mohd Rafie Bin and Bhargava, Suresh K. 2020. "Bifunctional rice husk-derived SiO2-Cu-Al-Mg nanohybrid catalyst for one-pot conversion of biomass-derived furfural to furfuryl acetate." *Fuel.* 275. Elsevier: 117953. doi: 10.1016/j.fuel.2020.117953.

Hronec, Milan and Fulajtarová, Katarina. 2012. "Selective transformation of furfural to cyclopentanone." *Catal. Commun.* 24. Elsevier: 100–104. doi: 10.1016/j.catcom.2012.03.020.

Hronec, Milan, Fulajtarová, Katarina and Liptaj, Tibor. 2012. "Effect of catalyst and solvent on the furan ring rearrangement to cyclopentanone." *Appl. Catal. A-Gen.* 26. Elsevier: 104–111. doi: 10.1016/j.apcata.2012.06.018.

Hronec, Milan,Fulajtárová, Katarína,Vávra, Ivo,Soták, Tomáš,Dobročka, Edmund and Mičušík, Matej. 2016. "Carbon supported Pd–Cu catalysts for highly selective rearrangement of furfural to cyclopentanone." *Appl. Catal. B-Environ.* 181. Elsevier: 210–219. doi: 10.1016/j.apcatb.2015.07.046.

Huang, Kefeng, Brentzel, Zachary J., Barnett, Kevin J., Dumesic, James A., Huber, George W. and Maravelias, Christos T. 2017. "Conversion of furfural to 1,5-pentanediol: Process synthesis and analysis." *ACS Sustain. Chem. Eng.* 5. American Chemical Society: 4699–4706. doi: 10.1021/acssuschemeng.7b00059.

Hutchings, Gregory S., Luc, Wesley, Lu, Qi, Zhou, Yang, Vlachos, Dionisios G. and Jiao, Feng. 2017. "Nanoporous Cu–Al–Co alloys for selective furfural hydrodeoxygenation to 2-methylfuran." *Ind. Eng. Chem. Res.* 56. American Chemical Society: 3866–3872. doi: 10.1021/acs.iecr.7b00316.

Huber, George W., Iborra, Sara and Corma, Avelino. 2006. "Synthesis of transportation fuels from biomass: Chemistry, catalysts, and engineering." *Chem. Rev. 106.* American Chemical Society: 4044–4098. doi: 10.1021/cr068360d.

Iqbal, Sarwat, Liu, Xi, Aldosari, Obaid F., Miedziak, Peter J., Edwards, Jennifer K., Brett, Gemma L., Akram, Adeeba, King, Gavin M., Davies, Thomas E., Morgan, David J., Knight, David K. and Hutchings, Graham J. 2014. "Conversion of furfuryl alcohol into

2-methylfuran at room temperature using Pd/TiO2 catalyst." *Catal. Sci. Technol.* 4. Royal Society of Chemistry: 2280–2286. doi: 10.1039/C4CY00184B.

Jaswal, Anurag, Singh, Piyush Pratap and Mondal, Tarak. 2022. "Furfural – A versatile, biomass-derived platform chemical for the production of renewable chemicals." *Green Chem.* 24. Royal Society of Chemistry: 510–551. doi: 10.1039/D1GC03278J.

Jithin, E. V., Raghuram, G. K. S., Keshavamurthy, T. V., Velamati, Ratna Kishore, Prathap, Chockalingam and Varghese, Robin John. 2021. "A review on fundamental combustion characteristics of syngas mixtures and feasibility in combustion devices." *Renew. Sust. Energ. Rev.* 146. Elsevier: 111178. doi: 10.1016/j.rser.2021.111178.

Jung, Hyun Seung, Kim, Byeong Gi and Bae, Jong Wook. 2024. "Synthetic routes of clean hydrocarbons fuels and oxygenates by catalytic conversions of carbon oxides." *Appl. Catal. B-Environ.* 343. Elsevier: 123477. doi: 10.1016/j.apcatb.2023.123477.

Khemthong, Pongtanawat, Yimsukanan, Chakrit, Narkkun, Thanitporn, Srifa, Atthapon, Witoon, Thongthai, Pongchaiphol, Suchat, Kiatphuengporn, Sirapassorn and Faungnawakij, Kajornsak. 2021. "Advances in catalytic production of value-added biochemicals and biofuels via furfural platform derived lignocellulosic biomass." *Biomass Bioenerg.* 148. Elsevier: 106033. doi: 10.1016/j.biombioe.2021.106033.

King, Gavin M., Iqbal, Sarwat, Miedziak, Peter J., Brett, Gemma L., Kondrat, Simon A., Yeo, Benjamin R., Liu, Xi, Edwards, Jennifer K., Morgan, David J., Knight, David K. and Hutchings, Graham J. 2015. "An investigation of the effect of the addition of tin to 5 %Pd/TiO2 for the hydrogenation of furfuryl alcohol." *ChemCatChem.* 7. Wiley-VCH Verlag GmbH & CO. KGaA: 2122–2129. doi: 10.1002/cctc.201500242.

Kobayashi, Hirokazu, Ohta, Hidetoshi and Fukuoka, Atsushi. 2012. "Conversion of lignocellulose into renewable chemicals by heterogeneous catalysis." *Catal. Sci. Technol.* 2. Royal Society of Chemistry: 869–883. doi: 10.1039/C2CY00500J.

Kosydar, Robert, Duraczyńska, Dorota, Gurgul, Jacek, Kryściak-Czerwenka, Joanna and Drelinkiewicz, Alicja. 2019. "Liquid phase hydrogenation of furfural under mild conditions over Pd/C catalysts of various acidity." *React. Kinet. Mech. Cat.* 126. Springer: 417–437. doi: 10.1007/s11144-018-1494-x.

Li, Joshua Q. 2000. "The Chemistry and Technology of Furfural and its Many By-products." *Chem. Eng. J.* 81. Elsevier: 338–339. doi: 10.1016/S1385-8947(00)00182-0.

Ma, Jiping, Shi, Song, Jia, Xiuquan, Xia, Fei, Ma, Hong, Gao, Jin and Xu, Jie. 2019. "Advances in catalytic conversion of lignocellulose to chemicals and liquid fuels." *J. Energy Chem.* 36. Elsevier: 74–86. doi: 10.1016/j.jechem.2019.04.026.

Mariscal, R., Maireles-Torres, P., Ojeda, M., Sádaba, I. and López Granados, M. 2016. "Furfural: A renewable and versatile platform molecule for the synthesis of chemicals and fuels." *Energ. Environ. Sci.* 9. Royal Society of Chemistry: 1144–1189. doi: 10.1039/C5EE02666K.

Meng, Ye, Jian, Yumei, Li, Jie, Wu, Hongguo, Zhang, Heng, Saravanamurugan, Shunmugavel, Yang, Song and Li, Hu. 2023. "Surface-active site engineering: Synergy of photo- and super molecular catalysis in hydrogen transfer enables biomass upgrading and H2 evolution." *Chem. Eng. J.* 452. Elsevier: 139477. doi: 10.1016/j.cej.2022.139477.

Merat, Nadine, Godawa, Christine and Gaset, Antoine. 1990. "High selective production of tetrahydrofurfuryl alcohol: Catalytic hydrogenation of furfural and furfuryl alcohol." *J Chem. Technol. Biot.* 48. John Wiley & Sons Ltd: 145–159. doi: 10.1002/jctb.280480205.

Mironenko, R. M., Belskaya, O. B. and Likholobov, V. A. 2022. "Solvent effect on the rate and direction of furfural transformations during hydrogenation over the Pd/C catalyst." *Russ. Chem. B.* 71. Springer: 64–69. doi: 10.1007/s11172-022-3377-6.

Morales, Gabriel, Iglesias, JoseandMelero, Juan A. 2020. "Sustainable catalytic conversion of biomass for the production of biofuels and bioproducts." *Catalysts.* 10. MDPI: 581. doi: 10.3390/catal10050581.

Nakagawa, Yoshinao, Takada, Kana, Tamura, Masazumi and Tomishige, Keiichi. 2014. "Total hydrogenation of furfural and 5-hydroxymethylfurfural over supported Pd–Ir alloy catalyst." *ACS Catal.* 4. American Chemical Society: 2718–2726. doi: 10.1021/cs500620b.

Nakagawa, Yoshinao, Nakazawa, Hiroya, Watanabe, Hideo and Tomishige, Keiichi. 2012. "Total hydrogenation of furfural over a silica-supported nickel catalyst prepared by the reduction of a nickel nitrate precursor." *ChemCatChem.* 4. Wiley-VCH Verlag GmbH & CO. KGaA: 1791–1797. doi: 10.1002/cctc.201200218.

Lesiak, M., Binczarski, M., Karski, S., Maniukiewicz, W., Rogowski, J., Szubiakiewicz, E., Berlowska, J., Dziugan, P. and Witońska, I. 2014. "Hydrogenation of furfural over Pd–Cu/Al2O3 catalysts. The role of interaction between palladium and copper on determining catalytic properties." *J Mol. Catal. A-Chem.* 395. 337–348. doi: 10.1016/j.molcata.2014.08.041.

Li, Chuang, Xu, Guangyue, Liu, Xiaohao, Zhang, Ying and Fu, Yao. 2017. "Hydrogenation of biomass-derived furfural to tetrahydrofurfuryl alcohol over hydroxyapatite-supported Pd catalyst under mild conditions." *Ind. Eng. Chem. Res.* 56. American Chemical Society: 8843–8849. doi: 10.1021/acs.iecr.7b02046.

Li, Hao, Liu, Jia, Cai, Chiliu, Wang, Haiyong, Huang, Youwang, Wang, Chenguang and Ma, Longlong. 2023. "Selectivity catalytic transfer hydrogenation of biomass-based furfural to cyclopentanone." *Fuel.* 332. Elsevier: 126057. doi: 10.1016/j.fuel.2022.126057.

Li, Hu, Fang, Zhen, Luo, Jia and Yang, Song. 2017. "Direct conversion of biomass components to the biofuel methyl levulinate catalyzed by acid-base bifunctional zirconia-zeolites." *Appl. Catal. B-Environ.* 200. Elsevier: 182–191. doi: 10.1016/j.apcatb.2016.07.007.

Li, Hu, He, Jian, Riisager, Anders, Saravanamurugan, Shunmugavel, Song, Baoan and Yang, Song. 2016. "Acid–base bifunctional zirconium N-alkyltriphosphate nanohybrid for hydrogen transfer of biomass-derived carboxides." *ACS Catal.* 6. American Chemical Society: 7722–7727. doi: 10.1021/acscatal.6b02431.

Liu, Lu-Jie, Guo, Hong-Mei, Xue, Bo, Lou, Hui and Chen, Min. 2015. "Hydrogenation in supercritical conditions catalyzed by palladium supported on modified activated carbon." *RSC Adv.* 5. Royal Society of Chemistry: 66704–66710. doi: 10.1039/C5RA14284A.

Liu, Sibao, Amada, Yasushi, Tamura, Masazumi, Nakagawa, Yoshinao and Tomishige, Keiichi. 2014. "One-pot selective conversion of furfural into 1,5-pentanediol over a Pd-added Ir–ReOx/SiO2 bifunctional catalyst." *Green Chem.* 16. Royal Society of Chemistry: 617–626. doi: 10.1039/C3GC41335G.

Liu, Ping, Sun, Luyang, Jia, Xinxin, Zhang, Chen, Zhang, Wei, Song, Yongji, Wang, Hong and Li, Cuiqing. 2020. "Efficient one-pot conversion of furfural into 2-methyltetrahydrofuran using non-precious metal catalysts." *Mol. Catal.* 490. Elsevier: 110951. doi: 10.1016/j.mcat.2020.110951.

Liu, Yanhua, Chen, Zhihao, Wang, Xiaofeng, Liang, Yu, Yang, Xiaomin and Wang, Zichen. 2017. "Highly selective and efficient rearrangement of biomass-derived furfural to cyclopentanone over interface-active Ru/carbon nanotubes catalyst in water." *ACS Sustain. Chem. Eng.* 5. American Chemical Society: 744–751. doi: 10.1021/acssuschemeng.6b02080.

Lucarelli, Carlo, Bonincontro, Danilo, Zhang, Yu, Grazia, Lorenzo, Renom-Carrasco, Marc, Thieuleux, Chloé, Quadrelli, Elsje A., Dimitratos, Nikolaos, Cavani, Fabrizio and Albonetti, Stefania. 2019. "Tandem hydrogenation/hydrogenolysis of furfural to

2-methylfuran over a Fe/Mg/O catalyst: Structure–activity relationship" *Catalysts.* 9. MDPI: 895. doi: 10.3390/catal9110895.

Luo, Jingjie, Cheng, Yuan, Niu, Hongyu, Wang, Tao and Liang, Changhai. 2022. "Efficient Cu/FeOx catalyst with developed structure for catalytic transfer hydrogenation of furfural." *J Catal.* 413. Academic Press Inc. 575–587. doi: 10.1016/j.jcat.2022.07.013.

Lv, Shuang, Liu, Huifang, Zhang, Jian, Wu, Qiang and Wang, Feng. 2022. "Water promoted photocatalytic transfer hydrogenation of furfural to furfural alcohol over ultralow loading metal supported on TiO2." *J. Energy Chem.* 73. Elsevier: 259–267. doi: 10.1016/j.jechem.2022.06.012.

Lv, Xianqing, Luo, Xiaolin, Cheng, Xin, Liu, Jing, Li, Changzhi and Shuai, Li. 2022. "Production of hydroxymethylfurfural derivatives from furfural derivatives via hydroxymethylation." *Front. Bioeng. Biotech.* 10. Frontiers Media Sa: 851668. doi: 10.3389/fbioe.2022.851668.

Rai, Mahendra, Ingle, Avinash P., Pandit, Raksha, Paralikar, Priti, Biswas, Jayanta Kumarandda Silva, and Silvio Silverio. 2019. "Emerging role of nanobiocatalysts in hydrolysis of lignocellulosic biomass leading to sustainable bioethanol production." *Catal. Rev.* 61. Taylor & Francis Inc: 1–26. doi: 10.1080/01614940.2018.1479503.

Rahimi, Abumuslim, Ghorbani-Vaghei, Ramin and Alavinia, Sedigheh. 2021. "Nickel nanoparticles anchored over porous triazine-thiourea-sulfonamide to explore the reduction of carbonyl compounds." *J Porous Mat.* 28. Springer: 1643–1653. doi: 10.1007/s10934-021-01104-1.

Ren, Puning, Zhou, Yue,Su, Kaiyi, Sun, Lulu, Luo, Nengchao and Wang, Feng. 2023. "Visible-light-driven furfural oxidation over CuOx/Nb2O5." *Chemistry-An Asian Journal.* 18. e202300732. doi: 10.1002/asia.202300732.

Rogers, Scott M., Catlow, C. Richard A., Chan-Thaw, Carine E., Chutia, Arunabhiram, Jian, Nan, Palmer, Richard E., Perdjon, Michal, Thetford, Adam, Dimitratos, Nikolaos, Villa, Alberto and Wells, Peter P. 2017. "Tandem site- and size-controlled Pd nanoparticles for the directed hydrogenation of furfural." *ACS Catal.* 7. American Chemical Society: 2266–2274. doi: 10.1021/acscatal.6b03190.

Tan, Jinyu, Liu, Yixuan, Li, Mingrui, Li, Hu and Yang, Song. 2022. "One-step catalytic upgrading of bio-based furfural to γ-valerolactone actuated by coordination organophosphate–Hf polymers." ACS *Sustain. Energ. Fuel.* 6. American Chemical Society: 484–501. doi: 10.1039/D1SE01476E.

Tarazanov, Sergey, Grigoreva, Katharine, Shipitcyna, Anastasiia, Repina, Olga, Ershov, Michael, Kuznetsova, Svetlana and Nikulshin, Pavel. 2020. "Assessment of the chemical stability of furfural derivatives and the mixtures as fuel components." *Fuel.* 271. Elsevier: 117594. doi: 10.1016/j.fuel.2020.117594.

Tukacs, József M., Bohus, Márton, Dibó, Gábor and Mika, László T. 2017. "Ruthenium-catalyzed solvent-free conversion of furfural to furfuryl alcohol." *RSC Adv.* 7. Royal Society of Chemistry: 3331–3335. doi: 10.1039/C6RA24723G.

Tuteja, Jaya, Choudhary, Hemant, Nishimura, Shun and Ebitani, Kohki. 2014. "Direct synthesis of 1,6-hexanediol from HMF over a heterogeneous Pd/ZrP catalyst using formic acid as hydrogen source." *ChemSusChem. 7.* John Wiley & Sons Ltd: 96–100. doi: 10.1002/cssc.201300832.

Salnikova, Ksenia E., Larichev, Yurii V., Sulman, Esther M., Bykov, Alexey V.,Sidorov, Alexander I., Demidenko, Galina N., Sulman, Mikhail G., Bronstein, Lyudmila M. and Matveeva, Valentina G. 2020. "Selective hydrogenation of biomass-derived furfural: Enhanced catalytic performance of Pd–Cu alloy nanoparticles in porous polymer." *ChemPlusChem.* 85. Wiley-VCH Verlag GmbH: 1697–1703. doi: 10.1002/cplu.202000383.

Salnikova, Ksenia E., Matveeva, Valentina G., Larichev, Yurii V., Bykov, Alexey V., Demidenko, Galina N., Shkileva, Irina P. and Sulman, Mikhail G. 2019. "The liquid phase catalytic hydrogenation of furfural to furfuryl alcohol." *Catal. Today.* 329. Elsevier: 142–148. doi: 10.1016/j.cattod.2018.12.036.

Scholz, David, Aellig, Christof and Hermans, Ive. 2014. "Catalytic transfer hydrogenation/hydrogenolysis for reductive upgrading of furfural and 5-(Hydroxymethyl)furfural." *ChemSusChem.* 7. John Wiley & Sons Ltd: 268–275. doi: 10.1002/cssc.201300774.

Šivec, Rok, Likozar, Blaž and Grilc, Miha. 2021. "Surface kinetics and transport phenomena modelling for furfural hydrotreatment over Pd/C in isopropanol and tetrahydrofuran." *Appl. Surf. Sci.* 541. Elsevier: 148485. doi: 10.1016/j.apsusc.2020.148485.

Shao, Yuewen, Wu, Jie, Zheng, Zhiyuan, Fan, Mengjiao, Sun, Kai, Bkangmo Kontchouo, Félix Mérimé, Zhang, Lijun, Zhang, Shu, Hu, Guangzhi and Hu, Xun. 2022. "Alloying cobalt in Co–Fe–Al catalyst for achieving the selective conversion of furfural to cyclopentanone." *Renew. Energ.* 195. Elsevier: 957–971. doi: 10.1016/j.renene.2022.06.068.

Sharma, Rajesh V., Das, Umashankar, Sammynaiken, Ramaswami and Dalai, Ajay K. 2013. "Liquid phase chemo-selective catalytic hydrogenation of furfural to furfuryl alcohol." *Appl. Catal. A-Gen.* 454. Elsevier: 127–136. doi: 10.1016/j.apcata.2012.12.010.

Shahbeik, Hossein, Shafizadeh, Alireza, Gupta, Vijai Kumar, Lam, Su Shiung, Rastegari, Hajar, Peng, Wanxi, Pan, Junting, Tabatabaei, Meisam and Aghbashlo, Mortaza. 2023. "Using nanocatalysts to upgrade pyrolysis bio-oil: A critical review." *J Clean. Prod.* 413. Elsevier: 137473. doi: 10.1016/j.jclepro.2023.137473.

Shivhare, Atal, Kumar, Abhinav and Srivastava, Rajendra. 2021. "An Account of the catalytic transfer hydrogenation and hydrogenolysis of carbohydrate-derived renewable platform chemicals over non-precious heterogeneous metal catalysts." *ChemCatChem.* 13. Wiley-VCH Verlag GmbH & CO. KGaA: 59–80. doi: 10.1002/cctc.202001415.

Shivhare, Atal, Kumar, Abhinav and Srivastava, Rajendra. 2021. "Metal phosphate catalysts to upgrade lignocellulose biomass into value-added chemicals and biofuels." *Green Chem.* 23. Royal Society of Chemistry: 3818–3841. doi: 10.1039/D1GC00376C.

Stevens, James G., Bourne, Richard A., Twigg, Martyn V. and Poliakoff, Martyn. 2010. "Real-time product switching using a twin catalyst system for the hydrogenation of furfural in supercritical CO2." *Angew. Chem. Int. Edit.* 49. John Wiley & Sons Ltd: 8856–8859. doi: 10.1002/anie.201005092.

Song, Heyuan, Jin, Fuxiang, Liu, Qiao and Liu, Hailong. 2021. "Zeolite-catalyzed acetalization reaction of furfural with alcohol under solvent-free conditions." *Mol. Catal.* 513. Elsevier: 111752. doi: 10.1016/j.mcat.2021.111752.

Sudarsanam, Putla, Li, Hu and Sagar, Tatiparthi Vikram. 2020. "TiO2-Based water-tolerant acid catalysis for biomass-based fuels and chemicals." *ACS Catal.* 10. American Chemical Society: 9555–9584. doi: 10.1021/acscatal.0c01680.

Sun, Daolai, Sato, Satoshi, Ueda, Wataru, Primo, Ana, Garcia, Hermenegildo and Corma, Avelino. 2016. "Production of C4 and C5 alcohols from biomass-derived materials." *Green Chem.* 18. Royal Society of Chemistry: 2579–2597. doi: 10.1039/C6GC00377J.

Sutton, Andrew D., Waldie, Fraser D., Wu, Ruilian, Schlaf, Marcel, 'Pete' Silks, Louis A. and Gordon, John C. 2013. "The hydrodeoxygenation of bioderived furans into alkanes." Nat. Chem. 5. Nature portfolio: 428–432. doi: 10.1038/nchem.1609.

Wang, Bing, Xiao, Rui and Zhang, Huiyan. 2019. "An overview of bio-oil upgrading with high hydrogen-containing feedstocks to produce transportation fuels: Chemistry, catalysts, and engineering." *Curr. Org. Chem.* 23. Bentham Science Publishers: 746–767. doi: 10.2174/1385272823666190405145007.

Wang, Chenguang, Zhang, Xinghua, Liu, Qiang, Zhang, Qi, Chen, Lungang and Ma, Longlong. 2020. "A review of conversion of lignocellulose biomass to liquid transport fuels by integrated refining strategies." *Fuel Process. Technol.* 208. Elsevier: 106485. doi: 10.1016/j.fuproc.2020.106485.

Wang, Hongliang, Yang, Bin, Zhang, Qian and Zhu, Wanbin. 2020. "Catalytic routes for the conversion of lignocellulosic biomass to aviation fuel range hydrocarbons." *Renew. Sust. Energ. Rev.* 120. Elsevier: 109612. doi: 10.1016/j.rser.2019.109612.

Wang, Keping, Guo, Zhenyan, Zhou, Min, Yang, Ying, Li, Lanyun, Li, Hu, Luque, Rafael and Saravanamurugan, Shunmugavel. 2024. "Biomass valorization via electrocatalytic carbon–carbon bond cleavage." *J Energy Chem.* 91. Elsevier: 542–578. doi: 10.1016/j.jechem.2023.12.041.

Wang, Yanling, Liu, Cun and Zhang, Xiongfu. 2020. "One-step encapsulation of bimetallic Pd–Co nanoparticles within UiO-66 for selective conversion of furfural to cyclopentanone." *Catal. Lett.* 150. Springer: 2158–2166. doi: 10.1007/s10562-020-03133-y.

Wang, Yi, Akbarzadeh, Abdolhamid, Chong, Li, Du, Jinyu, Tahir, NadeemandAwasthi, Mukesh Kumar. 2022. "Catalytic pyrolysis of lignocellulosic biomass for bio-oil production: A review." *Chemosphere.* 297. Elsevier: 134181. doi: 10.1016/j.chemosphere.2022.134181.

Xu, Nan, Gong, Jing and Huang, Zuohua. 2016. "Review on the production methods and fundamental combustion characteristics of furan derivatives." *Renew. Sust. Energ. Rev.* 54. Elsevier: 1189–1211. doi: 10.1016/j.rser.2015.10.118.

Xu, Ying, Qiu, Songbai, Long, Jinxing, Wang, Chenguang, Chang, Jiamin, Tan, Jin, Liu, Qiying, Ma, Longlong, Wang, Tiejun and Zhang, Qi. 2015. "In situ hydrogenation of furfural with additives over a RANEY® Ni catalyst." *RSC Adv.* 5. Royal Society of Chemistry: 91190–91195. doi: 10.1039/C5RA12844G.

Xu, Yufei, Long, Jingxuan, He, Jian and Li, Hu. 2019. "Alcohol-mediated reduction of biomass-derived furanic aldehydes via catalytic hydrogen transfer." *Curr. Org. Chem.* 23. Bentham Science Publishers: 2168–2179. doi: 10.2174/1385272823666190723141955.

Yan, Kai, Jarvis, Cody, Lafleur, Todd, Qiao, Yunxiang and Xie, Xianmei. 2013. "Novel synthesis of Pd nanoparticles for hydrogenation of biomass-derived platform chemicals showing enhanced catalytic performance." *RSC Adv.* 3. Royal Society of Chemistry: 25865–25871. doi: 10.1039/C3RA43619E.

Yan, Kai, Wu, Guosheng, Lafleur, Todd and Jarvis, Cody. 2014. "Production, properties and catalytic hydrogenation of furfural to fuel additives and value-added chemicals." *Renew. Sust. Energ. Rev.* 38. Elsevier: 663–676. doi: 10.1016/j.rser.2014.07.003.

Yang, Weiran and Sen, Ayusman. 2010. "One-step catalytic transformation of carbohydrates and cellulosic biomass to 2,5-dimethyltetrahydrofuran for liquid fuels." *ChemSusChem.* 3. John Wiley & Sons Ltd: 597–603. doi: 10.1002/cssc.200900285.

Yoo, Chang G. and Kim, Tae H. 2020. "Advancements in catalytic conversion of biomass into biofuels and chemicals." *Energies.* 13. MDPI: 5438. doi: 10.3390/en13205438.

Zhang, Bin, Zhu, Yulei, Ding, Guoqiang, Zheng, Hongyan and Li, Yongwang. 2012. "Selective conversion of furfuryl alcohol to 1,2-pentanediol over a Ru/MnOx catalyst in aqueous phase." *Green Chem.* 14. Royal Society of Chemistry: 3402–3409. doi: 10.1039/C2GC36270H.

Zhao, Yuying. 2014. "Facile synthesis of Pd nanoparticles on SiO2 for hydrogenation of biomass-derived furfural." *Environ. Chem. Lett.* 12. Springer: 185–190. doi: 10.1007/s10311-013-0424-4.

Zhou, Ziyuan, Liu, Dehua and Zhao, Xuebing. 2021. "Conversion of lignocellulose to biofuels and chemicals via sugar platform: An updated review on chemistry and mechanisms of acid hydrolysis of lignocellulose." *Renew. Sust. Energ. Rev.* 146. Elsevier: 111169. doi: 10.1016/j.rser.2021.111169.

Zhou, Peng, Li, Linbo, Mosali, Venkata Sai Sriram, Chen, Yu, Luan, Peng, Gu, Qinfen, Turner, David R., Huang, Liang and Zhang, Jie. 2022. "Electrochemical hydrogenation of furfural in aqueous acetic acid media with enhanced 2-methylfuran selectivity using CuPd bimetallic catalysts." *Angew. Chem. Int. Edit.* 61. John Wiley & Sons Ltd: e202117809. doi: 10.1002/anie.202117809.

6 Catalytic Hydrodeoxygenation of Furfural

Dinesh Kumar Mishra, Yeongin Jo, Cong Chien Truong, and Young-Woong Suh

6.1 INTRODUCTION

In the past few decades, lignocellulosic biomass (e.g., biodegradable waste materials such as crops, grasses, wood, and other plant sources) has emerged as an excellent alternative to traditional fossil fuel resources due to its low-cost and easily available in nature (Cherubini and Strømman 2011; Zhou et al. 2011; Yang et al. 2019; Sengupta and Pike 2014). It is composed of three major components such as cellulose (35–50%), hemicellulose (20–35%), and lignin (10–25%), as shown in Scheme 6.1 (Peng et al. 2010; De et al. 2015; Alonso et al. 2017; Mika et al. 2018). The composition of lignocellulosic biomass may vary depending on the source of bio-based feedstock materials (Mankar et al. 2021; Ruppert et al. 2012; Budzinski and Nitzsche 2016; Ma et al. 2020; Periyasamy et al. 2023). Thanks to the catalytic chemical processes, various bio-based platform chemicals from lignocellulosic biomass used as renewable feedstock were produced. Among them, one of the best examples of bio-based chemicals is furfural (FAL) (Jaswal et al. 2022; Yong et al. 2022). FAL is regarded as a vital platform chemical that has been listed as one of the "Top 12" platform chemicals recognized by the U.S. Department of Energy (DOE) (Werpy and Petersen 2004). In fact, FAL contains a heteroaromatic furan ring with highly reactive an aldehyde group at C2 position. Due to its unique structure and properties, FAL could further be transformed into a plethora of high-value-added chemicals via various catalytic processes such as hydrodeoxygenation (Lin et al. 2018; Lan et al. 2021; Taylor et al. 2017), decarbonylation (Lejemble et al. 1984; Ishida et al. 2016; Chen et al. 2023), hydrogenation (Chen et al. 2016; Matsheku et al. 2023; Nope et al. 2023; Mishra et al. 2020), hydrogenolysis (Yamaguchi et al. 2021; Nakagawa et al. 2015; Fu et al. 2021), and reductive-amination (Chatterjee et al. 2016; Nishimura et al. 2016; Saini et al. 2022). Of these, hydrodeoxygenation of bio-based FAL is known to be an attractive approach to produce 2-methylfuran (2-MF) (Lin et al. 2018; Lan et al. 2021). 2-MF has great potential to reduce greenhouse gas emission compared to fossil fuels. Despite the facts, 2-MF could further be used as a potential precursor for

DOI: 10.1201/9781003464839-6

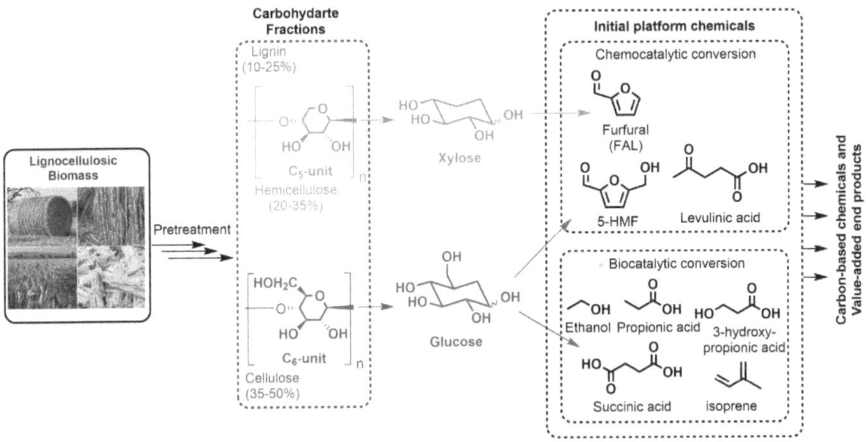

SCHEME 6.1 Transformation of lignocellulosic biomass to various bio-based chemicals. (Originally drawn based on Mika et al. 2018.)

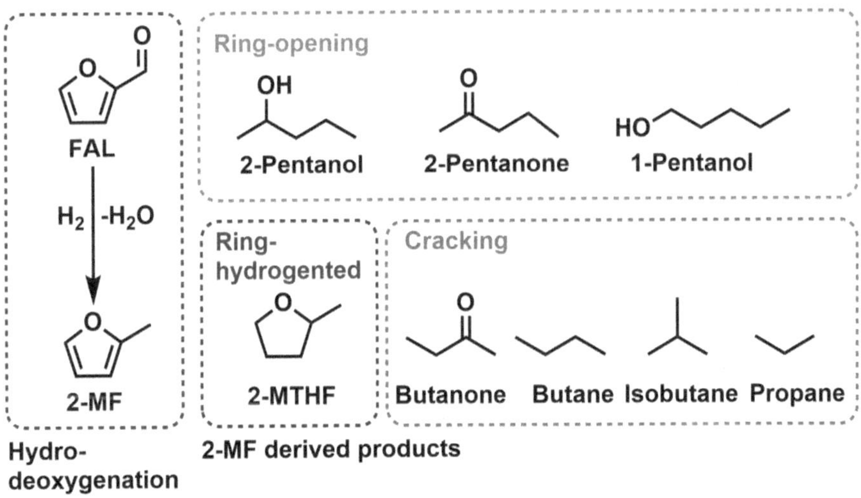

SCHEME 6.2 Catalytic hydrodeoxygenation of FAL to 2-MF and its derived products. (Originally drawn based on Biswas et al. 2014.)

the production of many industrial chemicals (e.g., ring opening, ring hydrogenated, and cracking products) (Scheme 6.2).

Herein, the production of 2-MF from catalytic hydrodeoxygenation of FAL is mainly focused. The catalytic hydrodeoxygenation of FAL has been extensively studied in both liquid and vapour phase. For this purpose, various noble and different non-noble metal-based catalysts are screened. Further, we described the merits and shortcomings of various catalytic systems used in catalytic hydrodeoxygenation of FAL. Based on the reaction mechanism, the reaction pathways in catalytic hydrodeoxygenation of FAL also delineated. At the end of the chapter, the environmental benefits and future opportunities of 2-MF are also described.

6.2 INDUSTRIAL RELEVANCE

This chapter focused on the production of 2-MF from catalytic hydrodeoxygenation of FAL derived from lignocellulosic biomass. It is distinguished that FAL had global market size of 662 million USD in 2023 and is anticipated to increase 767 USD million by 2028 at compound annual growth rate (CAGR) of 2.9% (www.marketsandmarkets.com/Market-Reports/furfural-market-101056456.html). As aforementioned, the 2-MF product is an important renewable biofuel and 2-MF could also offer many other fine chemicals (as depicted in Scheme 6.2) (Biswas et al. 2014). Due to increased demand of 2-MF, the global market size of 2-MF is anticipated to reach multimillion USD in the upcoming year. Top global companies and manufactures such as Capot chemical, Angene Chemical, Amadis Chemical, Guangzhu Chen Yu Chemicals, Finetech Chemical, Shandong YINIO Biologic Materials, AN Pharma Tech, Tractus company, and Haihang Industry are involved in the production, distribution, and sale of 2-MF. 2-MF is supposed to grow at a CAGR of 5.5% in future (www.precisionbusinessinsights.com/market-reports/2-methylfuran-market). Although the remarkable efforts have been made in the production of 2-MF from catalytic hydrodeoxygenation of FAL. However, the economic feasibility of producing 2-MF is not still reported yet. Therefore, technological advancements are indeed needed by using inexpensive non-noble-based catalysts, cost-effective support, and the process operation under mild reactions conditions. The strategies can enhance the feasibility of producing 2-MF from catalytic hydrogenation of FAL and facilitate considerable scale up of the commercialization process of 2-MF production.

6.3 SOURCE OF FAL AS A PRECURSOR

To the best of our knowledge, FAL was first isolated (in 1832) as a byproduct from formic acid production. These days, though, it is mainly produced from the most easily available lignocellulosic biomass as aforementioned (Jaswal et al. 2022; Yong et al. 2022). The physicochemical properties of lignocellulosic biomass are summarized in Table 6.1 (Periyasamy et al. 2023). In fact, many pretreatment methods such as physical, chemical, and biological are developed to fractionate the lignocellulosic biomass into their constituents (cellulose, hemicellulose, and lignin) (Tassinari et al. 1980; Sun and Cheng 2002; Cybulska et al. 2013; Millett et al. 1976; Sindhu et al. 2016). For the commercial production of FAL, xylan-rich hemicellulose is used as the main feedstock. In this process, xylan (C_5-unit) is first hydrolyzed into xylose (pentose unit), followed by dehydration of xylose into FAL, as shown Scheme 6.1 (Binder et al. 2010; Mariscal et al. 2016; Choudhary et al. 2012; Choudhary et al. 2011).

TABLE 6.1
Physicochemical Properties of Lignocellulosic Biomass

Feature	Cellulose	Hemicellulose	Lignin
General formula	$(C_6H_{10}O_5)n$	$(C_5H_8O_5)n$	$[C_9H_{10}O_3(OCH_3)_{0.9\text{-}1.7})]n$
Structure	Branched-chain, crystalline, and amorphous	Branched-chain, less crystalline, and highly amorphous	Branched-chain, amorphous, and non-uniform
Binding force	Hydrogen linkage	Chemical linkage	Chemical linkage
Structural unit	D-glucose	Xylose, mannose, galactose, L-arabinose, and glucuronic acid	p-hydroxyphenyl, guaiacyl, and syringyl
Pyrolysis	300–550 °C	220–315 °C	200–900 °C
Bonding	β-1,4-glycosidic acid linkage	β-(1,4) or (1,3)-glycosidic acid linkage	Aryl ether linkage (a-O-4 and a-O-4), C-C linkage
Solubility	Insoluble in water, Soluble in ionic liquid	Insoluble in water, Soluble in acid, alkali, ionic liquid, and organic solvents	Insoluble in water; soluble in acid, alkali, ionic liquid, and organic solvents
Degree of polymerization	1000–10000	Low (>200)	High (~4000)
Color	White	Brown	Yellow, light to dark brown
Density at 20°C (kg/m^3)	1559	1520	1300–1500
Applications	Use as fibers, composites and film derivatives, Resource for fuel and chemicals production	Use as drug carriers and preparation of hydrogels and cosmetics	Uses for natural binders, adhesives, coating and agricultural chemical production, and used in the form of polymer such as resins, polyurethane, epoxy, rubber, polyesters and polyolefins

Source: Periyasamy et al. (2023).

6.4 CATALYTIC HYDRODEOXYGENATION OF FAL

Catalytic hydrodeoxygenation of FAL has been extensively studied using various noble and non-noble metal-based catalysts. The results obtained from both liquid and vapor phases catalytic hydrodeoxygenation of FAL are discussed in the following sections.

6.4.1 Liquid Phase Catalytic hydrodeoxygenation of FAL

Liquid phase catalytic hydrodeoxygenation of FAL to the corresponding product 2-MF could be accomplished in batch-wise operation mode, which is the most preferable approach. The results obtained from liquid phase catalytic hydrodeoxygenation of FAL using various noble and non-metal-based catalysts are summarized in Table 6.2. In this context, Yu and co-workers described acetic acid-mediated catalytic transformation of FAL using Pd-based catalysts (Yu et al. 2011). This 5 wt% Pd/C catalyst showed 8.9% yield of hydrodeoxygenated 2-MF product. In addition, FOL as an intermediate, furfuryl acetate, and other byproducts were also observed. As compared to the 5 wt% Pd/C catalyst, a physical mixture of 5 wt% Pd/C and $Al_2(SiO_3)_3$ [5wt% Pd+ $Al_2(SiO_3)_3$] showed 18% yield of the 2-MF product. In this continuation, Hermans and co-workers described liquid phase hydrogenation of FAL using hematite (Fe_2O_3) supported different metal-based catalysts (Scholz et al. 2014). Among various metal-based catalysts, the 5 wt% Pd/Fe_2O_3 catalyst enabled generation of 10% 2-MF product at 180 °C after 7.5 h. This could be attributed to the morphology adopted by Pd particles (1–6 nm size) and their interaction of hematite (Fe_2O_3) support. Along with 2-MF product, there was 57% yield of FOL. In parallel to hydrogenation, decarbonylation of FAL was undertaken to form furan (8%). In another example, Biradar et al. screened various metals (Pd, Ru, and Re-based catalysts) in liquid phase hydrogenation of FAL (Biradar et al. 2014). Among all, Pd-based catalyst 3 wt% Pd/C catalyst showed total yield (32%) of hydrodeoxygenated products (14% yield of 2-MF and 18% yield of 2-MTHF) at complete conversion of FAL (100%). While, the other 3 wt% Pd/$CaCO_3$ and 3 wt% Ru/C catalysts showed 28.4% and 16.6% yield of hydrogenated products (2-MF and 2-MTHF), respectively. Interestingly, 3 wt% Re/C catalyst could give a total yield 39.5% of hydrogenated products (32.3% yield of 2-MF and 7.2% yield of 2-MTHF).

Based on Pt-based catalyst, Hronec and Fulajtarová performed liquid phase transformation of FAL in different solvents using 5 wt% Pt/C catalyst (Hronec and Fulajtarová 2012). Among various used solvents such as water, n-butanol, n-butanol/water (1:1 vol.), n-decanol, and tetrahydrofuran (THF), n-butanol was considered to be the best solvent for hydrodeoxygenated 2-MF and 2-MTHF viewpoints. The maximum yield of 2-MF (40.4%) was accomplished as displayed in Figure 6.1(a). Simultaneously, 2-MF could further be hydrogenated to form 2-MTHF with 1.9% yield under the applied reaction conditions (at 170 °C and 8.0 MPa H_2). Total yield of hydrogenated products (2-MF and 2-MTHF) was 42.3%. Unfortunately, targeted cyclopentanone product and cyclopentanol 0.08% and 0.2% were obtained. Meanwhile, other 5 wt% Pt/C catalyst, 5 wt% Pd/C, 5 wt% Ru/C, CoMnCr, and Raney Ni Actimet C catalysts were also tested in aqueous liquid phase transformation of FAL. H_2O used as solvent favors the formation of cyclopentanone, thereby lower yield of hydrodeoxygenated 2-MF and 2-MTHF products could be obtained (Figure 6.1(b)).

Investing further to improve the yield of 2-MF from catalytic hydrodeoxygenation of FAL, Panagiotopoulou and Vlachos performed transfer hydrogenation of FAL in IPA solution (as H_2 donor) using 5 wt% Ru/C catalyst (Panagiotopoulou and Vlachos

TABLE 6.2
Liquid Phase Catalytic Hydrodeoxygenation of FAL

Catalysts	Reaction conditions	Conv. (%)	Yield (%) 2-MF	Yield (%) 2-MTHF	Ref.
5 wt% Pd/C	FAL (100 mmol), acetic acid (100 mmol), cat. (400 mg), H_2 (2.0 MPa), toluene (10 mL), 150 °C, 4 h	41.2	8.9	-	(Yu et al. 2011)
5 wt% Pd/C+ $Al_2(SiO_3)_3$	FAL (100 mmol), acetic acid (100 mmol), Pd/C: $Al_2(SiO_3)$ ratio (1: 1 wt/wt), H_2 (2.0 MPa), toluene (10 mL), 150 °C, 4 h	69.4	18.0	-	(Yu et al. 2011)
5 wt% Pd/Fe_2O_3	FAL (400 mmol), cat. (200 mg), N_2 (0.1MPa), IPA (40 mL), 180 °C, 7.5 h	87	10	-	(Scholz et al. 2014)
3 wt% Pd/C	FAL (52 mmol), cat. (500 mg), H_2 (3.5 MPa), IPA (95 mL), 220 °C, 5 h	100	14	18	(Biradar et al. 2014)
3 wt% Pd/$CaCO_3$	FAL (52 mmol), cat. (500 mg), H_2 (3.5 MPa), IPA (95 mL), 220 °C, 5 h	91.5	13.7	14.6	(Biradar et al. 2014)
3 wt% Ru/C	FAL (52 mmol), cat. (500 mg), H_2 (3.5 MPa), IPA (95 mL), 220 °C, 5 h	57.3	9.7	6.9	(Biradar et al. 2014)
3 wt% Re/C	FAL (52 mmol), cat. (500 mg), H_2 (3.5 MPa), IPA (95 mL), 220 °C, 5 h	71.8	32.3	7.2	(Biradar et al. 2014)
5 wt% Pt/C	FAL (10.4 mmol), cat. (100 mg), H_2 (8.0 MPa), n-butanol (20 mL), 175 °C, 0.5 h	99.3	40.4	1.9	(Hronec and Fulajtarová 2012)
5 wt% Ru/C	FAL (2.6 mmol), cat. (100 mg), N_2 (2.04 MPa), IPA (24 mL), 180 °C, 10 h	95	61	-	(Panagiotopoulou and Vlachos 2014)
8 wt% Ru/$NiFe_2O_3$	FAL (2.6 mmol), cat. (100 mg), N_2 (2.04 MPa), IPA (24 mL), 180 °C, 6 h	>97	83	-	(Wang et al. 2017)
Cu-Fe catalyst	FAL (25.4 mmol), cat. (200 mg), H_2 (9.0 MPa), octane (5 mL), 220 °C, 14 h	99.4	51.1	41.7[a]	(Yan and Chen 2014)
$Cu_{2.5}$Zn-Al-600	FAL (6.2 mmol), cat. (150 mg), N_2 (0.1 MPa), IPA (24 mL), 180 °C, 4 h	99%	72	-	(Niu et al. 2019)
Cu-Pd/ZrO_2 (Cu: Pd ratio= 2:1)	FAL (1 mmol), cat. (120 mg), IPA (14 mL), 220 °C, 4 h	100	5.1	78.8	(Chang et al. 2016)
Ni-Cu/Al_2O_3 (Ni: Cu ratio= 1:1)	FAL (10.4 mmol), cat. (150 mg), FAL: HCOOH ratio (1: 2), IPA (12.7 mL), 210 °C, 4 h	97.6	79.0	13.1[a]	(Fu et al. 2017)
RuNiSB-16 (Ru: Ni ratio= 4:1)	FAL (12.1 mmol), cat. (200 mg), IPA (3.0 mL), H_2 (0.1 MPa) 180 °C, 4 h	100	88	12[a]	(Kumaravel et al. 2024)
Cu-Co (Cu: Co ratio= 1:1)	FAL (5.2 mmol), cat. (250 mg), H_2 (3 MPa), 200 °C, 3.5 h	100	93.8	4.9[a]	(Barranca et al. 2023)

Note: [a] Yield of FOL.

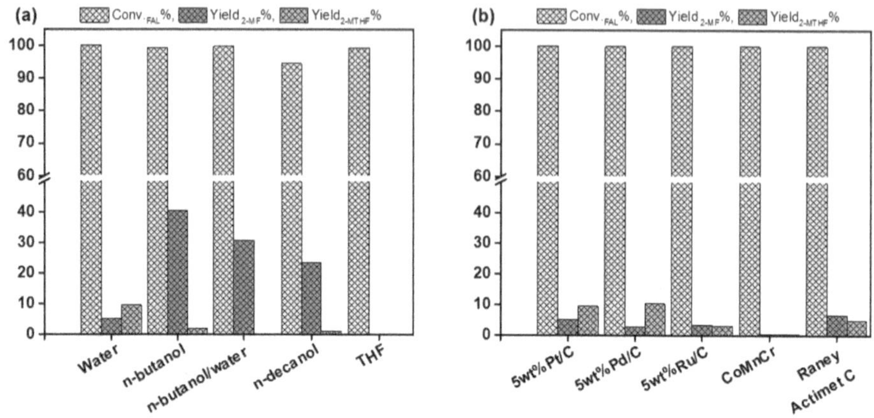

FIGURE 6.1 (a) Effect of solvents [FAL (10.4 mmol), 5 wt% Pt/C cat. (100 mg), H_2 (8.0 MPa), n-butanol (20 mL), 175 °C, 0.5 h]; (b) Effect of catalysts [FAL (5.2–10.4 mmol), cat. (100–200 mg), H_2 (3.0–8.0 MPa), water (20 mL), 160–175 °C, 0.5–1 h].

2014). Under the applied reaction conditions (at 180 °C and after 10 h), a high yield of hydrodeoxygenated 2-MF product (61%) was accomplished. The yield of 2-MF (61%), obtained from liquid phase catalytic hydrogenation of FAL below 200 °C, was highest. Another example of highly active and selective Ru-based catalyst in hydrodeoxygenation of FAL is associated with the formulation of Ru/$NiFe_2O_4$ catalyst (Wang et al. 2017). This Ru/$NiFe_2O_4$ catalyst was prepared by one-step sol-gel method and calcined at different temperatures (300, 400, and 500 °C). 4 wt% Ru/$NiFe_2O_4$-300 catalyst (calcined at 300 °C) showed lowest yield of hydrodeoxygenated 2-MF product, i.e., 62.9%. In fact, RuOx is reduced easily to Ru^0 and the remaining RuOx lead to lower hydrogenolysis of FOL. The yield of FOL intermediate was 22.1%. 4 wt% Ru/$NiFe_2O_4$-400 catalyst (calcined at 400 °C) showed 75.1% of hydrodeoxygenated 2-MF product. While, 4 wt% Ru/$NiFe_2O_4$-500 (calcined at 500 °C) showed 68.7% yield of 2-MF. This phenomenon ascribed due to low calcination temperature that could offer the formation of specific surface area and the smaller Ru particle size. Further, the effect of Ru loading was investigated in hydrogenation of FAL. Interestingly, Ru/$NiFe_2O_4$-400 (with 8 wt Ru%) catalyst exhibited 83% yield of 2-MF product, which is highest in the liquid-phase hydrogenation of FAL below 200 °C. Moreover, Ru/$NiFe_2O_4$-400 was found to be reusable 5 times, the activity and yield of 2-MF remained almost unchanged.

Many noble metal-based catalysts have been developed for the catalytic hydrodeoxygenation of FAL. However, high cost and sensitive surface of noble metal-based catalysts limit their practical use. Therefore, it is crucial to develop effective, cheap, stable, and ecofriendly catalysts. In this context, Yan and Chen et al. synthesized Cu-Fe catalyst (Cu^{2+}/Fe^{3+} ratio= 2) and used as catalyst in hydrodeoxygenation of FAL (Yan and Chen 2014). This Cu-Fe catalyst showed 51% yield of hydrodeoxygenated 2-MF and 41.7% yield of FOL as intermediate. This suggested that the amount of

Cu^{2+} could drastically influence the extent of hydrogenation. It turned out that copper sites controlled the dissociative activation of H_2 and regulated the catalytic performance. Recently, Niu et al. reported catalytic transfer hydrogenation of FAL using a series of CuZnAl catalysts in IPA as hydrogen donor (Niu et al. 2019). For this purpose, different ZnAl-600, $Cu_{2.5}$Zn-600, Cu-AL-600, and $Cu_{2.5}$Zn-Al-600 catalysts were prepared and tested. ZnAl-600 and $Cu_{2.5}$Zn-600 catalysts showed 89% and 96% conversion of FAL. However, the yield of desired 2-MF product from both catalysts was as low as <1%. FOL as main product was detected (85.7% for ZnAl-600 and 72% $Cu_{2.5}$Zn-600, respectively). On the contrary, Cu-AL-600 showed better performance and gave 55.1% yield of 2-MF, suggesting that specific ability of Cu-AL-600 catalyst for deoxygenation in depth. Cu-AL-600 catalyst, as compared to ZnAl-600 catalyst and $Cu_{2.5}$Zn-600 catalyst, could enhance the hydrogenolysis rate of FOL. Finally, $Cu_{2.5}$Zn-Al-600 demonstrated high activity in catalytic hydrogenation of FAL and 72% yield of 2-MF was accomplished under the identical reaction conditions (at 180 °C and after 4 h). This could be attributed to the complexed chemical composition/structure of $Cu_{2.5}$Zn-Al-600 catalyst and multi-combined effects of it increased yield of 2-MF product. It is worth mentioning that the catalyst was reused six times. The conversion of FAL was maintained with 7% loss of 2-MF yield.

In contrast to monometallic catalysts, bimetallic catalysts were found to have synergetic effect that shows superior catalytic performance. For instance, Chang et al. described the catalytic transfer hydrogenation of FAL over various bimetallic catalysts. In this study, various commercially available supports (ZrO_2, Al_2O_3, SiO_2, and TiO_2) were used to prepare bimetallic Cu-Pd catalysts. Among all, Cu-Pd/ZrO_2 bimetallic catalyst (Cu: Pd ratio= 2:1) was found to be highly active towards the total yield of 2-MF and 2-MTHF products (Chang et al. 2016). At the complete conversion of FAL (100%), a total yield of hydrodeoxygenated (2-MF and 2-MTHF) products was 83.9% (5.1% yield for 2-MF and 78.8% yield of 2-MTHF). This result could be possible due to the synergetic effect between Cu and Pd that enhanced the rate of hydrodeoxygenation and suppressed the possible side reactions. Further, this Cu-Pd/ZrO_2 bimetallic catalyst was found to be stable only after five catalyst recycling tests. In another typical example, alumina support (Al_2O_3), supporting other Ni-Cu bimetallic catalysts was prepared by the incipient wetness impregnation method and used as bimetallic catalyst in hydrogenation of FAL (Fu et al. 2017). Ni-Cu/Al_2O_3 catalyst gave 79% yield of 2-MF under the applied reaction conditions (at 210 °C after 4 h). Another example of bimetallic catalyst in catalytic hydrogenation of FAL is associated with the fabrication of RuNiSB-16 (Kumaravel et al. 2024). In a typical procedure, pluronic F-127 served as surfactant and tetraethyl orthosillicate (TEOS) as silica precursor were applied. Interestingly, this RuNiSB-16 catalyst exhibited a satisfactory yield of 2-MF (88%). This satisfactory yield of 2-MF is attributed to the coordination between metal and acid sites on the catalyst surface. More recently, Barranca et al. performed catalytic transformation of FAL in the presence of a γ-Al_2O_3-supported Cu-Co-based bimetallic catalyst (Barranca et al. 2023). During preparation of the Cu-Co-based catalyst, surface chlorine for metal precursor drastically affect both the activity and the yield of 2-MF product. Hydrogenation of furan ring was inhibited in the presence of chlorine on the surface of the catalyst, which could increase the selectivity of 2-MF. As a result, an excellent yield of 2-MF (93.8%) was achieved (at 200 °C after 3.5 h).

6.4.2 Liquid Phase Catalytic Hydrodeoxygenation of FOL

In another approach, FOL was used as the substrate to produce 2-MF and the results are summarized in Table 6.3. In this context, Iqbal et al. described selective transformation of FOL using two different 5 wt% Pd/Al_2O_3 and 2.58 wt% Pd/TiO_2 catalysts (Iqbal et al. 2014). At room temperature, the 5 wt% Pd/Al_2O_3 catalyst exhibited only 17% yield of 2-MF at complete conversion of FAL (100%) (at 25 °C and under 0.1 MPa of hydrogen pressure). Under the similar reaction conditions, the 2.58 wt% Pd/TiO_2 catalyst showed a better yield of 2-MF (55.5%) because it increased the rate of hydrogenolysis. Interestingly, increasing pressure improved both the conversion of FOL and yield of 2-MF. Conversion of FOL and yield of 2-MF were 93.7% and 87.7%, respectively. A high yield of 2-MF could be possible because the hydrodeoxygenation rate of C=O was much faster than that of reduction of C=C in the furan ring. Only a small amount of tetrahydrofurfuryl alcohol (THFOL) (< 6%) as the byproduct was detected. Recently, catalytic hydrodeoxygenation of FOL to 2-MF was performed using a SiO_2-supported Au: Pd bimetallic catalyst under very mild reaction conditions (at 25 °C and 0.1 MPa of hydrogen pressure) (Aldosari 2019). However, yield of 2-MF was quite low as 33.1%.

TABLE 6.3
Liquid Phase Catalytic Hydrodeoxygenation of FOL

Catalysts	Reaction Conditions	Conv. (%)	Yield (%) 2-MF	Yield (%) 2-MTHF	Ref.
5 wt% Pd/Al_2O_3	FOL (10.2 mmol), cat. (100 mg), H_2 (0.1 MPa), $C_2H_2Cl_2$ (20 mL), 25 °C, 3 h	100	17	-	(Iqbal et al. 2014)
2.58 wt% Pd/TiO_2	FOL (10.2 mmol), cat. (100 mg), H_2 (0.1 MPa), $C_2H_2Cl_2$ (20 mL), 25 °C, 3 h	65.1	55.5	9[a]	(Iqbal et al. 2014)
2.58 wt% Pd/TiO_2	FOL (10.2 mmol), cat. (100 mg), H_2 (0.3 MPa), $C_2H_2Cl_2$ (20 mL), 25 °C, 0.5 h	93.7	87.7	<6[a]	(Iqbal et al. 2014)
Au: Pd/SiO_2	FOL (10.2 mmol), cat. (100 mg), H_2 (0.1 MPa), $C_2H_2Cl_2$ (20 mL), 25 °C, 3 h	36	33.1	2.2[a]	(Aldosari 2019)

Note: [a] Yield of THFOL.

6.4.3 Vapor Phase Catalytic Hydrodeoxygenation of FAL

Vapor phase catalytic hydrodeoxygenation of FAL could also be possible over various noble and non-metal based catalysts and the results are summarized in Table 6.4. In the mid-19th century, Bremner and Keeys (in 1947) described vapor phase hydrogenation of FAL at atmospheric pressure using various Cu-based catalysts (Bremner and Keeys 1947). The Cu-Al catalyst gave a satisfactory yield (~89%) of hydrodeoxygenated 2-MF product at 253 °C. Other Cu-Cr catalysts showed similar behavior towards the product distribution from vapor hydrogenation of FAL. 87% yield of 2-MF at almost complete conversion of FAL. It could be noticed that the suitable temperature range for FOL production was found in the range of (≤200 °C), while 2-MF product formation could be possible at high temperature (>250 °C). Meanwhile, Burnett et al. reported vapor-phase hydrogenation of FAL at atmospheric pressure using various using catalyst carriers and Cu-Cr catalysts. Of these, Cu-Cr on activated charcoal (C) used as the catalyst was found to give excellent yield of 2-MF (90–95%) (Burnett et al. 1948). There is no doubt that Cu-Cr catalysts are highly active for the production of hydrodeoxygenated 2-MF product, but these catalysts could be deactivated. Although, these catalysts could be regenerated via coke burn off at 400 °C. However, the use of Cu-Cr-based catalysts were limited by heavy environmental pollution and energy crisis.

For vapor phase hydrodeoxygenation of FAL, it was necessary to develop highly stable and Cr-free catalysts. In this context, noble metal-based catalysts (especially Pt metal) were tested in vapor phase hydrogenation of FA (Kijeński et al. 2002). Unfortunately, 2 wt% Pt/TiO_2/SiO_2 and 2 wt% Pt/TiO_2/MgO could give 18.9% and 5.9% yield of 2-MF, respectively. Further, Sittisa et al. employed three different other metal-based catalysts (Cu, Pd, and Ni) supported on SiO_2 in vapor phase hydrodeoxygenation of FAL in the temperature range from 210–290 °C (Sitthisa et al. 2011; Sitthisa and Resasco 2011). Based on the results, it could be suggested that distribution of products depends on the nature of metal used. As a result, 10 wt% Cu/SiO_2 produced mainly FOL (63% yield), only a small amount of 2-MF (8.2%) obtained from the hydrogenolysis of the FOL product. While, the other 1 wt% Pd/SiO_2 and 5 wt% Ni/SiO_2 catalysts did not show hydrodeoxygenated products (neither 2-MF nor 2-MTHF). Recently, Yang et al. performed hydrogenation of FAL using Cu/ZnO-Al_2O_3 (PB) (PB: pseudo-boehmite) catalyst (Yang et al. 2018). As a result, a high yield of FOL (96%) was obtained. However, a trace amount of 2-MF (3.2%) was also formed. Vargas-Hernández et al. investigated hydrogenation of FAL using the Cu/SBA-15 catalyst (Vargas-Hernández et al. 2014). 37.5% yield of 2-MF could be observed under the applied reaction conditions (at 270 °C after 1 h). Further, Sittisa et al. reported vapor phase hydrodeoxygenation of FAL over SiO_2 supported Ni-Fe alloy used as a bimetallic catalyst (Sitthisa et al. 2011). The product selectivity was shifted to 2-MF, while decarbonylation products and their derivatives were suppressed. As a result, 39% yield was achieved at 250 °C. This reduction in decarbonylation could be attributed to an increase of the stability of η^2 (C, O)-surface species. The increase in the interaction of the carbonyl O with surfaces through Fe–O interaction hinders the formation of acyl species, which is responsible

TABLE 6.4
Vapor Phase Catalytic Hydrodeoxygenation of FAL

Catalysts	Reaction Conditions	Conv. (%)	Yield (%) 2-MF	Yield (%) FOL	Ref.
Cu-Al alloy	LHSV (0.16 h^{-1}), H_2/FAL ratio (7.3) H_2 (0.1 MPa), 253 °C	90.5	89	2	(Bremner and Keeys 1947)
Cu-Cr catalyst	LHSV (0.13 h^{-1}), H_2/FAL ratio (7.8) H_2 (0.1 MPa), 250 °C	99	87	7	(Bremner and Keeys 1947)
Cu-Cr/C catalyst	LHSV (0.81 h^{-1}), H_2/FAL ratio (11) H_2 (0.1 MPa), 220-230 °C	-	90–95	-	(Burnett et al. 1948)
2 wt% Pt/TiO_2/SiO_2	LHSV (2.0 h^{-1}), H_2/FAL ratio (2) H_2 (0.1 MPa), 250 °C	69.4	18.9	-	(Kijeński et al. 2002)
2 wt% Pt/TiO_2/MgO	LHSV (2.0 h^{-1}), H_2/FAL ratio (2) H_2 (0.1 MPa), 250 °C	50.4	5.9	-	(Kijeński et al. 2002)
10 wt% Cu/SiO_2	LHSV (2.0 h^{-1}), H_2/FAL ratio (25) H_2 (0.1 MPa), 290 °C	71	8.2	63	(Sitthisa et al. 2011; Sitthisa and Resasco 2011)
Cu/ZnO-Al_2O_3 (PB)	LHSV (0.5 h^{-1}), H_2/FAL ratio (25) H_2 (0.1 MPa), 120 °C	88.6	3.2	96.0	(Yang et al. 2018)
Cu/SBA-15	LHSV (2.0 h^{-1}), H_2/FAL ratio (11) H_2 (0.1 MPa), 270 °C	66.5	37.5	28.9	(Vargas-Hernández et al. 2014)
5 wt% Ni-2 wt% Fe/ SiO_2	LHSV (10.0 h^{-1}), H_2/FAL ratio (25) H_2 (0.1 MPa), 250 °C	96	39	10	(Sitthisa et al. 2011)
Co/Al_2O_3	LHSV (1.0 h^{-1}), H_2/FAL ratio (8.4) H_2 (1 MPa), 180 °C	100	93.4	3.2	(Liu et al. 2020)
Cu-Mn-Si	LHSV (0.49 h^{-1}), H_2/FAL ratio (50) H_2 (0.1 MPa), 279 °C	99.8	93.7	-	(Zheng et al. 2008)
Cu-Zn-Al	LHSV (0.7 h^{-1}), H_2/FAL ratio (2) H_2 (0.1 MPa), 225 °C	99.9	92.3	-	(Yang et al. 2004)

Note: LHSV: Liquid hour space velocity.

TABLE 6.5
Total Concentration of Acid Sites and Brønsted and Lewis Acidity Ratio (B/L) of Co-Based Catalysts

Catalyst	Total Acid Sites[a](μmol/g)	B/L[b]	Lewis acidity (μmol/g)
Co/Al_2O_3	466.8	0.110	420
Co/SiO_2	235.4	0.171	201
Co/TiO_2	101.7	0.078	94
Co/CeO_2	Not detected	0.150	Not detected

Notes: [a] Determined by ammonia (NH_3) temperature-programmed desorption (TPD) technique.
[b] Determined based on the integral of peak area obtained from pyridine-probed FTIR technique.

for decarbonylation. In addition, the C-O hydrogenolysis of the adsorbed FOL is improved by this interaction.

In order to achieve high yield of 2-MF, Liu et al. performed vapor phase catalytic hydrodeoxygenation of FAL using various metal oxide supported Co-based catalysts (Liu et al. 2020). Interestingly, Co-based catalysts showed distinct yield of 2-MF from hydrodeoxygenation of FAL. Of these, the Co/Al_2O_3 catalyst demonstrated an excellent yield of 2-MF (93.4%) with 3.2% yield of FOL (Table 6.4). This high yield of 2-MF product could be attributed to the availability of Lewis acid sites of support used and the high dispersion (8.5%) of Co metals. The concentration of acid sites and relative Brønsted and Lewis acidity ratio of Co-based catalysts are summarized in Table 6.5. In another approach, Zheng et al. performed coupling of cylcohexanol (CHN) dehydrogenation and FAL hydrogenation using a series of Cu-Mn-Si catalysts prepared by the co-precipitation method (Zheng et al. 2008). Of these, the binary Cu-Si catalyst could give 80% yield of 2-MF. While, the ternary (Cu-Mn-Si) catalyst was found to give 93.7% yield of 2-MF and 99.8% conversion of FAL (Table 6.4). The addition of Mn to the Cu-Si catalyst enhanced the ability of hydrogenation of FAL, resulting in a high yield of 2-MF. By following the similar approach of coupling process, Yang et al. performed coupling of 1,4-butanediol (BDO) dehydrogenation and FAL hydrogenation using a series of Cu-Zn-Al catalysts (Yang et al. 2004). From this study, it could be observed that the calcination temperature of the catalyst drastically influenced the catalytic activity and the selectivity of products (or product yields). A high yield of 2-MF (92.3%) could also be accomplished under the applied reaction conditions (at 225 °C and after 6 h) (Table 6.4).

6.5 REACTION MECHANISM

In fact, catalytic hydrogenation of FAL has been extensively studied with different metal-based catalysts. Except a few examples, no systematic studies could reveal the origin of products and the reaction mechanism involved in catalytic hydrogenation of FAL. Xu et al. investigated the complete reaction mechanism of catalytic hydrodeoxygenation of FAL (Xu et al. 2011). Based on several experiments, it could be suggested that 2-MF product is formed via the formation of FOL as an intermediate.

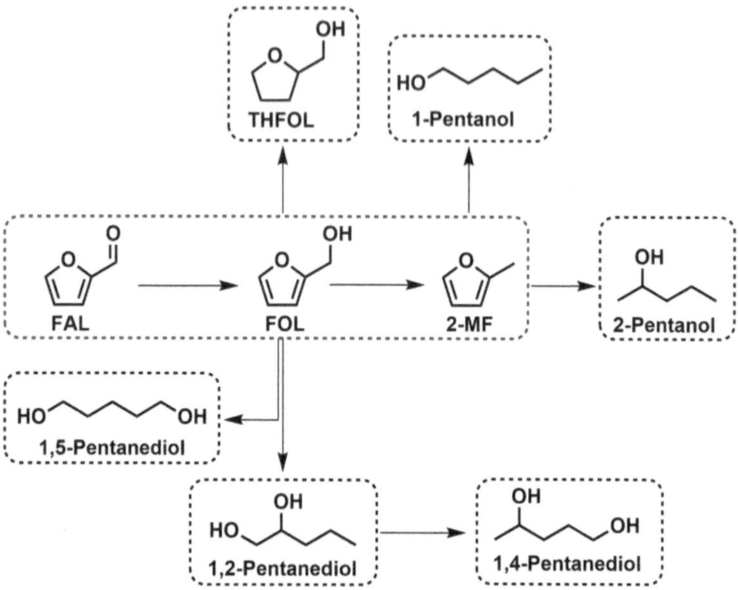

SCHEME 6.3 Reaction pathways for all possible products from catalytic hydrodeoxygenation of FAL. (Originally drawn based on Xu et al. 2011.)

Firstly, FOL is formed from the hydrogenation of C=O bond, which is most kinetically favored process. Then, the 2-MF product is formed from hydrogenolysis (dehydrogenation and subsequent hydrogenation) of FOL. It could further be suggested that FOL and 2-MF can further be hydrogenated to form THFOL and 2-MTHF products, respectively. Under the different reaction conditions, pentanols (e.g., 1-pentanol and 2-pentanols) and pentanediols (e.g., 1,2-pentanediols, 1,4-pentanediols, and 1,5-pentanediols,) are also formed. The reaction pathways for all possible products from catalytic hydrodeoxygenation of FAL is shown in Scheme 6.3.

6.6 CONCLUSIONS AND PERSPECTIVES

Catalytic hydrodeoxygenation of FAL to produce 2-MF has recently attracted great attention due to the potential of 2-MF as a renewable biofuel and precursor for fine industrial chemical practices. Various noble and non-noble metal-based catalysts were explored under the harsh reaction conditions (at high temperatures and long time). Several factors such as size and dispersion of metal nanoparticles, synergism between metal and support materials, nature of support materials (acidic or basic), and the reaction conditions significantly affect the distribution of products from catalytic hydrogenation of FAL. Undoubtedly, countless efforts have been put forward for the production of 2-MF with excellent yield (~94%).

However, several recommendations are to be taken into consideration to enhance the overall value and viability of the whole process. Despite the achieving excellent yield of 2-MF, low-cost and Cr-free catalysts are crucial. Particular attention must be paid to the use transition metals such as nickel, cobalt, iron, and molybdenum that can offer a promising alternative to noble metals for 2-MF production from catalytic hydrodeoxygenation of FAL. Adding promoters such as alkali and alkaline earth metals can enhance the activity of non-noble metal-based catalysts in catalytic hydrodeoxygenation of FAL. In fact, support materials also play a vital role in enhancing the affordability and scalability of the process. Thus, the use of inexpensive mixed metal oxides with high surface area and Lewis acid sites is highly recommended. These kinds of mixed metal oxide supported metal nano particles can also execute catalytic hydrodeoxygenation of FAL with high selectivity. Except a few examples, most of the catalysts applied catalytic hydrodeoxygenation of FAL were found to be deactivated. Implementation in the catalyst formulation is needed to exhibit high activity over prolonged reaction times. Catalysts with robust structure and resistance to sintering, leaching, or agglomeration can avoid the catalyst deactivation and ensure long-term stability. From the energy consumptions' point of view, it is also necessary to develop new strategies for modified reaction conditions (preferably green solvent, low temperatures, and shorter reaction times). 2-MF can mitigate the carbon footprints and can put bright impacts on the environment in eco-friendly manner. Thus, inexpensive non-noble-based catalysts and cost-effective support, the strategies for catalyst preparation, and its stability are still essential to promote 2-MF production from catalytic hydrogenation of FAL. With the advancement of 2-MF production technologies in the future, it is believed that both environmentally friendly and economically feasible 2-MF production approaches can be realized and commercialized.

REFERENCES

Aldosari, Obaid Fahad. 2019. "Selective conversion of furfuryl alcohol to 2-methylfuran over nanosilica supported Au:Pd bimetallic catalysts at room temperature." *Journal of Saudi Chemical Society* 23 (7). Elsevier B.V. 938–946. doi: 10.1016/j.jscs.2019.04.004.

Alonso, David Martin, Sikander H. Hakim, Shengfei Zhou, Wangyun Won, Omid Hosseinaei, Jingming Tao, Valerie Garcia-Negron, Ali Hussain Motagamwala, Max A. Mellmer, Kefeng Huang, Carl J. Houtman, Nicole Labbé, David P. Harper, Christos T. Maravelias, Troy Runge, and James A. Dumesic. 2017. "Increasing the revenue from lignocellulosic biomass: Maximizing feedstock utilization." *Science Advances* 3 (5). American Association for the Advancement of Science's (AAAS). e1603301. doi: 10.1126/sciadv.1603301.

Barranca, Alberto, Iker Agirrezabal-Tellería, Marcos Rellán-Piñeiro, Manuel A. Ortuño, and Inaki Gandarias. 2023. "Selective furfural hydrogenolysis towards 2-methylfuran by controlled poisoning of Cu–Co catalysts with chlorine." *Reaction Chemistry & Engineering* 8 (3). The Royal Society of Chemistry. 687–698. doi: 10.1039/D2RE00414C.

Binder, J. B., J. J. Blank, A. V. Cefali, and R. T. Raines. 2010. "Synthesis of furfural from xylose and xylan." *ChemSusChem* 3 (11). John Wiley & Sons, Ltd. 1268–1272. doi: 10.1002/cssc.201000181.

Biradar, Narayan S., Amol A. Hengne, Shobha N. Birajdar, Rameshwar Swami, and Chandrashekhar V. Rode. 2014. "Tailoring the product distribution with batch and continuous process options in catalytic hydrogenation of fFurfural." *Organic Process*

Research & Development 18(11). American Chemical Society. 1434–1442. doi: 10.1021/op500196x.

Biswas, Prakash, Jiann-Horng Lin, Jungshik Kang, and Vadim V. Guliants. 2014. "Vapor phase hydrogenation of 2-methylfuran over noble and base metal catalysts." *Applied Catalysis A: General* 475. Elsevier B.V. 379–385. doi: 10.1016/j.apcata.2014.01.054.

Bremner, John G. M., and Richard K. F. Keeys. 1947. "202. The hydrogenation of furfuraldehyde to furfuryl alcohol and sylvan (2-methylfuran)." *Journal of the Chemical Society (Resumed)* 1947. The Royal Society of Chemistry. 1068–1080. doi: 10.1039/JR9470001068.

Budzinski, Maik, and Roy Nitzsche. 2016. "Comparative economic and environmental assessment of four beech wood based biorefinery concepts." *Bioresource Technology* 216. Elsevier B.V. 613–621. doi: 10.1016/j.biortech.2016.05.111.

Burnett, L. W., I. B. Johns, R. F. Holdren, and R. M. Hixon. 1948. "Production of 2-methylfuran by vapor-phase hydrogenation of furfural." *Industrial & Engineering Chemistry* 40 (3). American Chemical Society. 502–505. doi: 10.1021/ie50459a034.

Chang, Xin, An-Feng Liu, Bo Cai, Jin-Yue Luo, Hui Pan, and Yao-Bing Huang. 2016. "Catalytic transfer hydrogenation of furfural to 2-methylfuran and 2-methyltetrahydrofuran over bimetallic copper–palladium catalysts." *ChemSusChem* 9 (23). John Wiley & Sons, Ltd. 3330–3337. doi: 10.1002/cssc.201601122.

Chatterjee, Maya, Takayuki Ishizaka, and Hajime Kawanami. 2016. "Reductive amination of furfural to furfurylamine using aqueous ammonia solution and molecular hydrogen: An environmentally friendly approach." *Green Chemistry* 18 (2). The Royal Society of Chemistry. 487–496. doi: 10.1039/C5GC01352F.

Chen, Hao-ze, Ji Liu, Wen-tao Li, Bin Hu, Xin-ru Liu, Yang-wen Wu, Bing Zhang, and Qiang Lu. 2023. "Mechanism insights into the decarbonylation of furfural to furan over Ni/MgO: A molecular simulation study." *Energy & Fuels* 37 (14). American Chemical Society. 10594–10602. doi: 10.1021/acs.energyfuels.3c01474.

Chen, Xiufang, Ligang Zhang, Bo Zhang, Xingcui Guo, and Xindong Mu. 2016. "Highly selective hydrogenation of furfural to furfuryl alcohol over Pt nanoparticles supported on g-C3N4 nanosheets catalysts in water." *Scientific Reports* 6 (1). Nature Portfolio. 28558. doi: 10.1038/srep28558.

Cherubini, Francesco, and Anders H. Strømman. 2011. "Chemicals from lignocellulosic biomass: opportunities, perspectives, and potential of biorefinery systems." *Biofuels, Bioproducts and Biorefining* 5 (5). John Wiley & Sons, Ltd. 548–561. doi: 10.1002/bbb.297.

Choudhary, Vinit, Ana B. Pinar, Stanley I. Sandler, Dionisios G. Vlachos, and Raul F. Lobo. 2011. "Xylose isomerization to xylulose and its dehydration to furfural in aqueous media." *ACS Catalysis* 1 (12). American Chemical Society. 1724–1728. doi: 10.1021/cs200461t.

Choudhary, Vinit, Stanley I. Sandler, and Dionisios G. Vlachos. 2012. "Conversion of xylose to furfural using Lewis and Brønsted acid catalysts in aqueous media." *ACS Catalysis* 2 (9). American Chemical Society. 2022–2028. doi: 10.1021/cs300265d.

Cybulska, Iwona, Grzegorz Brudecki, and Hanwu Lei. 2013. "Hydrothermal pretreatment of lignocellulosic biomass." In Tingyue Gu (Ed.), *Green Biomass Pretreatment for Biofuels Production*, 87–106. Springer Netherlands. doi: 10.1007/978-94-007-6052-3_4.

De, S., B. Saha, and R. Luque. 2015. "Hydrodeoxygenation processes: Advances on catalytic transformations of biomass-derived platform chemicals into hydrocarbon fuels." *Bioresource Technology* 178. Elsevier B.V. 108–118. doi: 10.1016/j.biortech.2014.09.065.

Fu, Xiaomin, Xiaoqian Ren, Jiecan Shen, Yong Jiang, Yuehua Wang, Yasin Orooji, Wenlong Xu, and Jinhua Liang. 2021. "Synergistic catalytic hydrogenation of furfural to

1,2-pentanediol and 1,5-pentanediol with LDO derived from CuMgAl hydrotalcite." *Molecular Catalysis* 499. Elsevier B.V. 111298. doi: 10.1016/j.mcat.2020.111298.

Fu, Zhaolin, Ze Wang, Weigang Lin, Wenli Song, and Songgeng Li. 2017. "High efficient conversion of furfural to 2-methylfuran over Ni-Cu/Al_2O_3 catalyst with formic acid as a hydrogen donor." *Applied Catalysis A: General* 547: Elsevier B.V. 248–255. doi: 10.1016/j.apcata.2017.09.011.

Hronec, Milan, and Katarina Fulajtarová. 2012. "Selective transformation of furfural to cyclopentanone." *Catalysis Communications* 24. Elsevier B.V. 100–104. doi: 10.1016/j.catcom.2012.03.020. https://www.marketsandmarkets.com/Market-Reports/furfural-market-101056456.html. https://www.precisionbusinessinsights.com/market-reports/2-methylfuran-market.

Iqbal, Sarwat, Xi Liu, Obaid F. Aldosari, Peter J. Miedziak, Jennifer K. Edwards, Gemma L. Brett, Adeeba Akram, Gavin M. King, Thomas E. Davies, David J. Morgan, David K. Knight, and Graham J. Hutchings. 2014. "Conversion of furfuryl alcohol into 2-methylfuran at room temperature using Pd/TiO2 catalyst." *Catalysis Science & Technology* 4 (8). The Royal Society of Chemistry. 2280–2286. doi: 10.1039/C4CY00184B.

Ishida, Tamao, Kurumi Kume, Kota Kinjo, Tetsuo Honma, Kengo Nakada, Hironori Ohashi, Takushi Yokoyama, Akiyuki Hamasaki, Haruno Murayama, Yusuke Izawa, Masaru Utsunomiya, and Makoto Tokunaga. 2016. "Efficient decarbonylation of furfural to furan catalyzed by zirconia-supported palladium clusters with low atomicity." *ChemSusChem* 9 (24). John Wiley & Sons, Ltd. 3441–3447.

Jaswal, Anurag, Piyush Pratap Singh, and Tarak Mondal. 2022. "Furfural – A versatile, biomass-derived platform chemical for the production of renewable chemicals." *Green Chemistry* 24 (2). The Royal Society of Chemistry. 510–551. doi: 10.1039/D1GC03278J.

Kijeński, Jacek, Piotr Winiarek, Tadeusz Paryjczak, Andrzej Lewicki, and Agnieszka Mikołajska. 2002. "Platinum deposited on monolayer supports in selective hydrogenation of furfural to furfuryl alcohol." *Applied Catalysis A: General* 233 (1). Elsevier B.V. 171–182. doi: 10.1016/S0926-860X(02)00140-0.

Kumaravel, Sakthivel, Balakrishna Avula, Jagadeesh Kumar Alagarasan, Moonyong Lee, Wahid Ali, Mohammad Ehtisham Khan, Syed Kashif Ali, Abdullateef H. Bashiri, Anwar Ulla Khan, and Krishnakumar Balu. 2024. "Development of bimetallic Ru/Ni/SBA-16 catalysts for catalytic hydrogenation of bio-based furfural to 2-methylfuran and furfuryl alcohol." *Materials Chemistry and Physics* 316. Elsevier B.V. 129083. doi: 10.1016/j.matchemphys.2024.129083.

Lan, Xuefang, Robert Pestman, Emiel J. M. Hensen, and Thomas Weber. 2021. "Furfural hydrodeoxygenation (HDO) over silica-supported metal phosphides – The influence of metal–phosphorus stoichiometry on catalytic properties." *Journal of Catalysis* 403. Elsevier B.V. 181–193. doi: 10.1016/j.jcat.2021.01.031.

Lejemble, Philippe, Antoine Gaset, and Philippe Kalck. 1984. "From biomass to furan through decarbonylation of furfural under mild conditions." *Biomass* 4 (4). MDPI. 263–274. doi: 10.1016/0144-4565(84)90039-8.

Lin, Zhexi, Weiming Wan, Siyu Yao, and Jingguang G. Chen. 2018. "Cobalt-modified molybdenum carbide as a selective catalyst for hydrodeoxygenation of furfural." *Applied Catalysis B: Environmental* 233. Elsevier B.V. 160–166. doi: 10.1016/j.apcatb.2018.03.113.

Liu, Ping, Luyang Sun, Xinxin Jia, Chen Zhang, Wei Zhang, Yongji Song, Hong Wang, and Cuiqing Li. 2020. 'Efficient one-pot conversion of furfural into 2-methyltetrahydrofuran using non-precious metal catalysts." *Molecular Catalysis* 490. Elsevier B.V. 110951. doi: 10.1016/j.mcat.2020.110951.

Ma, Yan, Weihong Tan, Jingxin Wang, Junming Xu, Kui Wang, and Jianchun Jiang. 2020. "Liquefaction of bamboo biomass and production of three fractions containing aromatic compounds." *Journal of Bioresources and Bioproducts* 5 (2). Elsevier B.V. 114–123. doi: 10.1016/j.jobab.2020.04.005.

Mankar, Akshay R., Ashish Pandey, Arindam Modak, and K. K. Pant. 2021. "Pretreatment of lignocellulosic biomass: A review on recent advances." *Bioresource Technology* 334. Elsevier B.V. 125235. doi: 10.1016/j.biortech.2021.125235.

Mariscal, R., P. Maireles-Torres, M. Ojeda, I. Sádaba, and M. López Granados. 2016. "Furfural: A renewable and versatile platform molecule for the synthesis of chemicals and fuels." *Energy & Environmental Science* 9 (4). The Royal Society of Chemistry. 1144–1189. doi: 10.1039/C5EE02666K.

Matsheku, Asanda C., Munaka Christopher Maumela, and Banothile C. E. Makhubela. 2023. "Convenient hydrogenation of furfural to furfuryl alcohol in metal-catalyzed and organo-catalyzed environments." *RSC Sustainability* 1 (6). The Royal Society of Chemistry. 1471–1483. doi: 10.1039/D3SU00056G.

Mika, László T., Edit Cséfalvay, and Áron Németh. 2018. "Catalytic conversion of carbohydrates to initial platform chemicals: Chemistry and sustainability." *Chemical Reviews* 118 (2). American Chemical Society 505–613. doi: 10.1021/acs.chemrev.7b00395.

Millett, M. A., A. J. Baker, and L. D. Satter. 1976. "Physical and chemical pretreatments for enhancing cellulose saccharification." In *Biotechnololgy and Bioengneering Symposium.* John Wiley & Sons. 125–153.

Mishra, Dinesh Kumar, Santosh Kumar, and Ram Sambhar Shukla. 2020. "Chapter 12- Furfuryl alcohol–A promising platform chemical." In Shunmugavel Saravanamurugan, Ashok Pandey, Hu Li, and Anders Riisager (Eds.), *Biomass, Biofuels, Biochemicals: Recent Advances in Development of Platform Chemicals*, 323–353. Elsevier Inc. doi: 10.1016/B978-0-444-64307-0.00012-3.

Nakagawa, Yoshinao, Masazumi Tamura, and Keiichi Tomishige. 2015. "Catalytic conversions of furfural to pentanediols." *Catalysis Surveys from Asia* 19 (4). Springer. 249–256. doi: 10.1007/s10563-015-9194-2.

Nishimura, Shun, Kunihiko Mizuhori, and Kohki Ebitani. 2016. "Reductive amination of furfural toward furfurylamine with aqueous ammonia under hydrogen over Ru-supported catalyst." *Research on Chemical Intermediates* 42(1). Springer. 19–30. doi: 10.1007/s11164-015-2334-5.

Niu, Hongyu, Jingjie Luo, Chuang Li, Baowei Wang, and Changhai Liang. 2019. "Transfer hydrogenation of biomass-derived furfural to 2-methylfuran over CuZnAl catalysts." *Industrial & Engineering Chemistry Research* 58 (16). American Chemical Society. 6298–6308. doi: 10.1021/acs.iecr.9b00408.

Nope, Eliana, Ángel G. Sathicq, José J. Martínez, Gustavo P. Romanelli, and Rafael Luque. 2023. "Transfer hydrogenation of furfural to furfuryl alcohol under microwave irradiation using mixed oxides." *ChemPlusChem* 88 (8). John Wiley & Sons, Ltd. e202300265. doi: 10.1002/cplu.202300265.

Panagiotopoulou, Paraskevi, and Dionisios G. Vlachos. 2014. "Liquid phase catalytic transfer hydrogenation of furfural over a Ru/C catalyst." *Applied Catalysis A: General* 480. Elsevier B.V. 17–24. doi: 10.1016/j.apcata.2014.04.018.

Peng, Feng, Jun-Li Ren, Feng Xu, Jing Bian, Pai Peng, and Run-Cang Sun. 2010. "Fractional study of alkali-soluble hemicelluloses obtained by graded ethanol precipitation from sugar cane bagasse." *Journal of Agricultural and Food Chemistry* 58 (3). American Chemical Society. 1768–1776. doi: 10.1021/jf9033255.

Periyasamy, Selvakumar, J. Beula Isabel, S. Kavitha, V. Karthik, Badr A. Mohamed, Desta Getachew Gizaw, P. Sivashanmugam, and Tejraj M. Aminabhavi. 2023. "Recent advances in consolidated bioprocessing for conversion of lignocellulosic biomass into bioethanol – A review" *Chemical Engineering Journal* 453. Elsevier B.V. 139783. doi: 10.1016/j.cej.2022.139783.

Ruppert, Agnieszka M., Kamil Weinberg, and Regina Palkovits. 2012. "Hydrogenolysis goes bio: From carbohydrates and sugar alcohols to platform chemicals." *Angewandte Chemie International Edition*, 51 (11). John Wiley & Sons, Ltd. 2564–2601. doi: 10.1002/anie.201105125.

Saini, Ms Kanika, Sahil Kumar, Hu Li, Srinivasarao Arulananda Babu, and Shunmugavel Saravanamurugan. 2022. "Advances in the catalytic reductive amination of furfural to furfural amine: The momentous role of active metal sites." *ChemSusChem* 15 (7). John Wiley & Sons, Ltd. e202200107. doi: 10.1002/cssc.202200107.

Scholz, David, Christof Aellig, and Ive Hermans. 2014. "Catalytic transfer hydrogenation/hydrogenolysis for reductive upgrading of furfural and 5-(Hydroxymethyl)furfural." *ChemSusChem* 7 (1). John Wiley & Sons, Ltd. 268–275. doi: 10.1002/cssc.201300774.

Sengupta, Debalina, and Ralph W. Pike. 2014. "Chemicals from biomass." In Maximilian Lackner, Baharak Sajjadi, and Wei-Yin Chen(Eds.), *Handbook of Climate Change Mitigation and Adaptation*, Springer New York. 1–38. doi: 10.1007/978-1-4614-6431-0_28-2.

Sindhu, Raveendran, Parameswaran Binod, and Ashok Pandey. 2016. "Biological pretreatment of lignocellulosic biomass–An overview." *Bioresource Technology* 199. Elsevier B.V. 76–82. doi: 10.1016/j.biortech.2015.08.030.

Sitthisa, Surapas, Wei An, and Daniel E. Resasco. 2011. "Selective conversion of furfural to methylfuran over silica-supported NiFe bimetallic catalysts." *Journal of Catalysis* 284 (1). Elsevier B.V. 90–101. doi: 10.1016/j.jcat.2011.09.005.

Sitthisa, Surapas, and Daniel E. Resasco. 2011. "Hydrodeoxygenation of furfural over supported metal catalysts: A comparative study of Cu, Pd and Ni." *Catalysis Letters* 141 (6). Springer 784–791. doi: 10.1007/s10562-011-0581-7.

Sitthisa, Surapas, Tawan Sooknoi, Yuguang Ma, Perla B. Balbuena, and Daniel E. Resasco. 2011. "Kinetics and mechanism of hydrogenation of furfural on Cu/SiO_2 catalysts." *Journal of Catalysis* 277 (1). Elsevier B.V. 1–13. doi: 10.1016/j.jcat.2010.10.005.

Sun, Ye, and Jiayang Cheng. 2002. "Hydrolysis of lignocellulosic materials for ethanol production: A review." *Bioresource Technology* 83 (1). Elsevier B.V. 1–11. doi: 10.1016/S0960-8524(01)00212-7.

Tassinari, Thomas, Charles Macy, Leo Spano, and Dewey D. Y. Ryu. 1980. "Energy requirements and process design considerations in compression-milling pretreatment of cellulosic wastes for enzymatic hydrolysis." *Biotechnology and Bioengineering* 22 (8). John Wiley & Sons, Ltd. 1689–1705. doi: 10.1002/bit.260220811.

Taylor, Martin J., Li Jiang, Joachim Reichert, Anthoula C. Papageorgiou, Simon K. Beaumont, Karen Wilson, Adam F. Lee, Johannes V. Barth, and Georgios Kyriakou. 2017. "Catalytic hydrogenation and hydrodeoxygenation of furfural over Pt (111): A model system for the rational design and operation of practical biomass conversion catalysts." *The Journal of Physical Chemistry C* 121 (15). American Chemical Society. 8490–8497. doi: 10.1021/acs.jpcc.7b01744.

Vargas-Hernández, D., J. M. Rubio-Caballero, J. Santamaría-González, R. Moreno-Tost, J. M. Mérida-Robles, M. A. Pérez-Cruz, A. Jiménez-López, R. Hernández-Huesca, and P. Maireles-Torres. 2014. "Furfuryl alcohol from furfural hydrogenation over copper supported on SBA-15 silica catalysts." *Journal of Molecular Catalysis A: Chemical* 383–384: Elsevier B.V. 106–113. doi: 10.1016/j.molcata.2013.11.034.

Wang, Baowei, Chuang Li, Bo He, Ji Qi, and Changhai Liang. 2017. "Highly stable and selective Ru/$NiFe_2O_4$ catalysts for transfer hydrogenation of biomass-derived furfural to

2-methylfuran." *Journal of Energy Chemistry* 26 (4). Elsevier B.V. 799–807. doi: 10.1016/j.jechem.2017.04.008.

Werpy, T, and Petersen, G. 2004. "Top Value Added Chemicals from Biomass: Volume I -- Results of Screening for Potential Candidates from Sugars and Synthesis Gas". United States. https://doi.org/10.2172/15008859. https://www.osti.gov/servlets/purl/15008859.

Xu, Wenjie, Haifeng Wang, Xiaohui Liu, Jiawen Ren, Yanqin Wang, and Guanzhong Lu. 2011. "Direct catalytic conversion of furfural to 1,5-pentanediol by hydrogenolysis of the furan ring under mild conditions over Pt/Co_2AlO_4 catalyst." *Chemical Communications* 47 (13). The Royal Society of Chemistry. 3924–3926. doi: 10.1039/C0CC05775D.

Yamaguchi, Aritomo, Yuka Murakami, Tomohiro Imura, and Kazuaki Wakita. 2021. "Hydrogenolysis of furfuryl alcohol to 1,2-pentanediol over supported ruthenium catalysts." *ChemistryOpen* 10 (8). John Wiley & Sons, Ltd. 731–736. doi: 10.1002/open.202100058.

Yan, Kai, and Aicheng Chen. 2014. "Selective hydrogenation of furfural and levulinic acid to biofuels on the ecofriendly Cu–Fe catalyst." *Fuel* 115. Elsevier B.V. 101–108. doi: 10.1016/j.fuel.2013.06.042.

Yang, J., H. Y. Zheng, Y. L. Zhu, G. W. Zhao, C. H. Zhang, B. T. Teng, H. W. Xiang, and Y. Li. 2004. "Effects of calcination temperature on performance of Cu-Zn-Al catalyst for synthesizing γ-Butyrolactone and 2-methylfuran through the coupling of dehydrogenation and hydrogenation." *Catalysis Communications* 5 (9). Elsevier B.V. 505–510. doi: 10.1016/j.catcom.2004.06.005.

Yang, Weisheng, Xu Du, Wei Liu, Zewei Wang, Hongqi Dai, and Yulin Deng. 2019. "Direct valorization of lignocellulosic biomass into value-added chemicals by polyoxometalate catalyzed oxidation under mild conditions." *Industrial & Engineering Chemistry Research* 58 (51). American Chemical Society. 22996–23004. doi: 10.1021/acs.iecr.9b05311.

Yang, Xiaohai, Qingwei Meng, Guoqiang Ding, Yueqing Wang, Huimin Chen, Yu lei Zhu, and Yong Wang Li. 2018. "Construction of novel $Cu/ZnO\text{-}Al_2O_3$ composites for furfural hydrogenation: The role of Al components." *Applied Catalysis A: General* 561. Elsevier B.V. 78–86. doi: 10.1016/j.apcata.2018.04.005.

Yong, Khai Jie, Ta Yeong Wu, Cornelius Basil Tien Loong Lee, Zhi Jin Lee, Qinpu Liu, Jamaliah Md Jahim, Qiaoqiao Zhou, and Lian Zhang. 2022. "Furfural production from biomass residues: Current technologies, challenges and future prospects." *Biomass and Bioenergy* 161. Elsevier B.V. 106458. doi: 10.1016/j.biombioe.2022.106458.

Yu, Wanjin, Yang Tang, Liuye Mo, Ping Chen, Hui Lou, and Xiaoming Zheng. 2011. "One-step hydrogenation–esterification of furfural and acetic acid over bifunctional Pd catalysts for bio-oil upgrading." *Bioresource Technology* 102 (17). Elsevier B.V. 8241–8246. doi: 10.1016/j.biortech.2011.06.015.

Zheng, H. Y., Y. L. Zhu, L. Huang, Z. Y. Zeng, H. J. Wan, and Y. W. Li. 2008. "Study on Cu-Mn-Si catalysts for synthesis of cyclohexanone and 2-methylfuran through the coupling process." *Catalysis Communications* 9 (3). Elsevier B.V. 342–348. doi: 10.1016/j.catcom.2007.06.026.

Zhou, Chun-Hui, Xi Xia, Chun-Xiang Lin, Dong-Shen Tong, and Jorge Beltramini. 2011. "Catalytic conversion of lignocellulosic biomass to fine chemicals and fuels." *Chemical Society Reviews* 40(11). The Royal Society of Chemistry. 5588–5617. doi: 10.1039/C1CS15124J.

7 Reductive Amination of Furfural to Furfurylamine

Kanika Saini and Shunmugavel Saravanamurugan

7.1 INTRODUCTION: BACKGROUND OF BIOMASS VALORIZATION

The increasing global energy, fuel, and chemical demands of humankind drive the need to reduce dependence on rapidly diminishing non-renewable fossil resources. Unlike fossil resources, sustainable alternatives that are renewable and environment-friendly are highly sought to meet future demands. Significant steps were taken worldwide to move towards net-zero emissions, such as the Paris Agreement, to have a safer future for humankind (Savaresi 2016). Fossil resources are not only the predominant source for deriving energy for our society but also chemicals to a large extent, as chemical products have become ubiquitous today and essential for our modern society (Lopez et al. 2023). In this regard, terrestrial lignocellulosic biomass (LCB) is the only carbon-neutral source with huge potential to replace fossil resources for chemical production. LCB is composed of mainly three components, i.e., cellulose (35–45 %), hemicellulose (20–25 %), and lignin (15–20 %), which can be disintegrated systematically via acid or base treatment (Figure 7.1). (Pal et al. 2022) Various platform chemicals, such as furfural (FUR), 5-hydroxymethylfurfural (5-HMF) and levulinic acid (LA), can be produced from the monomers, xylose and glucose of the cellulosic components mentioned above via acid-catalysed depolymerization followed by dehydration (Wu et al. 2019). In particular, xylose via catalytic dehydration can be used to produce FUR. FUR has been identified as one of the top 10+4 value-added chemicals by the US Department of Energy (DOE) (Bozell et al. 2010) and is a valuable platform chemical with a global production of around ~300000 tons per year. (Pirmoradi et al. 2020)

7.2 INDUSTRIAL RELEVANCE AND APPLICATIONS

The FUR market size was estimated at USD 520 million in 2021 and is expected to grow with a compound annual growth rate (CAGR) of 6.5% from 2021 to 2030 (Jorqueira et al. 2023) Further, the derivatives of FUR, such as furfuryl alcohol (FUL), furfurylamine (FUA), furan, and 5-methylfurfural (MFUR), hold a considerable market demand because of numerous applications in pharmaceuticals, polymer

DOI: 10.1201/9781003464839-7

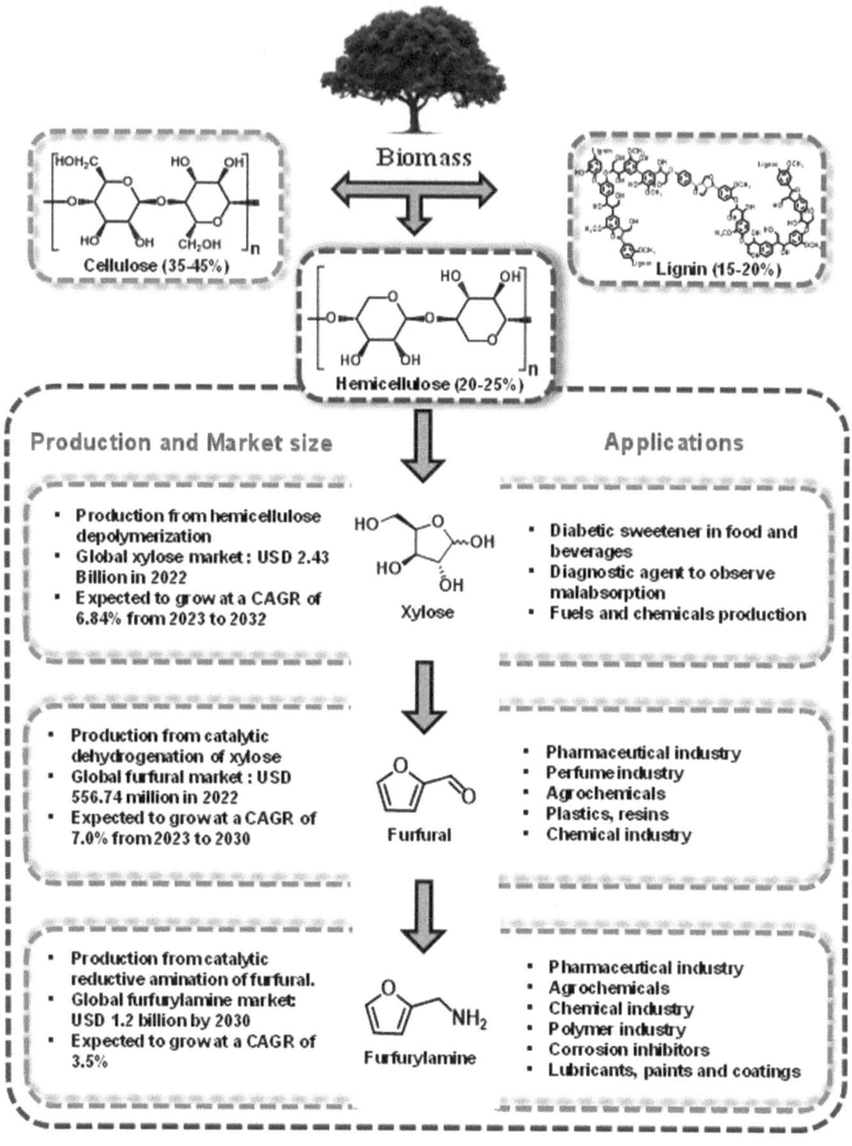

FIGURE 7.1 Production pathway, market size and applications of furfural and furfuryl amine.

industry, agrochemicals, as an industrial solvent, paints, coatings and textile industry (Figure 7.1).

Notably, the catalytic conversion of FUR to value-added N-containing compounds has attracted a great deal of attention as it holds more than US $50 billion of annual market size and is majorly reliant on fossil resources (Xu et al. 2020). The search for other sustainable routes is thus highly required to produce N-containing compounds.

The catalytic reductive amination of bio-based FUR to corresponding primary amines, especially FUA, has recently been widely studied. FUA have various applications in pharmaceuticals, agrochemicals, polymer industry, corrosion inhibitors, lubricants, paints and coatings (Figure 7.1). The global FUA market is anticipated to reach USD 1.2 billion by 2030 at a CAGR of 3.5% due to its potential applications. ("https://Industrygrowthinsights.Com/Report/Furfurylamine-Market/" n.d.) According to the different market reports, the major industry players of FUA are Alkyl Amines Chemicals Ltd., MINAFIN, Changzhou Huayang Technology Co. Ltd., Galaxy Laboratories, Hubei Jiangyan Tianxiang Chemical Co., Ltd. Liyang Yutian Chemical Co. Ltd., Pingquan Fengsheng Chemical Co. Ltd., Shandong Yuexing Chemical Co., Ltd., and Pennakem ("https://Industrygrowthinsights.Com/Report/Furfurylamine-Market/").

7.3 CHALLENGES, DRAWBACKS AND GENERAL REACTION MECHANISM OF CATALYTIC REDUCTIVE AMINATION OF FUR

For any process to be industrially viable, it should be simple, practical and operational at a low cost. Developing highly efficient, safer, low-waste-producing, and atom economy processes are also key points of the principle of green chemistry. The production of FUA via reductive amination using NH_3 and H_2 as nitrogen and hydrogen sources is the most atom-economical and cost-effective process (Saini et al. 2022). The reductive amination of FUA from FUR entails NH_3, and the process involves the simultaneous activation of intermediates and NH_3 along with the dissociative adsorption of H_2. Although this reaction involves the hydrogenation step, entailing H_2 as a reductant, but the reaction is somewhat complex and comparatively more challenging than a simple hydrogenation reaction due to the various steps involved. The presence of NH_3 in the reaction mixture could prevent the activation of H_2 and other substrates by suppressing the active species on the catalyst surface. Therefore, understanding the reaction mechanism is crucial before comprehending the role of the catalyst in the reductive amination of carbonyl compounds to produce primary amines.

Generally, the reductive amination reactions to primary amines involving NH_3 and H_2 as nitrogen and hydrogen sources occur in two steps (Chatterjee et al. 2016). The first step is the catalytic hydrogenation of the imine intermediate formed via condensation of carbonyl moiety with NH_3 along with H_2 activation on the catalyst surface to produce primary amine. Next, the reaction between a highly nucleophilic *in situ* produced primary amine with an imine intermediate takes place, resulting in inevitable Schiff's base formation. Further, the transformation of this Schiff's base back to primary amine is the crucial step and generally found to be the rate-determining step. It is usually known that two main pathways are proposed for converting Schiff's base to primary amine, i.e., hydrogenolysis and/or ammonolysis. The formation of primary amine via hydrogenolysis of Schiff's base results in the formation of primary amine and furfural in a 1:1 ratio. Meanwhile, the ammonolysis step directly results in the formation of two molecules of the required primary amine, possibly contributing to a faster reaction rate than the hydrogenolysis step. However, no direct evidence has

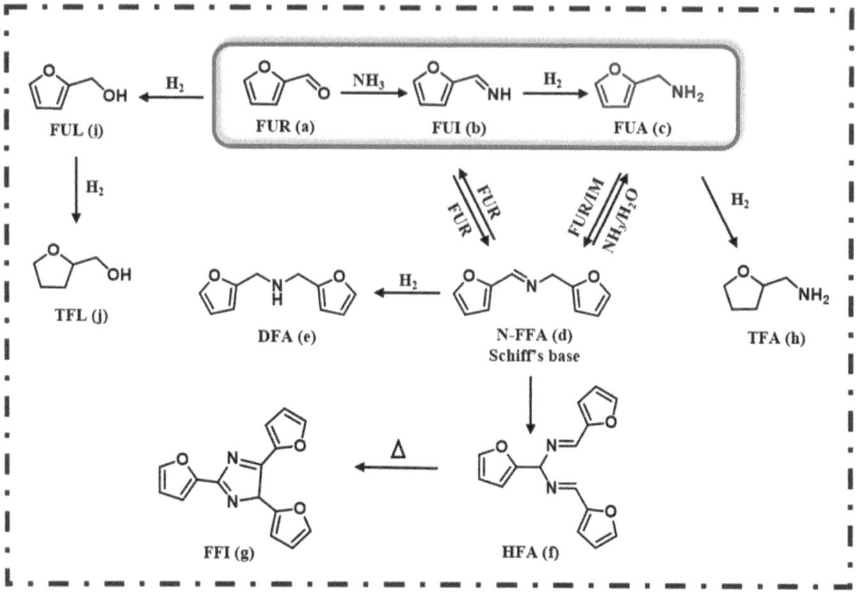

SCHEME 7.1 General reaction pathway for reductive amination of FUR to FUA (FUR – furfural, FUI – furfurimine, FUA – furfurylamine, FUL – furfuryl alcohol, TFL – tetrahydro furfuryl alcohol, TFA – tetrahydro furfurylamine, DFA – difurfurylamine, FDA – N-furfurylidenefurfurylamine, HFA – hydrofuramide, FFI – furfurine).

been reported so far for these two steps. In addition, depending on the catalyst properties, other side reactions can also take place, as shown in Scheme 7.1, leading to the decline in FUA selectivity. In a nutshell, the following points are considered to play a decisive role during the catalytic reductive amination of FUR.

(a) The hydrogenation capability of the catalyst in the presence of NH_3.
(b) The activation of NH_3 on the catalyst surface.
(c) The activation of intermediates on alternate sites on the catalyst surface in the presence of excess NH_3.

In this regard, the catalyst plays an essential role in the selective production of FUA. FUA is currently produced via the reductive amination pathway by reacting FUR with NH_3 and H_2 gas using Raney nickel as a hydrogenation catalyst (Ayedi et al. 2012; Ayusawa et al. 1986). However, this reaction is carried out under high pressure in an anhydrous solvent, indicating harsh reaction conditions with cumbersome catalytic systems. Thus, this chapter focuses on the recently developed catalytic system for the reductive amination of FUR to FUA. Significantly, the nature of the supports and role of metals (precious and non-precious) in the reductive amination of FUA are delineated in detail to identify the potential catalytic systems.

7.4 SIGNIFICANCE OF HETEROGENOUS CATALYST AND ROLE OF SUPPORT ON THE REDUCTIVE AMINATION OF FUR

Catalysts play a pivotal role in the chemical and fuel industries (Fechete et al. 2012). The forecasts indicate that the worldwide catalyst market will expand from an initial 2020 valuation of $35.5 billion to a whopping $57.5 billion in 2030, with a CAGR of 4.9% ("www.Alliedmarketresearch.Com/Catalysts-Market," n.d.). The use of catalysts in chemical reactions, whether continuous flow or batch and in liquid or gas flow reactions, has been the subject of extensive research (Munnik et al. 2015). It has been found that catalysis contributes to >35% of the global GDP and depicted that out of 85% of industrial catalytic processes, approximately 80% are heterogeneously catalysed (Juhász et al. 2021; Taseska et al. 2023) owing to the various advantages of heterogenous catalysts such as reusability, easy tunability, high thermal stability and easy separation. The surface properties of heterogeneous catalysts play an important role in achieving high catalytic activity and selectivity towards any product, which can be tuned systematically. Materials like metal oxides, zeolites, mesoporous materials, carbon-based materials, metal-organic frameworks, metal sulphides and phosphides offer a significant surface area to tune their surface properties. Employing these materials as support for precious and non-precious metals has been widely explored towards the reductive amination of bio-based FUR to FUA (Saini et al. 2022).

The nature of support in supported metal catalysts plays a significant role in any catalytic reaction. Support materials generally contribute to generating specific active sites together with the employed metal, thereby influencing the structural and electronic properties of the final material (Nian et al. 2023). The various physicochemical properties/phenomena, such as oxygen spillover, electronic interaction, strong metal support interaction (SMSI), and hydrogen spillover, contribute to the catalytic activity towards specific reaction types and are decisively controlled by the support properties (Karim et al. 2017; Pan et al. 2017; Su et al. 2021). Also, the construction or placement of supported metals as single atoms, clusters or nanoparticles and their dispersion, depends on the interaction between the support and the doped metal, and thus, the nature of support becomes very crucial (Liu et al. 2018). As for the industrial application, the stability of the active catalytic system is the primary concern that can be ensured by the stability of the support material and its capability to stabilize the employed metal structurally, morphologically, and electronically. Therefore, studying the support effect becomes crucial for understanding and modifying the materials to meet industrial needs.

In line with this, several studies have been reported on developing various catalytic systems using both precious and non-precious metals on multiple supports to achieve selective production of FUA via catalytic reductive amination of FUR. Initially, metals supported on acidic supports were found to be playing a pivotal role in the reductive amination of FUR to FUA as the high number of acidic sites on the catalyst surface contributes to the facile -C=N bond activation, consequently forming the corresponding amine product. However, acidic supports generally require harsh reaction conditions like high NH_3 concentration and high H_2 pressure. Also, the spent catalyst(s) after recycling (up to five) studies have not been characterized in-depth to pinpoint the retainment of active sites and characteristic features. Although not significantly explored, in contrast to acidic-supported systems, metal supported on basic supports were found to be more active for FUA formation under comparatively milder reaction conditions than metal on acidic-supported systems.

Basic supported metal catalytic systems also exhibited recyclability (up to six cycles), but only a few studies indicated better retainment of the surface properties after use, which could be because of milder reaction conditions. However, in-depth characterization of the catalysts before and after use studies is still lacking. Further, in contrast to acidic and basic supports, metal supported on neutral supports was found to be relatively more stable due to the encapsulation of metals, avoiding leaching in a highly basic reaction medium. On the other hand, metals on neutral supports encounter inaccessibility of active metal species due to the coverage of layers of support material, indicating the requirement for high metal loading and further development. Based on the above viewpoints, this section thus provides insight into how the nature of support (acidic, basic and neutral) contributes to the catalytic active surface properties like the electronic environment of supported metal and the material stability. Further, the different roles of support for precious and non-precious metal-supported catalyst systems towards their stability are discussed.

7.4.1 The Acidic Support for FUR Reductive Amination

The acidic sites of the support play an important role in different reactions, like hydrogenation, oxidation, isomerization, and dehydration, by assisting facile substrate activation on the catalyst surface (Ali et al. 2023; Holm et al. 2010; Pal et al. 2020; Saravanamurugan et al. 2013; Saini et al. 2023). The strength and nature of acidic sites play different roles for various reactions; for example, in a typical reaction of glucose to HMF, isomerization reactions are primarily assisted via Lewis acidic sites, and dehydration happens at Bronsted acidic sites (Song et al. 2021). Similarly, the activation of substrates and intermediates to produce FUA via reductive amination of FUR occurs at Lewis acidic sites on the catalyst surface, contributing to high catalytic activity (Saini et al. 2023). Along with the substrate activation on the catalyst surface, the nature of support can modify the electronic properties of the supported metal, consequently contributing to the catalytic activity. The electronic characteristics of the supported metal are crucial in reductive amination reactions as they involve activation and dissociative adsorption of H_2 in addition to substrate activation (Saini et al. 2022). In this regard, the role of acidic supports in modulating the catalytic activity of supported precious and non-precious metals was widely explored for the reductive amination of FUR (Table 7.1).

7.4.1.1 The Role of Precious Metal on Acidic Support on the Reductive Amination of FUR

The properties of supported metals, like particle size, morphology, and electronic environment, significantly contribute to the reductive amination reaction that is directly or indirectly influenced by the nature of the support material. In addition, as the reductive amination reaction involves the hydrogenation step to produce the required primary amine, the type of supported metal plays a vital role in the

TABLE 7.1
Various Acidic and Basic Supports for Metals for the Reductive Amination of FUR

Entry	Catalyst	Reaction Conditions	FUR Con. (%)	FUA Yield (%)	DFA Yield (%)	Ref.
1	Rh/Al_2O_3	0.03 $FUR:NH_3$ (mol.), 80 °C, 20 bar H_2, 2 h.	100.0	91.5	-	(Chatterjee et al. 2016)
	Ru/Al_2O_3		33.5	-	-	
	Pd/Al_2O_3		100	18.4	77.2	
	Pt/Al_2O_3		55.6	12.4	18.8	
2	Pt/SiO_2	1:5 $FUR:NH_3$ (mol.), 130 °C, 51.7 bar H_2, 10 h, 40:1 Dioxane:Water (wt.)	100	41	23	(Gould et al. 2020)
	Pd/SiO_2		100	10	60	
	Rh/SiO_2		97	30	21	
	Ru/SiO_2		100	72	4	
	Ni/SiO_2		98	51	1	
	Co/SiO_2		100	70	2	
3	Pd/HZSM-5(46)	6 ml 7 M NH_3 in MeOH, 100 °C, 30 bar H_2, 15 min	>99	44	-	(Dong et al. 2020)
	Pt/HZSM-5(46)		>99	47	-	
	Rh/HZSM-5(46)		>99	49	-	
	Ru/HZSM-5(46)		>99	76	-	
4	Ru/Nb_2O_5	8 mmol NH_3, 90 °C, 40 bar H_2, 4 h	100	89	-	(Komanoya et al. 2017)
	Ru/SiO_2		100	84	-	
	Ru/TiO_2		100	72	-	
	Rh/Nb_2O_5		100	75	-	
	Pd/Nb_2O_5		100	26	-	
	Ru/Nb_2O_5	10 bar NH_3, 40 bar H_2, 90 °C, 4 h	100	99	-	
5	Ru/ $Nb_2O_5{\cdot}nH_2O$-300	5 ml 1.6 M NH_3 in MeOH, 70 °C, 40 bar H_2, 4 h	>99	89	-	(Deng et al. 2019)
	Ru/ $Nb_2O_5{\cdot}nH_2O$-400		>99	73	-	
6	Ru/Nb_2O_5-L	5 mL 2 M NH_3 in MeOH, 90 °C, 20 bar H_2, 8 h	100	60	-	(Guo et al. 2019)
7	Ni/SiO_2-I-DP	8 bar NH_3, 90 °C, 40 bar H_2, 1.5 h	100	94.6	-	(Yang, Zhou, et al. 2023)
8	Ru-PVP/HAP	Aq. NH_3, 100 °C, 4 atm H_2, 2 h	100	60	-	(Nishimura et al. 2016)
	Ru/Al_2O_3		100	33	-	
	Ru/TiO_2		100	27	-	
	Ru/CeO_2		100	9	-	
	Ru/ZrO_2		100	9	-	

(continued)

TABLE 7.1 (Continued)
Various Acidic and Basic Supports for Metals for the Reductive Amination of FUR

Entry	Catalyst	Reaction Conditions	FUR Con. (%)	FUA Yield (%)	DFA Yield (%)	Ref.
9	Ru/MMT	4 ml 25% aq. NH_3, 90 °C, 10 bar H_2, 3 h	97	89	-	(Gokhale et al. 2021)
	Ni/MMT	4 ml 25% aq. NH_3, 90 °C, 10 bar H_2, 7 h	96	84	-	
10	Ru/TiP-100	3 bar NH_3, 30 °C, 17 bar H_2, 24 h	>99	91	-	(Xie et al. 2020)
11	Ru/Ni_1MgAlO_x	4.5 mmol NH_3, 90 °C, 20 bar H_2, 5 h	>99	91.3	-	(Gao et al. 2022)
12	Pd/MoO_{3-x}	16 ml 7M NH_3 in MeOH, 80 °C, 20 bar H_2, 4 h	>99	84	~10	(Wang et al. 2023)
13	Ni/Al_2O_3-0.5LaO_x	8 bar NH_3, 90 °C, 20 bar H_2, 2h	>99	90.9	-	(Yang et al. 2023)
14	Ni/Al_2O_3	4 bar NH_3, 60 °C, 30 bar H_2, 4 h	>99	97.7	-	(Wan et al. 2024)
15	NiAl-Zr	NH_3/furfural (mol/mol) = 12/1, 70 °C,10 bar H_2	>99	98.8	-	(Xu et al. 2024)

dissociative adsorption of H_2. Various precious metals, such as Pt, Pd, Ru, and Rh, on different supports like Nb_2O_5, Al_2O_3, TiO_2, carbon, MCM-41, and H-ZSM-5 have been investigated to understand their catalytic role on the reductive amination of FUR. Chatterjee et al. found Rh the most suitable metal when supported on Al_2O_3, giving 92% FUA selectivity (Chatterjee et al. 2016). Whereas Pt/Al_2O_3 and Pd/Al_2O_3 resulted in the formation of over-hydrogenated products, i.e., FUL and DFA formations, respectively, considering Pt and Pd are known to have higher hydrogenation capability compared to other precious metals like Ru (Qi et al. 2020). In line with this, Gould et al. reported the hydrogenation capabilities of the different metals supported on SiO_2, where Pt/SiO_2, Pd/SiO_2, and Rh/SiO_2 exhibited poor catalytic activity towards primary amine (FUA) to secondary amine (DFA) selectivity (Gould et al. 2020).

In contrast, Ru/SiO_2 exhibited comparatively better selectivity towards primary amine, owing to the moderate hydrogenation capability of Ru, which is required for primary amine synthesis. A similar trend has been observed by Dong et al. when employing Pd and Pt on H-ZSM-5 (Dong et al. 2020). The role of different precious and non-precious metals was also investigated, and it was found that Ru showed better activity towards primary amine with high selectivity than Rh, Pt, Pd, Co, and Ni-based catalysts under identical conditions (Wang et al. 2023). These studies clearly demonstrated the high catalytic activity of Ru metal compared to other precious metals for the synthesis of primary amines. However, in a recent study (Wang et al. 2023), Pd-supported on MoO_x also exhibited high catalytic activity, affording a maximum FUA yield of 84% owing to the formation of Mo^V species contributing to the Lewis acidity, and thus, facilitating the facile carbonyl activation. Concurrently, the Pd^0 species facilitated the hydrogenolysis of FDA to form corresponding FUA along with 10% DFA, even at high NH_3 concentrations. One can clearly understand from the studies mentioned above that irrespective of the nature of the support, Pt and Pd are unsuitable for the selective reductive amination of FUR to FUA due to their very high hydrogenation capability. Surprisingly, as Ru is found to be suitable for the reductive amination of FUR to FUA; however, Ru supported on Al_2O_3 exhibited no catalytic activity towards FUA formation (Chatterjee et al. 2016), inferring that along with metal, the nature of support is crucial that also contributes to the selective product formation.

Numerous studies have thoroughly examined the function of support in the reductive amination reactions of FUR. The role of support nature can be realized from the study reported by Komanoya et al., investigating Ru on various supports, such as TiO_2, SiO_2, Nb_2O_5, ZrO_2, and Al_2O_3. Ru/Nb_2O_5 afforded the highest yield of 99% of FUA at 90 °C in 4 h (Komanoya et al. 2017). However, Ru/TiO_2 and Ru/SiO_2 gave good FUA yield under similar conditions, but with a prolonged reaction time, FUA converted to tetrahydrofurfurylamine (THFUA), i.e., the over-hydrogenated product of FUA. The rationale behind this discrepancy in catalytic activity was the acidic nature of Nb_2O_5 that reduced the electron density around Ru, preventing the ring hydrogenation of FUA by affecting the furan-ring adsorption on the catalyst surface as observed with Ru/TiO_2 and Ru/SiO_2. This demonstrates how the nature of the support plays a crucial role in selectively producing FUA by regulating the electron density around the supported metal. Although in the case of Ru/Nb_2O_5, the controlled electron density around Ru prevented the over-hydrogenation, it is essential to understand first the surface-active sites that primarily contributed to the formation of primary amine. Therefore, if the Ru electronic environment is identical on various acidic supports (e.g., Nb_2O_5, TiO_2), it thus becomes imperative to get insight into other surface properties, especially the number of acidic sites and their type, strength, and density, which could significantly influence the selective catalytic activity, as the surface acidic sites contribute to reductive amination reaction via facile C=N bond activation of FUI and FDA.

In line with that, Deng et al. studied the role of acidic sites by strategically varying the reduction temperature of Ru/$Nb_2O_5 \cdot nH_2O$ and found that the Ru/$Nb_2O_5 \cdot nH_2O$ reduced at 300 °C exhibited higher catalytic activity with 89% FUA yield at 70 °C in 4 h, attributed to the higher density of acidic sites on Ru/$Nb_2O_5 \cdot nH_2O$-300 compared to Ru/$Nb_2O_5 \cdot nH_2O$-400 (Deng et al. 2019). However, the acidic strength and Ru electronic environment were similar in both cases. Interestingly, the large-scale applicability of the $Ru/Nb_2O_5 \cdot nH_2O$-300 was also studied using 72 times (1.44 g) higher amount of FUR as the starting amount and obtained 87% yield of FUA under optimized conditions. The morphology effect of acidic Nb_2O_5 support was also explored by Guo et al. (Guo et al. 2019). The Ru/Nb_2O_5-L (L-layered) having layered morphology with high surface area resulted in high Ru dispersion, contributing to its catalytic activity. In comparison, Ru/Nb_2O_5-H (spherical morphology) and Ru/Nb_2O_5-F (nanoflower-like morphology) with low surface area and poor Ru dispersion resulted in undesired product formation like FDA and others. These studies have nicely demonstrated how the electronic interaction between metal and acidic support, the support morphology and surface acidic sites contribute to the formation of primary amine (cyclopentylamine from cyclopentanone, similar to FUR to FUA) selectively. As the studies mentioned above concluded the decisive role of acidic sites towards reductive amination reaction, in line with that of Gao et al. who studied the surface-active acidic sites on Ru/Ni_1MgAlO_x that can selectively activate Schiff's base intermediate in the presence of highly basic NH_3 (Gao et al. 2022). The proximity and appropriate proportion of Al_{4c} (4 co-ordinated Al) and Al_{6c} (6 co-ordinated Al) sites provided a facile activation of NH_3 and Schiff's intermediate activation, respectively, and subsequently facilitated hydrogenolysis to produce primary amine. The generation of alternative active sites to activate Schiff's base and NH_3 could be the reason for the high FUA yield (91.3%) at 90 °C in 4 h, even at low NH_3 concentration (3 times that of the substrate). Therefore, in addition to controlling the electronic environment around supported metal, acidic supports contribute to the availability of acidic sites, facilitating the production of selective FUA under mild reaction conditions.

7.4.1.2 The Role of Non-Precious Metal on Acidic Support on the Reductive Amination of FUR

The high abundance, low toxicity, and unique catalytic properties of non-precious metals are of significant interest in developing environment-friendly processes. Only Ni and Co, because of their excellent hydrogenation capability, have been significantly explored for reductive amination reactions among various non-precious metals. Figure 7.2 shows the comparative catalytic activity of different metals supported on SiO_2 towards the selective production of FUA from Schiff's base intermediate (Gould et al. 2020). Ni and Co displayed better catalytic activity than other metals under similar reaction conditions. In the previous section, electron density around supported metal, metal dispersion, and support morphology predominantly contributed to selective FUA formation. In order to develop efficient non-precious metal-based catalysts, it will be interesting to find out whether non-precious metal-containing materials show a similar trend of properties as observed in the previous section.

In this regard, a study by Dong et al. disclosed that the metal support interaction and good metal dispersion contributed towards the higher catalytic activity of $10Ni/Al_2O_3$, giving a 99% FUA yield (Dong et al. 2021). A moderate FUA yield (45–70%) was obtained with other supports like ZrO_2, CeO_2, SiO_2, H-ZSM-5 and Nb_2O_5 because of a lack of metal support interaction. Surprisingly, 10Ni/H-ZSM-5, with a high number of acidic sites that help in substrate activation, gave only 48% of FUA yield at 100 °C

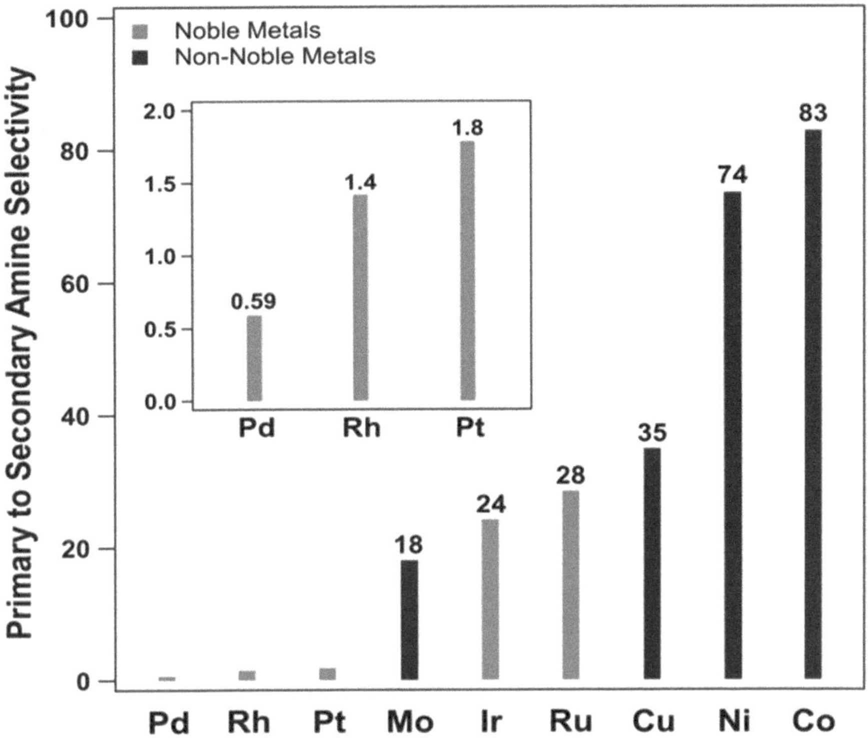

FIGURE 7.2 The effect of the metal identity on the ratio of primary amine (FUA) to secondary amine (DFA) selectivity at full conversion of the secondary imine. (Adapted from Gould et al. 2020.)

in 2 h. XPS analysis revealed that the Ni^0 peak in the case of 10Ni/H-ZSM-5 shifted towards higher binding energy compared to 10Ni/Al_2O_3, inferring the decreased electron density around Ni, hence reducing its hydrogenation capability that could be one of the reasons for its lower catalytic activity. Further, Ni/Al_2O_3 modified with La to enhance the surface acidity, metal dispersion, and electronic properties of supported metal was also reported (Yang et al. 2023). It was observed that enhanced adsorption of FDA was promoted by the strong acidic sites and the electron-deficient Ni^0 species, and small Ni particles simultaneously favored the H_2 activation. However, no clear picture of the role of the electronic environment of Ni in H_2 activation has been disclosed, that could have been done to understand the role of acidic sites to alter the electronic environment around Ni, hence altering its hydrogenation capability. Further, Ni supported on two different exposed facets {(111) and (011)} of Al_2O_3 was also explored (Wan et al. 2024). Ni/Al_2O_3(111) possessing stronger acidic sites than Ni/Al_2O_3(011) caused a reduction in the electron density around Ni, resulting in facile hydrogen migration from Ni to Al_2O_3(111) with enhanced H affinity. The H migration from Ni to Al_2O_3 resulted in promotional adsorption and activation of NH_3 and Schiff's base over Ni sites, hence decreasing the activation energy of Schiff base reductive ammonolysis to produce the final product, i.e., FUA. Thus, the role of alternative active sites for the activation of Schiff's base and for reactive H species can be clearly understood.

Non-precious metals possess a comparatively limited hydrogenation capacity, which can be reduced further by the acidic supports; however, it is still ambiguous whether acidic supports are better suited to precious metals than non-precious metals. Yang et al. investigated the role of Ni nanoparticles and their shape supported on SiO_2 by varying the Ni doping method (Ni/SiO_2-I-DP, Ni/SiO_2-II-DP (DP: deposition-precipitation method) and Ni/SiO_2-I-IM (IM: impregnation method)) (Yang et al. 2023). It has been disclosed that small Ni flat-shaped particles (2.8 nm) and the surface Lewis acidic sites contributed to facile H_2 activation and intermediate adsorption to produce FUA. The higher number of acidic sites on SiO_2 support in the case of Ni/SiO_2-I-DP was found to contribute significantly to reductive amination. Although the CO-DRIFT studies indicated a redshift of the Ni^0-CO band in the case of Ni/SiO_2-I-DP compared to Ni/SiO_2-I-IM and Ni/SiO_2-II-DP, suggesting comparatively high electron density around Ni^0 in Ni/SiO_2-I-DP that might be contributed to its high catalytic activity; however, authors disregarded as an electronic effect but as a geometric effect. Yolk-shell structured layered double hydroxide (LDH)-derived Zr-assisted NiAl (NiAl-Zr) was recently reported, exhibiting three-fold higher catalytic activity towards FUA formation than simple NiAl catalyst (Xu et al. 2024). The introduction of Zr in NiAl resulted in increased electron density around Ni^0, lowering the adsorption energy of hydrogen on the catalyst surface and enhancing the hydrogenation capability of NiAl-Zr.

To sum up, the studies mentioned above suggest that the electron density surrounding the supported metal is crucial to produce primary amines via facile H_2 activation. An optimized electron density surrounding a supported metal can help to achieve a moderate hydrogenation capacity of the catalyst. Precious metals are known to have high electron density around them and, hence, are well-known to exhibit high hydrogenation capacity. Therefore, selective FUA formation is achieved when the electron density around them is reduced using acidic supports. On the other hand, non-precious metals are more active when supported on less acidic supports to maintain significant electron density around metal; hence, a moderate hydrogenation capability can be achieved. Along with controlling the electron density around metal, the acidic support provides suitable acidic sites, facilitating the selective product formation via intermediate adsorption on the catalyst surface. Apart from metal support interaction, it was observed that the metal particle size, dispersion, and morphology can also affect the catalytic activity significantly.

7.4.2 The Basic Support for FUR Reductive Amination

Basic supports are not explored much compared to acidic supports but can play a significant role in the catalytic reductive amination of FUR to FUA. The basic sites on the catalyst surface are well known to promote the facile hydrogen activation (Saini et al. 2023). The basic supports can regulate the electron density around supported metal, which helps lower the reaction temperature due to the activation of H_2 at low temperatures and selectively produces primary amines. Metal with high electron density around it can activate H_2 by donating electrons to its antibonding orbitals, weakening the H-H bond and resulting in dissociative adsorption of hydrogen, hence contributing to high hydrogenation activity (Saini et al. 2022). However, the highly basic nature of the catalyst can also restrict the adsorption of the substrate and intermediates on its surface, reducing the catalytic activity. Also, the presence of NH_3 in the reaction mixture can hinder the facile H_2 activation on the catalyst surface, altering the hydrogenation capability. Therefore, the H_2 activation under ambient conditions is challenging in reductive amination reactions. This section, therefore, covers how basic supports, for both precious and non-precious metals, affect the catalytic properties of the supported metal.

7.4.2.1 The Role of Precious Metal on Basic Support on the Reductive Amination of FUR

Nishimura et al. screened different basic (HAP, HT) and acidic supports (Al_2O_3 and TiO_2, CeO_2, and ZrO_2) for Ru metal and found that the former contributed more towards catalytic activity than the latter (Nishimura et al. 2016). 4 wt% Ru-PVP/HAP with an average diameter of 1.7 ± 0.3 nm of Ru clusters yielded 60 % FUA. However, no role of metal support interaction was revealed, reaching no firm conclusion that rather than the role of support, Ru particle size and loading could have significantly influenced the catalytic activity, as observed in previous studies. Further, Gokhale et al. synthesized Ru/MMT and achieved 96% FUA selectivity at 90 °C in 3 h (Gokhale et al. 2021). A large-scale study with approximately 42 times higher substrate amount (2 g FUR) was done over Ru/MMT, producing 69% FUA yield at 120 °C in 14 h. However, no significant conclusions have been drawn to pinpoint the role of support.

Further, the role of metal interaction with the basic sites of supports towards the reductive amination of carbonyls was well studied by Xie et al. (Xie et al. 2020). The presence of low-temperature reduction peaks in the H_2-TPR profile of synthesized RuO_2/TiP demonstrated that electron transfer from support to metal contributes to facile Ru reduction. The high acidity of TiP, along with a suitable reduction degree of RuO_2 to Ru^0 (52%) in Ru/TiP-100 that controls the electron density around Ru^0 contributed to enhanced catalytic activity, giving a 91% FUA yield under ambient conditions. XPS and CO-DRIFT analysis indicate that the increased electron density around Ru^0 contributes to the room temperature H_2 activation compared to Ru/TiO_2. In addition, the presence of acidic sites promotes the C=N bond activation of *in situ* generated Schiff's base. However, as carbon is highly electrophilic in the C=N moiety of FDA, a very high electron density surrounding metal can also lead to strong adsorption, requiring a high reaction temperature to desorb the intermediate from the catalyst surface. Thus, an optimized amount of electron density around supported metal and the electronic metal support interaction (EMSI) plays a crucial role in the reductive amination of carbonyls to primary amines.

7.4.2.2 The Role of Non-Precious Metal on Basic Support on the Reductive Amination of FUR

As indicated in the previous section, a comparison study between Ru/MMT and Ni/MMT was conducted for the reductive amination of FUR (Gokhale et al. 2021). Ni/MMT afforded 85% FUA selectivity, but no suitable discussion has been provided to reveal the role of the support. It should be noted that a higher loading of Ni (12.5 wt%) compared to Ru (2 wt%) was required to achieve a high yield of FUA with a longer reaction time at a relatively high reaction temperature. Also, the Ni/MMT showed poor recyclability under the employed reaction conditions. Though the role of basic support for non-precious metals is scarcely reported, an in-depth study is required to investigate the effect of basic support in non-precious metal-supported catalysts towards understanding the characteristic features of the material.

7.4.3 The Neutral Support for FUR Reductive Amination

Apart from employing acidic and basic supports for precious and non-precious metals for reductive amination, neutral supports have also been considerably explored in relation to their contribution to producing FUA selectively from FUR (Table 7.2). Thus, understanding the type of active species generated on the metal using neutral supports contributing to catalytic activity is important.

7.4.3.1 The Role of Precious Metal on Neutral Support on the Reductive Amination of FUR

Carbon-based materials are significantly explored for different biomass transformations in the category of neutral supports. In contrast to metal oxide systems, these carbon-based materials have significant advantages, such as high availability, inexpensiveness, and environmental friendliness. The electronic properties of the metal can be easily modified by creating defective sites by introducing heteroatoms or by thermal treatment at different pyrolysis temperatures (Rangraz et al. 2021). Also, introducing heteroatoms like N in carbon support enhances the stability of supported metals because of the strong interaction between heteroatoms and supported metals. In line with that, Qi et al. fabricated highly active Ru SACs supported on N-doped carbon support pyrolysed at various temperatures (700–900 °C) (Qi et al. 2021). The increased electron density around Ru, with the change in Ru coordination structure with N from RuN_5 to RuN_3, was found to play a critical role in FUA formation. This increased the electron density around Ru, facilitating H_2 activation in the Ru_1/NC-900-800NH_3 (Ru_1/NC-900 treated with NH_3 at 800 °C) catalyst and enhancing the hydrogenation capability, as observed with basic supports.

The activation of H_2 in the presence of NH_3 in reductive amination is challenging compared to a simple hydrogenation reaction due to the competitive adsorption of NH_3 and H_2 on the catalyst surface. This was proved by calculating the change in H_2 uptake from 7.9 to 4.1 μmol/g, with and without pre-adsorption of NH_3, respectively, on Ru_1/NC-900–800NH_3, indicating the competitive adsorption of the two reactants on the active sites playing a crucial role. This could be the reason for the requirement of high pressure of NH_3 and H_2 to achieve high FUA yield compared to the catalytic

TABLE 7.2
Various Metals on Neutral Supports and Nanoparticle-Based Catalysts for the Reductive Amination of FUR

Entry	Catalyst	Reaction Conditions	FUR Con. (%)	FUA Yield (%)	Ref.
1	Ru_1/NC-900-800NH_3	50 bar NH_3, 100 °C, 20 bar H_2, 10 h	100	97 (170.7)	(Qi et al. 2021)
	Ru_1/NC/Nb_2O_5		>99	97 (227.6)	
2	4Ru1Co/AC	2 bar H_2, water/ammonia solution (2:3), 100 °C, 2 h.	>99	92	(Gou et al. 2022)
3	Ru/BN-e	2 equiv. of 25% aqueous NH_3, 90 °C,15 bar H_2, 2 h.	>99	90.2	(Gao et al. 2021)
	Ru/BN-b		>99	2.4	
4	Co@C-700-EtOH	5 ml 7 M NH_3 in MeOH, 90 °C, 20 bar H_2, 2 h.	>99	71.5	(Zhuang et al. 2022)
5	Co/NC-700	1 ml aq. NH_3, 120 °C, 20 bar H_2 pressure, 1 h	100	99	(Yogita et al. 2022)
6	Co@NC-800	2 ml 26.5 wt% aq. NH_3, 130 °C, 10 bar H_2, 12 h.	>99	81.8	(Yuan et al. 2019)
7	MC/Ni	2.8 mmol NH_3, 80 °C, 1 bar H_2, 12 h.	>99	88.4	(Zhang et al. 2019)
8	Ni/pNC	4 bar NH_3, 60 °C, 30 bar H_2, 6 h.	>99	92.3	(Song et al. 2022)
9	Ni_6AlO_x	49:1 NH_3:aldehyde, 100 °C, 4 bar H_2, 5h	>99	90 (isolated)	(Yuan et al. 2019)
10	NiAl-10	5 bar NH_3, 80 °C, 20 bar H_2, 1 h.	100	99	(Pan et al. 2022)
	NiAl-120		>99	-	
11`	Ru NP (10)	8 mmol NH_3, 90 °C, 20 bar H_2, 2 h	100	99	(Chandra et al. 2018)
12	Pd NP	H_2 filled in a balloon, RT, 1 h, pH set at 2.0	100	99	(Jv et al. 2020)
13	Raney Co	1 bar NH_3, 120 °C, 10 bar H_2, 2 h	100	98.9	(Zhou et al. 2019)
	Raney Ni	1 bar NH_3, 120 °C, 10 bar H_2, 2 h	100	65.1	
14	Co_2P NR	3 ml aq. NH_3 25%, 100 °C, 5 bar H_2, 10 h.	93	91	(Sheng et al. 2021)

systems like Ru/Ni_1MgAlO_x, which provided a high FUA yield even under low NH_3 concentration due to the existence of alternative active sites for H_2, NH_3, and intermediate activation.

Guo et al. synthesized RuCo alloy stabilized on activated carbon (4Ru1Co/AC) catalyst and obtained a high FUA yield (92.1%) at 80 °C in 1 h (Gou et al. 2022). The metallic Ru facilitated the Co reduction to the low-valence state due to the formation of RuCo alloy, resulting in increased surface Co^0 species and contributing to significant hydrogenation ability, accomplishing a high FUA yield. However, no insightful information has been given regarding the role of support. Also, the carbon-based support catalysts with anisotropic locations of Ru on hexagonal boron nitride (Ru/BN) were synthesised to produce FUA selectively (Gao et al. 2021). Ru^0 located on the edges of h-BN (Ru/BN-e) was found to be electron-rich compared to that on basal (Ru/BN-b), contributing to the high hydrogenation capability of Ru/BN-e and hence affording a high FUA yield.

7.4.3.2 The Role of Non-precious Metal on Neutral Support on the Reductive Amination of FUR

The use of non-precious metal-based catalytic systems is considerably attractive due to their advantages of sufficient abundance and cost-effectiveness; however, their stability is a major challenge due to highly basic reaction conditions required for reductive amination to primary amines as it involves the use of NH_3 as a nitrogen source. As mentioned before, the heteroatoms in the carbon support can stabilize the supported metals. Also, encapsulating non-precious metal inside nanotubes and spheres is another interesting approach, preventing metal leaching in the reaction mixture and thereby increasing the stability of the material during the reaction (Yu et al. 2020). In line with that, several studies have explored the activity and stability of carbon-supported non-precious metal-based catalysts for reductive amination of FUR to FUA. Zhuang et al. synthesized graphene-shelled metallic Co nanoparticles (Co@C-600-EtOH) as a highly stable and selective catalyst for the reductive amination of FUR, giving 86.9% FUA yield at 90 °C in 2 h (Zhuang et al. 2022). The presence of metallic Co promotes imine formation by activating the C=O bond of FUR by the electron transfer from metallic Co species to outer graphitic shells, thus activating the substrate molecules, i.e., FUA followed by dehydration to form imine intermediate (FUI).

At the same time, the acid sites, possibly generated from CoO species on the particle surface, were speculated to be responsible for the hydrogenation of FUI to FUA via FUI activation. However, a more in-depth study of the catalytic properties of the materials synthesized at different pyrolysis temperatures could have been done to understand each step of the reaction mechanism along with H_2 activation. Further, the catalyst maintained its catalytic activity for up to eight cycles because the cobalt nanoparticles are embedded in the carbon layer as the armour to effectively protect it from leaching. Hence, this chainmail property contributed to high catalyst stability even in the highly basic environment of reductive amination reaction. Also, the role of N doped in carbon supports (NC) in stabilizing the non-precious metals was also significantly explored. Yogita et al. reported a similar role of N-stabilized metallic

Co in graphitic layers (Co@NC-700), yielding 99% of FUA yield in 1 h at 120 °C (Yogita et al. 2022). Co@NC-700 retained its catalytic activity up to the fifth run with no significant decline in FUA yield. H_2-TPD studies showed the facile desorption of hydrogen from the surface of Co@NC-700 compared to Co@NC-400, Co@NC-500, and Co@NC-600, resulting in its higher catalytic activity. Also, Co@NC-700 gave 68.7% of FUA yield using crude FUR obtained from xylose.

Moreover, Yaun et al. reported that the metallic Co with a relatively smaller particle size (13.1 nm) (Co@NC-800) compared to Co@NC-900 (32.3 nm) by systematically varying the pyrolysis temperature yielded 81.8% of FUA yield at 130 °C in 12 h and maintained its activity up to the five runs (Yuan et al. 2019). These studies indicate that the binding of Co with N, as discussed above, not only played a significant role in the catalytic activity but also stabilized the supported non-precious metal by encapsulating them in the carbon layers, making them more robust. However, as few studies indicated, the presence of different types of N species, pyridinic, pyrrolic and graphitic N, might have contributed to the catalytic activity, but the role of each N species was not unambiguously untangled for in-depth understanding (Yogita et al. 2022; Yuan et al. 2019).

Ni supported on mesoporous carbon (Ni/MC) was also reported, yielding 88.4% FUA at 80 °C in 12 h (Zhang et al. 2019). It was realized that N-doping happened unexpectedly because of the use of Ni ammonium salt during the synthesis; however, the catalytic role of N-doping was not discussed. Later, Ni supported with and without N-doped carbon, i.e., Ni/pNC and Ni/pC, was also reported to understand the effect of metal-N bonding (Song et al. 2022). Ni/pNC exhibited excellent catalytic activity (92.3 % FUA) at 60 °C in 6 h, whereas Ni/pC showed no transformation of Schiff's intermediate to FUA. The formation of Ni–N_x bonding resulted in increased electron density around Ni, as confirmed by CO-DRIFT analysis. As a result, the formation of Ni–N_x species significantly promoted the reductive amination of C=O bonds in carbonyl compounds along with easy H_2 activation.

The above studies related to non-precious metal supported on neutral supports conclude that, in contrast to basic and acidic supports, stabilizing supported metal on neutral supports is simple and can be achieved by introducing heteroatoms (e.g., N) into the support or by encasing the metal in carbon-based materials. Furthermore, the pyrolysis temperature can be systematically changed, or a heteroatom can be introduced into the support to easily control the electron density surrounding the supported metal, thereby enhancing catalytic activity. Despite this, the encapsulation of active metal by carbon material could limit the effective utilization of actual metal content used during the synthesis of these catalysts.

7.4.4 Reductive Amination of FUR with Metal Nanoparticles

Apart from the metal-oxide-supported catalytic systems, the metal nanoparticles were also employed as catalysts for the reductive amination of FUR. The type (precious/non-precious), size, and morphology of different nanoparticles can decisively contribute to their catalytic activity (Ishida et al. 2020). With regard to this, HCP (hexagonal close-packed) and flat-shaped pristine FCC (face-centred cubic) ruthenium nanoparticles with {111} exposed facets were reported and found that the latter exhibited an outstanding activity (99% FUA) at 90 °C in 2 h. The presence of a large fraction of metallic Ru and weak electron donating ability was responsible for the high catalytic activity. The study showed Ru nanoparticles superior to other Ru-based catalysts (Ru/Nb_2O_5 and Ru/SiO_2), giving the highest TOF of 1850 h^{-1} (Chandra et al. 2018). Pd nanoparticles was also tested for the reductive amination of FUR, leading to the formation of undesired over hydrogenated products, such as FUL and DFA (Jv et al. 2020). As described in Section 7.4.1.1 and due to their high hydrogenation capability, Pt and Pd are generally unsuitable for producing primary amines via reductive amination of carbonyls. However, to obtain a high yield of primary amine, the pH of the reaction mixture was maintained at 2.0, and a 98% benzylamine yield was obtained at room temperature at 3 h. Further, 97% of the FUA yield was obtained under similar optimized conditions at pH 2.0.

The reason behind the high primary amine yield using Pd nanoparticles could be the protonation of the final product, i.e., primary amine, which inhibits the formation of Schiff's base intermediate, which is inevitable at a neutral pH. Overall, Ru can be a suitable candidate among precious metals for synthesizing primary amines via reductive amination of carbonyls using NH_3 and H_2 as nitrogen sources and reductants, respectively. The catalytic activity of commercially available catalysts like Raney Ni, Raney Co, Ru/C, Pt/C, Pd/C and Rh/C were also investigated, among which, Raney Co exhibited the best catalytic reductive amination performance, giving 98.9% of FUA yield but at relatively high temperature of 120 °C (Zhou et al. 2019). On the other hand, Raney Co yielded 99.5% of 5-hydroxymethyl furfurylamine (HMFA) at 120 °C in 2 h and 82.3% 2,5-bis(aminomethyl)furan (BAMF) was obtained with Raney Ni as the catalyst at 160 °C for 12 h from HMF (Wei et al. 2021). In contrast to Raney Co, Raney Ni can also be employed for the reductive amination of furfuryl alcohol via the hydrogen borrowing approach due to the formation of BAMF from HMF, as reported in another study by Liu et al. (Liu et al. 2017).

7.4.5 Other Miscellaneous Catalytic Reductive Amination Systems

Apart from metal oxide, carbon-based materials and metal nanoparticles as catalysts, other catalytic systems were also explored for reductive amination reactions. One of them is metal phosphides; these materials have been explored significantly for hydrogenation, hydrodesulfurization, hydrodenitrogenation, and hydrodeoxygenation reactions for biomass transformation to value-added chemicals (Prins et al. 2012). The P-alloying in metal phosphides can stabilize metal in a low valent state and can also modulate the electronic state of the metal species and increase its d-electron density. Cobalt phosphide (Co_2P) nanorods were mainly tested for the reductive amination of FUR, achieving 90% FUA yield at 100 °C in 10 h (Sheng et al. 2021). Single-crystal Co_2P NRs with coordinatively unsaturated Co-Co as active sites were air-stable, highly active, and reusable catalysts for the reductive amination of a wide range of carbonyl compounds.

Ni-AlO_x-type materials were also explored for reductive amination reaction. Ni_yAlO_x with different Ni/Al ratios were tested, among which Ni_6AlO_x is highly active for the reductive amination of HMF and FUR at 100 °C (Yuan et al. 2019). The synthesized material was also tested at a gram scale, using 3 g HMF, and an 81% isolated yield of 5-aminomethyl-2-furylmethanol (AFM) was obtained. In contrast to carbon-based materials, the recrystallization of Al_2O_3 from an amorphous to γ-phase and Ni leaching after the fifth run indicated the instability of the material, hence limiting its application towards large-scale studies and suggesting to improve the stability of the material. Although the previous study (Yuan et al. 2019) exhibited a quantitative yield of primary amines and is applicable at the gram scale, poor recyclability hinders its further development. In line with that, a similar kind of Ni-Al (metallic Ni particle size of 7.5 nm; NiAl-10) layered double hydroxides (NiAl-LDH) grafted on hollow polymer nanosphere catalysts was found to be active towards FUA formation and able to recycle up to 10 times with no catalytic loss (Pan et al. 2022). Unlike Ni_6AlO_x, NiAl-10 displayed superior catalytic activity due to the formation of surface Ni-C_x species during the synthesis, preventing the oxidation of Ni nanoparticles as observed with carbon-based materials. However, the above studies did not rationalize the role of Al towards their catalytic contribution, as discussed in another study (Gao et al. 2022).

7.5 CONCLUSION AND PERSPECTIVES

In summary, this chapter provided an overview of the reductive amination of FUR to FUA, a potential N-containing furanic having multiple applications, with a wide range of metals (Ru, Pt, Pd, Ni, and Co) on various supports (acidic: TiO_2, Nb_2O_5, basic: HAP, TiP and neutral: carbon) and metal nanoparticles, especially recent development on the catalytic materials, based on the previously published research articles. It has been exhibited how a supported metal primarily contributes to the dissociative adsorption of H_2 on the catalyst surface by donating its electrons to the antibonding orbitals of H_2. Ru, Co, and Ni are found to be suitable for the reductive amination of FUR to FUA due to their moderate hydrogenation capability compared to other precious and non-precious metals. However, non-precious metal-based catalytic systems require relatively harsh reaction conditions to achieve higher FUA selectivity than precious metal-based ones. It has also been pinpointed in this chapter that the role of acidic support, especially for Ru among the metals employed, reduces the electron density around Ru, thereby enhancing its catalytic activity. It has also been emphasized that employing basic supports could alter the degree of reduction of the supported metal; eventually, the catalytic performance could be improved at lower reaction temperatures. Incorporating heteroatoms such as N on neutral support (carbon) could enhance the electron density around the supported metal's (e.g., Co) vicinity to N and consequently improve its catalytic activity. Though significant progress has been recently made in the catalytic reductive amination of FUR to FUA, almost all studies indicate academic interest in lab scale, typically technology readiness level (TRL) between 3–4, and there is still a long way to go towards commercialization. The following suggestions should be considered for further improvement in the future.

i. None of the studies present the synthesis of material on a large scale (at least >10 g) to understand the retainment of active species that is achieved in typical lab-scale synthesis.
ii. None of the studies exhibits long-term material stability in a continuous flow experiment, though a high yield of FUA was achieved under batch experimental conditions.
iii. An unambiguous characterization study is required to determine the exact number of active sites that contribute to the catalytic activity so that the efficiency of the catalytic materials can be found in terms of TON/TOF/productivity.
iv. None of the studies demonstrates a life-cycle assessment (LCA) for understanding the economic viability of the process.

ACKNOWLEDGEMENTS

We sincerely thank the Department of Biotechnology (DBT), New Delhi, India, for the financial support and infrastructural facilities provided through CIAB. KS gratefully thanks UGC for financial support through the senior research fellowship and the Indian Institute of Science Education and Research (IISER) Mohali for the PhD registration.

REFERENCES

Ali, Hadi, Tom Vandevyvere, Jeroen Lauwaert, Sushil Kumar Kansal, Maarten K. Sabbe, Shunmugavel Saravanamurugan, and Joris W. Thybaut. 2023. "Enhancing the Anisole Hydrodeoxygenation Activity over Ni/Nb2O5−x by Tuning the Oxophilicity of the Support." *Catalysis Science and Technology* 13 (4). Royal Society of Chemistry: 1140–53. doi:10.1039/d2cy01745h.

Ayedi, Mohamed Ali, Ammar Houcine, and Souhir Abid. 2012. "Simple, Novel Synthesis of Furfurtlamine from Furfural by One-Pot Reductive Amination in Water using Zinc Metal." *Journal de la Société Chimique de Tunisie* 14, 109–16.

Ayusawa, T., Mori, S., Aoki, T., Hamana, R. 1986. "Process for producing Furfurylamine and /or Tetrahydrofurfurylamine." U.S. Patent 4,598,159.

Bozell, Joseph J., and Gene R. Petersen. 2010. "Technology Development for the Production of Biobased Products from Biorefinery Carbohydrates—the US Department of Energy's 'Top 10' Revisited." *Green Chemistry* 12 (4): 539–55. doi:10.1039/b922014c.

Chandra, Debraj, Yasunori Inoue, Masato Sasase, Masaaki Kitano, Asim Bhaumik, Keigo Kamata, Hideo Hosono, and Michikazu Hara. 2018. "A High Performance Catalyst of Shape-Specific Ruthenium Nanoparticles for Production of Primary Amines by Reductive Amination of Carbonyl Compounds." *Chemical Science* 9 (27). Royal Society of Chemistry: 5949–56. doi:10.1039/c8sc01197d.

Chatterjee, Maya, Takayuki Ishizaka, and Hajime Kawanami. 2016. "Reductive Amination of Furfural to Furfurylamine Using Aqueous Ammonia Solution and Molecular Hydrogen: An Environmentally Friendly Approach." *Green Chemistry* 18 (2). Royal Society of Chemistry: 487–96. doi:10.1039/c5gc01352f.

Deng, Dian, Yusuke Kita, Keigo Kamata, and Michikazu Hara. 2019. "Low-Temperature Reductive Amination of Carbonyl Compounds over Ru Deposited on Nb 2 O 5 · NH 2 O." *ACS Sustainable Chemistry and Engineering* 7 (5). American Chemical Society: 4692–98. doi:10.1021/acssuschemeng.8b04324.

Dong, Chenglong, Hongtao Wang, Haochen Du, Jiebang Peng, Yang Cai, Shuai Guo, Jianli Zhang, Chanatip Samart, and Mingyue Ding. 2020. "Ru/HZSM-5 as an Efficient and Recyclable Catalyst for Reductive Amination of Furfural to Furfurylamine." *Molecular Catalysis* 482 (February). Elsevier B.V. doi:10.1016/j.mcat.2019.110755.

Dong, Chenglong, Yushan Wu, Hongtao Wang, Jiebang Peng, Yanzhi Li, Chanatip Samart, and Mingyue Ding. 2021. "Facile and Efficient Synthesis of Primary Amines via Reductive Amination over a Ni/Al2O3Catalyst." *ACS Sustainable Chemistry and Engineering* 9 (21). American Chemical Society: 7318–27. doi:10.1021/acssuschemeng.1c01456.

Fechete, Ioana, Ye Wang, and Jacques C. Védrine. 2012. "The Past, Present and Future of Heterogeneous Catalysis." *Catalysis Today* 189: 2–27. doi:10.1016/j.cattod.2012.04.003.

Gao, Mingxia, Xiuquan Jia, Jiping Ma, Xiaomeng Fan, Jin Gao, and Jie Xu. 2021. "Self-Regulated Catalysis for the Selective Synthesis of Primary Amines from Carbonyl Compounds." *Green Chemistry* 23 (18). Royal Society of Chemistry: 7115–21. doi:10.1039/d1gc01275d.

Gao, Zhexi, Xinxin Zhao, Xiumin Li, Huifang Wu, Mingyu Gao, Qian Wang, Dianqing Li, and Junting Feng. 2022. "Rational Regulation of Spatially Adjacent Al4c and Al6c Sites Assisted Ru Catalysts for Low-NH3 Furfural Tandem Reductive Amination." *Chemical Engineering Science* 258 (August). Elsevier Ltd. doi:10.1016/j.ces.2022.117777.

Gokhale, Tejas A., Amol B. Raut, and Bhalchandra M. Bhanage. 2021a. "Comparative Account of Catalytic Activity of Ru- and Ni-Based Nanocomposites towards Reductive Amination of Biomass Derived Molecules." *Molecular Catalysis* 510 (June). Elsevier B.V. doi:10.1016/j.mcat.2021.111667. doi:10.1016/j.mcat.2021.111667

Gou, Le, Longfei Xie, Yuanyuan Wang, and Liyi Dai. 2022. "Efficient Conversion of Furfural to Furfural Amine over 4Ru1Co/AC Catalyst." *Applied Catalysis A: General* 647 (October). Elsevier B.V. doi:10.1016/j.apcata.2022.118902.

Gould, Nicholas S., Harrison Landfield, Brian Dinkelacker, Casper Brady, Xuan Yang, and Bingjun Xu. 2020. "Selectivity Control in Catalytic Reductive Amination of Furfural to Furfurylamine on Supported Catalysts." *ChemCatChem* 12 (7). Wiley Blackwell: 2106–15. doi:10.1002/cctc.201901662.

Guo, Wanjun, Tao Tong, Xiaohui Liu, Yong Guo, and Yanqin Wang. 2019. "Morphology-Tuned Activity of Ru/Nb2O5 Catalysts for Ketone Reductive Amination." *ChemCatChem* 11 (16). Wiley Blackwell: 4130–38. doi:10.1002/cctc.201900335.

Holm, Martin Spangsberg, Shunmugavel Saravanamurugan, and Esben Taarning. 2010. "Conversion of Sugars to Lactic Acid Derivatives Using Heterogeneous Zeotype Catalysts." *Science* 328 (5978): 602–5. doi:10.1126/science.1183990.

Ishida, Tamao, Toru Murayama, Ayako Taketoshi, and Masatake Haruta. 2020. "Importance of Size and Contact Structure of Gold Nanoparticles for the Genesis of Unique Catalytic Processes." *Chemical Reviews* 120(2). American Chemical Society: 464–525. doi:10.1021/acs.chemrev.9b00551.

Jorqueira, Diogo Silva Sanches, Letícia Franzo de Lima, Silvia Fernanda Moya, Léa Vilcocq, Dominique Richard, Marco A. Fraga, and Raphael Soeiro Suppino. 2023. "Critical review of furfural and furfuryl alcohol production: Past, present, and future on heterogeneous catalysis." *Applied Catalysis A: General* 665 (September). Elsevier Ltd: 119360. doi:10.1016/j.apcata.2023.119360.

Juhász, Kinga, Krisztina Lévay, László Hegedűs, Diána Balogh-Weiser, Laurence Pirault-Roy, and Zoltán Hell. 2021. "Application of Supported Lanthanum Catalysts in the Hydrogenation of Nitriles." *Reaction Kinetics, Mechanisms and Catalysis* 133 (2). Springer Science and Business Media B.V.: 687–98. doi:10.1007/s11144-021-02028-2.

Jv, Xinchun, Shuting Sun, Qun Zhang, Muyao Du, Lan Wang, and Bo Wang. 2020. "Efficient and Mild Reductive Amination of Carbonyl Compounds Catalyzed by Dual-Function

Palladium Nanoparticles." *ACS Sustainable Chemistry and Engineering* 8 (3). American Chemical Society: 1618–26. doi:10.1021/acssuschemeng.9b06464.

Karim, Waiz, Clelia Spreafico, Armin Kleibert, Jens Gobrecht, Joost Vandevondele, Yasin Ekinci, and Jeroen A. Van Bokhoven. 2017. "Catalyst Support Effects on Hydrogen Spillover." *Nature* 541 (7635). Nature Publishing Group: 68–71. doi:10.1038/nature20782.

Komanoya, Tasuku, Takashi Kinemura, Yusuke Kita, Keigo Kamata, and Michikazu Hara. 2017. "Electronic Effect of Ruthenium Nanoparticles on Efficient Reductive Amination of Carbonyl Compounds." *Journal of the American Chemical Society* 139 (33). American Chemical Society: 11493–99. doi:10.1021/jacs.7b04481.

Liu, Lichen, and Avelino Corma. 2018. "Metal Catalysts for Heterogeneous Catalysis: From Single Atoms to Nanoclusters and Nanoparticles." *Chemical Reviews* 118(10). American Chemical Society: 4981–5079. doi:10.1021/acs.chemrev.7b00776.

Liu, Yingxin, Kuo Zhou, Huimin Shu, Haiyan Liu, Jiongtao Lou, Dechao Guo, Zuojun Wei, and Xiaonian Li. 2017. "Switchable Synthesis of Furfurylamine and Tetrahydrofurfurylamine from Furfuryl Alcohol over RANEY® Nickel." *Catalysis Science and Technology* 7 (18). Royal Society of Chemistry: 4129–35. doi:10.1039/c7cy00981j.

Lopez, Gabriel, Dominik Keiner, Mahdi Fasihi, Tuomas Koiranen, and Christian Breyer. 2023. "From Fossil to Green Chemicals: Sustainable Pathways and New Carbon Feedstocks for the Global Chemical Industry." *Energy and Environmental Science* 16 (7). Royal Society of Chemistry: 2879–909. doi:10.1039/d3ee00478c.

Munnik, Peter, Petra E. De Jongh, and Krijn P. De Jong. 2015. "Recent Developments in the Synthesis of Supported Catalysts." *Chemical Reviews* 115 (14). American Chemical Society: 6687–718. doi:10.1021/cr500486u.

Nian, Yao, Xinyuan Huang, Menghui Liu, Jinli Zhang, and You Han. 2023. "Insight into the Dynamic Evolution of Supported Metal Catalysts by in situ/Operando Techniques and Theoretical Simulations." *ACS Catalysis* 13 (16). American Chemical Society: 11164–71. doi:10.1021/acscatal.3c02370.

Nishimura, Shun, Kunihiko Mizuhori, and Kohki Ebitani. 2016. "Reductive Amination of Furfural toward Furfurylamine with Aqueous Ammonia under Hydrogen over Ru-Supported Catalyst." *Research on Chemical Intermediates* 42 (1). Springer Netherlands: 19–30. doi:10.1007/s11164-015-2334-5.

Pal, Priyanka, Hu Li, and Shunmugavel Saravanamurugan. 2022. "Removal of Lignin and Silica from Rice Straw for Enhanced Accessibility of Holocellulose for the Production of High-Value Chemicals." *Bioresource Technology* 361 (October). Elsevier Ltd. doi:10.1016/j.biortech.2022.127661.

Pal, Priyanka, and Shunmugavel Saravanamurugan. 2020. "Heterostructured Manganese Catalysts for the Selective Oxidation of 5-Hydroxymethylfurfural to 2,5-Diformylfuran." *ChemCatChem* 12 (8). Wiley Blackwell: 2324–32. doi:10.1002/cctc.202000086.

Pan, Chun Jern, Meng Che Tsai, Wei Nien Su, John Rick, Nibret Gebeyehu Akalework, Abiye Kebede Agegnehu, Shou Yi Cheng, and Bing Joe Hwang. 2017. "Tuning/Exploiting Strong Metal-Support Interaction (SMSI) in Heterogeneous Catalysis." *Journal of the Taiwan Institute of Chemical Engineers* 74 (May). Taiwan Institute of Chemical Engineers: 154–86. doi:10.1016/j.jtice.2017.02.012.

Pan, Zhengyi, Qian Zhang, Wenquan Wang, Lulin Wang, and Guang Hui Wang. 2022. "Size-Tunable Carbon-Doped Ni Nanoparticles for Switchable Reductive Amination of Biomass-Derived Carbonyl Compounds to Primary Amines and Secondary Imines." *ACS Sustainable Chemistry and Engineering* 10 (11). American Chemical Society: 3777–86. doi:10.1021/acssuschemeng.2c00800.

Pirmoradi, Maryam, Nida Janulaitis, Robert J. Gulotty, and James R. Kastner. 2020. "Continuous Hydrogenation of Aqueous Furfural Using a Metal-Supported Activated Carbon Monolith." *ACS Omega* 5 (14). American Chemical Society: 7836–49. doi:10.1021/acsomega.9b04010.

Prins, Roel, and Mark E. Bussell. 2012. "Metal Phosphides: Preparation, Characterization and Catalytic Reactivity." *Catalysis Letters* 142: 1413–36. doi:10.1007/s10562-012-0929-7.

Qi, Haifeng, Fei Liu, Leilei Zhang, Lin Li, Yang Su, Jingyi Yang, Rui Hao, Aiqin Wang, and Tao Zhang. 2020. "Modulating: Trans -Imination and Hydrogenation towards the Highly Selective Production of Primary Diamines from Dialdehydes." *Green Chemistry* 22 (20). Royal Society of Chemistry: 6897–901. doi:10.1039/d0gc02280b.

Qi, Haifeng, Ji Yang, Fei Liu, Lei Lei Zhang, Jingyi Yang, Xiaoyan Liu, Lin Li, et al. 2021. "Highly Selective and Robust Single-Atom Catalyst Ru1/NC for Reductive Amination of Aldehydes/Ketones." *Nature Communications* 12 (1). Nature Research: 3295. doi:10.1038/s41467-021-23429-w.

Rangraz, Yalda, and Majid M. Heravi. 2021. "Recent Advances in Metal-Free Heteroatom-Doped Carbon Heterogonous Catalysts." *RSC Advances* 11 (38). Royal Society of Chemistry: 23725–78. doi:10.1039/d1ra03446d.

Saini, Kanika, Sahil Kumar, Ramandeep Kaur, Srinivasarao Arulananda Babu, and Shunmugavel Saravanamurugan. 2023. "Accelerated H2 Activation over Pt/M-ZrO2 for the Reductive Amination of Levulinic Acid Esters under Benign Conditions." *Catalysis Science and Technology* 13 (6). Royal Society of Chemistry: 1666–76. doi:10.1039/d2cy01550a.

Saini, Ms Kanika, Sahil Kumar, Hu Li, Srinivasarao Arulananda Babu, and Shunmugavel Saravanamurugan. 2022. "Advances in the Catalytic Reductive Amination of Furfural to Furfural Amine: The Momentous Role of Active Metal Sites." *ChemSusChem* 15(7). John Wiley and Sons Inc. doi:10.1002/cssc.202200107.

Saravanamurugan, Shunmugavel, Marta Paniagua, Juan A. Melero, and Anders Riisager. 2013. "Efficient Isomerization of Glucose to Fructose over Zeolites in Consecutive Reactions in Alcohol and Aqueous Media." *Journal of the American Chemical Society* 135 (14): 5246–49. doi:10.1021/ja400097f.

Savaresi, Annalisa. 2016. "The Paris Agreement: A New Beginning?" *Journal of Energy and Natural Resources Law* 34 (1). Taylor and Francis Ltd.: 16–26. doi:10.1080/02646811.2016.1133983.

Sheng, Min, Shu Fujita, Sho Yamaguchi, Jun Yamasaki, Kiyotaka Nakajima, Seiji Yamazoe, Tomoo Mizugaki, and Takato Mitsudome. 2021. "Single-Crystal Cobalt Phosphide Nanorods as a High-Performance Catalyst for Reductive Amination of Carbonyl Compounds." *JACS Au* 1 (4): 501–7. doi:10.1021/jacsau.1c00125.

Song, Wenjing, Yujie Wan, Yuefeng Li, Xin Luo, Weiping Fang, Quanxing Zheng, Pengfei Ma, Jianping Zhang, and Weikun Lai. 2022. "Electronic Ni-N Interaction Enhanced Reductive Amination on an N-Doped Porous Carbon Supported Ni Catalyst." *Catalysis Science and Technology* 106. Royal Society of Chemistry: 7208–18. doi:10.1039/d2cy01551j.

Song, Xiangbo, Jun Yue, Yuting Zhu, Chengyan Wen, Lungang Chen, Qiying Liu, Longlong Ma, and Chenguang Wang. 2021. "Efficient Conversion of Glucose to 5-Hydroxymethylfurfural over a Sn-Modified SAPO-34 Zeolite Catalyst." *Industrial and Engineering Chemistry Research* 60 (16). American Chemical Society: 5838–51. doi:10.1021/acs.iecr.1c01121.

Su, Ya Qiong, Guang Jie Xia, Yanyang Qin, Shujiang Ding, and Yang Gang Wang. 2021. "Lattice Oxygen Self-Spillover on Reducible Oxide Supported Metal Cluster: The Water-Gas Shift Reaction on Cu/CeO2catalyst." *Chemical Science* 12 (23). Royal Society of Chemistry: 8260–67. doi:10.1039/d1sc01201k.

Taseska, Teona, Wanqing Yu, Madeleine K. Wilsey, Connor P. Cox, Ziyi Meng, Soraya S. Ngarnim, and Astrid M. Müller. 2023. "Analysis of the Scale of Global Human Needs and Opportunities for Sustainable Catalytic Technologies." *Topics in Catalysis* 66 (5–8). Springer: 338–74. doi:10.1007/s11244-023-01799-3.

Wan, Yujie, Pengfei Ma, Hongliang Lu, Jianping Zhang, Jie Wang, Weiping Fang, Wenjing Song, Quanxing Zheng, and Weikun Lai. 2024. "External [1 1 1] Facets on Nanorod Gamma Alumina Boosts Catalytic Reductive Amination of Carbonyl Compound to Primary Amine." *Journal of Catalysis* 429 (January). Academic Press Inc. doi:10.1016/j.jcat.2023.115285.

Wang, Huanjiang, Yuchen Zhang, Dan Luo, He Wang, Yurong He, Fei Wang, and Xiaodong Wen. 2023. "Active Metal Dependent Side Reactions for the Reductive Amination of Furfural." *Molecular Catalysis* 536 (February). Elsevier B.V. doi:10.1016/j.mcat.2023.112914.

Wang, Zhiyuan, Yinjian Zheng, Jiye Feng, Wenbiao Zhang, and Qingsheng Gao. 2023. "Promoting Amination of Furfural to Furfurylamine by Metal-Support Interactions on Pd/MoO3−x Catalysts." *Chemistry – A European Journal* 29 (47). John Wiley and Sons Inc. doi:10.1002/chem.202300947.

Wei, Zuojun, Yuran Cheng, Kuo Zhou, Yue Zeng, En Yao, Qing Li, Yingxin Liu, and Yong Sun. 2021. "One-Step Reductive Amination of 5-Hydroxymethylfurfural into 2,5-Bis(Aminomethyl)Furan over Raney Ni." *ChemSusChem* 14 (11). John Wiley and Sons Inc: 2308–12. doi:10.1002/cssc.202100564.

Wu, Ping, Jianmei Li, Ting He, and Changwei Hu. 2019. "The Direct Conversion of Hemicelluloses to Selectively Produce Xylose from Corn Stover Catalysed by Maleic Acid." *Bioresource Technology* 14 (1). Elsevier Ltd: 816–41. doi:10.15376/biores.14.1.

Xie, Chao, Jinliang Song, Manli Hua, Yue Hu, Xin Huang, Haoran Wu, Guanying Yang, and Buxing Han. 2020. "Ambient-Temperature Synthesis of Primary Amines via Reductive Amination of Carbonyl Compounds." *ACS Catalysis* 10 (14). American Chemical Society: 7763–72. doi:10.1021/acscatal.0c01872.

Xu, Lujiang, Chenchen Shi, Zijian He, Huan Zhang, Mingying Chen, Zhen Fang, and Ying Zhang. 2020. "Recent Advances of Producing Biobased N-Containing Compounds via Thermo-Chemical Conversion with Ammonia Process." *Energy and Fuels* 34 (9). American Chemical Society: 10441–58. doi:10.1021/acs.energyfuels.0c01993.

Xu, Yanlin, Jiahao Zhang, Peixia Wang, Yiwei Fan, Yifan Jiang, Xiaonan Duan, Yingying Li, and Jisong Zhang. 2024. "Highly Efficient and Stable Ni-Based Catalyst for Selective Conversion of Biomass-Derived Aldehydes and Ketones to Primary Amines." *Chemical Engineering Journal* 480 (January). Elsevier B.V. doi:10.1016/j.cej.2023.148175.

Yang, Yinze, Liyan Zhang, Leilei Zhou, Haiyang Cheng, and Fengyu Zhao. 2023. "A Highly Efficient Ni/Al2O3-LaOx Catalyst for the Reductive Amination of Furfural to Furfurylamine: The Promoting Effect of La." *Chemical Research in Chinese Universities*, February 40 (1). Springer Science and Business Media Deutschland GmbH: 36–46. doi:10.1007/s40242-023-3216-9.

Yang, Yinze, Leilei Zhou, Xinchao Wang, Liyan Zhang, Haiyang Cheng, and Fengyu Zhao. 2023. "Catalytic Reductive Amination of Furfural to Furfurylamine on Robust Ultra-Small Ni Nanoparticles." *Nano Research* 16 (3). Tsinghua University: 3719–29. doi:10.1007/s12274-022-4923-0.

Yogita, None, K. T. Venkateshwara Rao, P. Mahesh Kumar, and N. Lingaiah. 2022. "Cobalt Nanoparticles Embedded in a Nitrogen-Doped Carbon Matrix for Reductive Amination of Biomass-Derived Furfural to Furfurylamine." *Sustainable Energy and Fuels* 6 (20). Royal Society of Chemistry: 4692–705. doi:10.1039/d2se00408a.

Yu, Liang, Dehui Deng, and Xinhe Bao. 2020. "Chain Mail for Catalysts." *Angewandte Chemie – International Edition* 59 (36). Wiley-VCH Verlag: 15294–7. doi:10.1002/anie.202007604.

Yuan, Hangkong, Jerry Peng Li, Fangzheng Su, Zhen Yan, Bright T. Kusema, Stéphane Streiff, Yongji Huang, Marc Pera-Titus, and Feng Shi. 2019. "Reductive Amination of Furanic Aldehydes in Aqueous Solution over Versatile NiyAlOx Catalysts." *ACS Omega* 4 (2). American Chemical Society: 2510–16. doi:10.1021/acsomega.8b03516.

Yuan, Ziliang, Bing Liu, Peng Zhou, Zehui Zhang, and Quan Chi. 2019. "Preparation of Nitrogen-Doped Carbon Supported Cobalt Catalysts and Its Application in the Reductive Amination." *Journal of Catalysis* 370 (February). Academic Press Inc.: 347–56. doi:10.1016/j.jcat.2019.01.004.

Zhang, Yangmin, Hanmin Yang, Quan Chi, and Zehui Zhang. 2019. "Nitrogen-Doped Carbon-Supported Nickel Nanoparticles: A Robust Catalyst to Bridge the Hydrogenation of Nitriles and the Reductive Amination of Carbonyl Compounds for the Synthesis of Primary Amines." *ChemSusChem* 12 (6). Wiley-VCH Verlag: 1246–55. doi:10.1002/cssc.201802459.

Zhou, Kuo, Bixian Chen, Xiaoting Zhou, Shimin Kang, Yongjun Xu, and Jinjia Wei. 2019. "Selective Synthesis of Furfurylamine by Reductive Amination of Furfural over Raney Cobalt." *ChemCatChem* 11 (22). Wiley Blackwell: 5562–69. doi:10.1002/cctc.201901269.

Zhuang, Xiuzheng, Jianguo Liu, Shurong Zhong, and Longlong Ma. 2022. "Selective Catalysis for the Reductive Amination of Furfural toward Furfurylamine by Graphene-Co-Shelled Cobalt Nanoparticles." *Green Chemistry* 24 (1). Royal Society of Chemistry: 271–84. doi:10.1039/d1gc03578a.

8 Catalytic Etherification of Furfurals to Furanic Ethers

Alkoxymethyl Furfurals

Xiaofang Liu, Zhuangzhuang Zhang, Zhengfei Pei, Pan Meng, and Lei Yu

8.1 INTRODUCTION

The main objective of this review is to compare the conversion yield of various feedstocks (e.g., CMF (5-chloromethylfurfural), HMF (5-hydroxymethylfurfural), fructose or glucose, carbohydrates or biomass) to EMF and the applied reaction conditions. Moreover, the type of acid catalysts (e.g., homogenous or heterogeneous) and their performance are discussed as another objective. Fossil fuels are the back bone of carbon-based economy. Fossil fuels and the chemicals derived from them have been produced from limited natural sources. Due to the ever-increasing demand for fossil fuels, their supply is often unstable. At the same time, the environment has been negatively affected by the carbon-based economy over the years. Thus, it is the consensus of the general opinion that fossil fuels, or part of them, should be replaced by alternative and sustainable energy sources (Hamid et al. 2015; Hamid et al. 2014). As one of the possible candidates, biomass is regarded as a very effective sustainable energy source. Indeed, biomass is the only sustainable source of organic compounds that has been proposed as the ideal equivalent to petroleum for the production of fuels and chemicals. Biomass is widely available as a proper feedstock, whose production estimates 1.0×10^{11} tons per year. Extensive research and studies have been conducted to produce biofuels and biodegradable products from biomass (Huber et al. 2006; Kamm et al. 2007).

Carbohydrates are the major source of carbon in biomass. Dehydration of carbohydrates results in furan formation, such as HMF (van Putten et al. 2013), the DOE's top 10 list of bio-based chemicals. There are many reports on HMF chemistry, synthesis, and applications available in the literature (van Putten et al. 2013; Jarosław et al. 2001; Rosatella et al. 2011; Kuster et al. 1990; Corma et al. 2007; Zakrzewska et al. 2010; Delidovich et al. 2014; Teong et al. 2014). HMF is an important platform molecule that connects carbohydrates to chemical and fuel precursors. HMF derivatives have various applications (van Putten et al. 2013) and some of them

DOI: 10.1201/9781003464839-8

have huge market potential. For instance, one of the HMF oxidation products, 2,5-furandicarboxylic acid (FDCA), is a viable replacement for terephthalic acid in the polyethylene terephthalate (PET) production and adipic acid production in the nylon industry (Patel et al. 2010; Buntara et al. 2011). In another case, hydrogenation of HMF results in 2,5-dimethylfuran (DMF) which is often used as a fuel additive (Román-Leshkov et al. 2007; Binder et al. 2009; Chidambaram et al. 2010). In addition, it is possible to produce drop-in fuel from hydro-deoxygenation of HMF condensation products (Avelino et al. 2012; Avelino et al. 2011; Huber et al. 2005; Shen et al. 2011; West et al. 2008; Yang et al. 2013).

The active molecule HMF reacts with alcohols in the presence of an acid catalyst and produces its corresponding ether (Moye et al. 1966). In other studies, HMF etherification products are reported when aliphatic alcohols are used as an extracting solvent during hexose sugars conversion to HMF in aqueous media (Tyrlik et al. 1999; Brown et al. 1982). A common etherification example is the HMF reaction with a simple alcohol such as methanol or ethanol that leads to the production of 5- (methoxymethyl)furfural (MMF) and EMF. Although alkoxymethyl furfurals have been known for a long time, and patented as a tobacco flavor in the early 1960s, they have remained largely unexplored until recently reported as diesel fuel additives (Gerardus et al. 2007). Among alkoxymethyl furfurals, EMF is recognized as an excellent additive for regular diesel with promising properties as follows:

- With a high boiling point (508 K) in comparison with diesel fuel.
- The high energy density (30.3 $MJ{\cdot}L^{-1}$) is similar to regular gasoline (31.9 $MJ{\cdot}L^{-1}$) and is compared to diesel fuel (33.6 $MJ{\cdot}L^{-1}$), which is notably higher than ethanol (21 $MJ{\cdot}L^{-1}$) (Murphy et al. 2004).
- Low toxicity. EMF can be used as a flavor and aroma ingredient in the food and beverage industries (Câmara et al. 2006; Vanderhaegen et al. 2003; Pérez-Coello et al. 2003; Silva et al. 2008; Cutzach et al. 1999).
- To blend EMF with diesel fuel in a diesel engine reducing the formation of particulate contamination, SO_2 emissions, and soot. Meanwhile, EMF blended fuel can make the engine run smoothly for hours (Hamid et al. 2015; Hamid et al. 2014; Gruter et al. 2014; Omidvarborna et al. 2016).
- Ethers including EMF have high cetane number (Murphy et al. 2004), which is a very important factor for combustion performance and emission.
- No hydrogenation step is required in EMF production, which is an advantage over other fuel additives (e.g., DMF) obtained from HMF production.

Due to these significant properties of EMF, its syntheses from renewable sources have been attracted by many researchers and commercial companies. Avantium company also filed a patent in 5-(alkoxymethyl)furfural production from either glucose or fructose (Gruter et al. 2010). The patent claimed the simultaneous production technology for 5-(alkoxymethyl)furfural and furfural from C5 and C6 sugars in the presence of acid catalysts. In the other patents (Gruter et al. 2010; Gruter et al. 2008; Gruter et al. 2012), they claimed ether production technologies from the etherification of branched C3-C20 alcohols with HMF. They proposed another pathway for HMF ether production through hydroxyl-alkoxy addition. In this regard, they conducted an HMF reaction with an olefin using an acid catalyst (Gruter et al. 2010).

Considering the fact that EMF plays an important role as a fuel candidate, diesel fuel additive (X. Liu et al. 2016) FDCA, or cyclopentenone precursor (Janka et al. 2012, Bredihhin et al. 2016), it is vital to review their different production aspects with detailed attention. Thus, recent developments in EMF synthesis are summarized in this review.

8.2 SYNTHESIS OF EMF

For the production of EMF, researchers pay much attention to HMF etherification with ethanol in the presence of acidic catalysts. Fructose or glucose also acted as an initiator feed for cascade conversion with HMF occurring in an acidic environment followed by HMF etherification reaction with ethanol. This "one-pot" process can convert disaccharides, polysaccharides, and biomass to EMF through intermediate HMF, which is economically attractive and easy to operate. However, the synthesized EMF yield decreases obviously when HMF or fructose has not been used as the feedstocks (Scheme 8.1).

Herein we reviewed and compared different acidic catalysts and their catalytic performance on the synthesis of EMF from multiple initial substrates. Optimum reaction conditions for the case were provided.

8.2.1 Nucleophilic Substitution of 5-Chloromethylfurfural

Synthesis and applications of CMF have been explored for a long time and can be translated to EMF following the Williamson reaction (Climent et al. 2014). In addition, Mascal and Nikitin (Mascal et al. 2008) published high yields of excess of >70% CMF derived from glucose, sucrose, and cellulose. The introduced system by Mascal and Niktin (Mascal et al. 2008) was a simple biphasic reactor where 1,

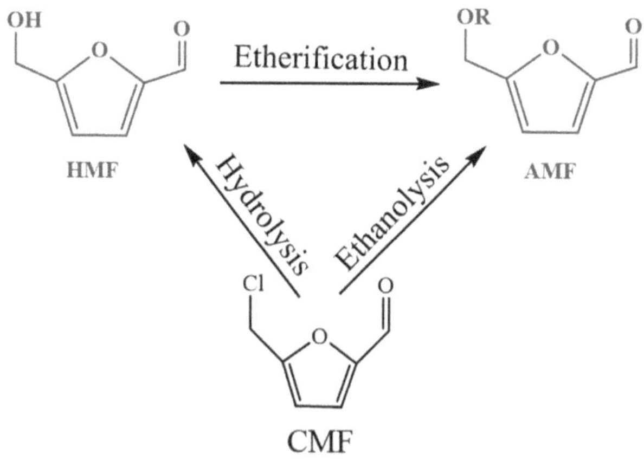

SCHEME 8.1 The scheme of the etherification reaction of HMF.

2-dichloroethane was chosen as the organic solvent, and the aqueous phase was highly concentrated with HCl solution containing LiCl. After CMF was separated, it was reacted with ethanol providing high up to 95% EMF yield at room temperature within 8 h. Bredihhin et al. (2013) also conducted halomethyl furfural conversion to EMF.

Viil et al. (2014) reported a high and clean EMF yield of up to 88% with 96% aqueous ethanol, which is obtained without byproducts such as EL benefited from the introduction of oil shale ash to the reaction medium. Notably, the study described a procedure suitable for large-scale synthesis for the cycle solvent ethanol and pure production EMF without further purification. It is important to note that the introduction of oil shale ash suppresses the hydrolysis of the furan ring and the research lays the foundation for the synthesis of biofuels on a large scale. Due to the promising result of CMF conversion to EMF, Mascal, and Nikitin (2009) also studied filter paper, cotton, wood, straw, newsprint, and corn stover as feedstocks for CMF production (Mascal et al. 2009).

Despite the high yields carried out for the EMF synthesis from halomethyl furfural, there exist some points that prevent large-scale production of EMF by this progress. Because of the high concentration of HCl, hazardous chlorinated organic products may potentially form (e.g., 1,2-dichloroethane) and the recycle is a problematic issue for sustainable development (Climent et al. 2014).

8.2.2 Etherification of 5-Hydroxymethylfurfural

EMF synthesis from HMF etherification has been investigated extensively in order to achieve the suitable catalysts and establish optimum reaction conditions.

8.2.2.1 Homogenous Mineral Salts

In the past several decades, investigators focused on homogenous catalysts because of the easily accessible substrates and the reactions carried out faster. However, their isolation and reuse require a huge amount of energy and do not meet the green and sustainable development. On the other hand, the target chemical catalyzed by heterogeneous catalysts purification is simple, and in addition, they cause lower corrosion for the environment and equipment than homogenous catalysts. Therefore, many researchers pay attention to appropriate solid acid catalysts and compare their performances and activities with common soluble acid catalysts (Balakrishnan et al. 2012; Lanzafame et al. 2011; Che et al. 2012). Furthermore, reaction temperature plays a key role in carbohydrate conversion and EMF preparation. The major byproduct that exists in the etherificationprogress is EL, in addition to 5,5′(oxy-bis(methylene))bis-2-furfural (OBMF) (Liu et al. 2013a) and 5-(ethoxymethyl)-furfural diethylacetal (EMFDEA) (Balakrishnan et al. 2012).

Initially, Balakrishnan et al. (2012) conducted HMF etherification with ethanol catalyzed by H_2SO_4, organic acid, and different acidic catalysts. The obtained highest EMF yield was 81% when H_2SO_4 was applied as the catalyst. Their findings also showed that resins with sulfonic acid group catalyzed HMF to EMF at promising yields. The main byproduct EMFDEA was detected in the reaction system at a high percentage compared to EL. High yields of diethylacetal in these reaction conditions could be attributed to the reactivity of HMF's aldehyde.

$AlCl_3$ is a commercially available Lewis acid with low toxicity and proved a high catalytic activity in the dehydration of hexoses. The highest EMF yield through single pot reaction from HMF was testified 92.9% under 100 °C for 5 h in Liu et al. (2013a). Alam et al. (2012) catalyzed HMF etherification by ILs in the presence of Brønsted acidic catalysts, which were synthesized from N,N-dimethylacetamide (DMA), N-methyl-2-pyrrolidinone (NMP) and methane sulfonate ($[NMP]^+[CH_3SO_3]^-$ or $[DMA]^+[CH_3SO_3]^-$). Starting with 0.3 g of HMF, 0.28 g, and 0.31 g EMF could obtained under the reaction condition at 120 °C, and 2 h, respectively. This work also afforded an extended conversion of weed biomass including foxtail and red nut sedge weed.

8.2.2.2 Heterogeneous Catalysts

Zeolites Act as Catalyst

Zeolite has a large surface area and strong acid sites, catalytic activity for HMF etherification is not desirable, the low activity of which can be attributed to the diffusion limitations of HMF and EMF because of their molecular sizes and small pore openings of zeolites (Alam et al. 2012; Saravanamurugan et al. 2012).

Lanzafame et al. (2011) reported the etherification of HMF catalyzed by the mesoporous silica catalysts, including Al-MCM-41 (with different Si/Al ratios), zirconia or sulfated zirconia supported on SBA-15. The data results illustrated that the presence of strong Lewis acid sites, such as ZrO_2 and extra-framework isolated Al^{3+} sites, are in favor of EMF production. Meanwhile, ethyl 4-oxopentanoate formation is observed at high yield when strong Brønsted acid sites exist. In addition, Lanzafame et al. (2011) utilized the catalytic activities of prepared series catalysts to compare to the two classical strong Brønsted acids, namely H_2SO_4 and Amberlyst-15, confirming the effective and influences of materials for the reaction progress.

Based on the perfect structure characteristic of MCM-41, Che et al. (2012) developed a nanosphere catalyst with highly dispersed ($H_4SiW_{12}O_{40}$/MCM-41) for the etherification of the hydroxyl group of HMF. Results exhibited that the $H_4SiW_{12}O_{40}$/MCM-41 had superior catalytic performance, and both the heteropolyacid activity and selectivity were better than those classic acidic catalysts of *p*-TSA, H_2SO_4, Amberlyst-15, and H_3PO_4. The newly $H_4SiW_{12}O_{40}$/MCM-41 exhibited 84.1% selectivity to EMF with 92.0% conversion of HMF reacting in ethanol under mild conditions. This work also proved that 5,5'(oxybis(methylene))bis-2-furfural as an intermediate can be further transformed into the main product EMF. The strength of acids and the effects of heteropoly anion determined the effective catalytic performance of the nanosphere $H_4SiW_{12}O_{40}$/MCM-41. Furthermore, Liu et al. (2014) developed a solid acid catalyst by immobilizing different dosages of phosphotungstic acid (HPW) on MCM-41. The optimum 40 wt% MCM-41-HPW exhibited that a high EMF yield of 83.4% derived from HMF was observed under the relative reaction condition 100 °C, 12 h. The new method for the production of EMF indicated that MCM-41-HPW has stable acidic sites for the EMF preparation with reusable activity and has potential utilization in multiple other important acid-catalyzed progresses for biomass conversion.

To gain insights into the relation between the zeolites and the etherification of HMF, Lewis et al. (2014) have explored a series of zeolites including Hf-, Zr-, Ti-, Ta-, Nb- and Sn-Beta promote the etherification reaction for the production of EMF with ethanol. The catalytic performance of these Lewis acids has been conducted under batch conditions, and the results data indicated that Ta-Beta and Al-Beta showed comparable selectivity for EMF preparation with 56% and 41% yields, respectively.

Recently, Lanzafame et al. (2015) highlighted the effect of the number of Lewis and Brønsted acid sites on EMF synthesis from HMF in the presence of silicalite-1 catalysts. Besides, the location of ZrO_2 in a sulfated ZrO_2 catalyst with SBA-15 support is an important parameter, and it should be investigated in detail (Barbera et al. 2015).

On the other hand, Barbera et al. (2016) introduced NH_4-exchanged zeolites for HMF etherification with ethanol. In order to prevent secondary reactions catalyzed by strong acid sites of zeolites, NH_4 was used to block the strong sites. Thus, the product selectivity increased. Here, Zeolite Socony Mobil-5 (ZSM-5) with two different frameworks, MFI and BEA, were used to prepare NH_4-BEA and NH_4-MFI catalysts. The catalytic performance of the zeolites in the corresponding NH^{4+} and H^+ forms are compared with the other catalytic reactions. The results indicated that the presence of NH^{4+} increases the product selectivity for BEA catalyst. However, NH_4-MFI cannot catalyze etherification reaction effectively. Consequently, the HMF conversion yield was 60% and EMF and EL space-time yields were lower than the case where the original zeolite was used.

Mesoporous aluminosilicates of the type Al-TUD-1 were also proposed by Neves et al. (2013) as a solid acid catalyst in EMF production from HMF. The main advantages of TUD-1 are tunable pore sizes and a 3D channel structure that prevents internal diffusion. It is also possible to produce them cost-effectively and to improve their acidic properties by introducing aluminum into the framework (Neves et al. 2013). Al-TUD-1 (Si/Al ratio: 21) showed promising results for HMF etherification. Decreasing the Si/Al ratio in Al-TUD-1 to 4 resulted in slightly low EMF and EL yields. This difference can be explained by considering the high acidic strength of Al-TUD-1 with the Si/Al ratio of 21.

Heteropolyacid-Based Catalysts

Heteropolyacid as a soluble acid catalyst, has good performance, its separation and recycling is not an easy task. Thus, it is beneficial to transform it into a heterogeneous acid catalyst. Grafting of it on a solid support is a method that has been used by different groups.

Considering the unique magnetic separation property of Fe_3O_4, Wang et al. (2013) prepared catalyst $Fe_3O_4@SiO_2$-HPW by supporting phosphotungstic acid (HPW) on silica-coated Fe_3O_4 nanoparticles ($Fe_3O_4@SiO_2$) and introduced for EMF production. The obtained yield was high up to 83.6% derived from the etherification of

HMF under the optimized reaction condition with relatively the same yield as Liu and Zhang (2013).

Organic–inorganic hybrid materials that act as catalysts have attracted much attention. For instance, an HPA-based IL hybrid catalyst, methylimidazolebutylsulfate phosphotungstate ($[MIMBS]_3PW_{12}O_{40}$) has been reported by Liu et al. (2012) as a catalyst or additive for the production of EMF derived from HMF. The results demonstrated that the IL combination with $H_3PW_{12}O_{40}$ not only enhanced the EMF yields but also provided a simple separation of the acidic catalyst. A high EMF yield of 90.7% with 98.1% conversion of HMF was achieved under the optimum reaction condition at 70 °C for 24 h.

Wang et al. (2017) by controlling the molar ratio of phosphotungstic acid (HPW) and pyridine (PY) or trimethylamine (TEA) self-assembly in ethanol during the self-assembly process, the uniform nanospheres hybrids with various acid capacities in different nano sizes were prepared without using template and surfactant. Particularly, for the etherification reaction of HMF to EMF, PY-PW-1 exhibited an excellent catalytic activity with a high yield of 90% at 80 °C for 24 h, which might be due to the relatively higher acidity and uniform particle size. The kinetic study further proved the superior performance of PY-PW-1 in HMF-to-EMF conversion.

Immobilizing soluble acids such as Keggin-type heteropolyacid $H_3PW_{12}O_{40}$ (HPW) with strong Brønsted acidity on an insoluble and high cation exchange capacity inorganic support K-10 clay is an effective way to prepare the heterogeneous solid catalysts. The synthesized K-10 clay-HPW can prevent the HPW dissolve in polar organic solvents and express high activity. Liu et al. (2014) utilized 30 wt% of K-10 clay-HPW for HMF etherification converted to EMF. The catalytic performance was excellent and the EMF yield was high up to 91.5% under the optimized reaction conditions (100 °C, 10 h).

Recently, magnetic nanoparticles (MNPs) received great attention as catalyst support because of their stability and efficient separation by magnetic force that prevents catalyst loss (Reddy et al. 2012). It was also possible to use a silica layer to modify MNPs and to prepare an inert surface. For instance, $Fe_3O_4@SiO_2$ could not catalyze the reaction of HMF with ethanol, and therefore, the production of EMF was negligible. However, after the addition of heteropolyacid, the activity of $Fe_3O_4@SiO_2$-HPW increased dramatically and the EMF yield reached 83.6%.

They reported 89.3% yields for EMF in the presence of this solid acid catalyst. Ren et al. (2015) applied a completely different method in this regard. They exchanged the H^+ ion of HPW with Ag^+ and prepared Ag_1H_2PW, which was insoluble in ethanol. The result of HMF conversion to EMF with it presented a high activity of the catalyst and EMF yield, 88.7%. It appears to be possible to replace another H^+ in the Ag_1H_2PW, but, the acid strength of the catalyst would be decreased, and as a result, the EMF yield may be reduced. In order to provide an insoluble acid catalyst from heteropolyacids (HPAs), Raveendra et al. (2016) prepared a cesium-exchanged silicotungstic acid (STA) catalyst. After establishing an optimum content of Cs in STA catalysts, reaction parameters, such as reaction time and temperature were investigated. Their results indicated that the EMF production yield was 91% in the presence of the Cs_2STA.

Recently, Wang and Chen (2016) examined the heterogeneous and recyclable acid catalyst MOF-based polyoxometalate [Cu-BTC][HPM] (NENU-5), which encapsulates the homogeneous phosphomolybdic acid hydrate (HPM) into the insolubility, large surface area and hierarchical pores of Cu-BTC (BTC= benzene-1,3,5-tricarboxylate). Optimized performance of NENU-5 in the reaction conditions, the etherification of HMF to EMF obtained high up to 68.4% yield under normal pressure conditions. The recycling experiments demonstrated the reusability of the catalyst and catalytic activity retained with 55% yield of EMF. Benefiting from the unique structure, MOFs can provide adsorption sites for HPM to avoid HPM dissolving in solvents, making the catalysts recyclable.

Sulfonic Acid Groups' Functionalized Catalyst

Over the last decades, sulfonic acid (-SO_3H) was very frequently selected as a functional acidic group for its efficient proton acid catalytic ability and to be easily introduced into many materials.

Zhang et al. (2014) immobilized of acidic functional group sulfonic acid (-SO_3H) on Fe_3O_4@SiO_2, giving a high up to 89.3% yield of EMF derived from HMF at 100 °C within 10 h. Yuan et al. (2015) reported a feasible method to prepare a new magnetic solid acid catalyst with magnetic nanoparticles (MNPs) as the catalyst support and sulfonic acid groups (-SO_3H) acting as the acid functional group for EMF production. The constructed acidic catalyst Fe_3O_4@C-SO_3H was encapsulated by amorphous carbon in order to avoid the aggregation of the cores. The heterogeneous acid catalyst was successfully used to convert HMF to EMF at a high yield of 88.4%, which is the etherification of HMF with ethanol.

With a wheat straw biomass as the raw material, Yao et al. (2016) prepared a novel magnetically recoverable carbonaceous heterogeneous catalyst Fe_3O_4@C-SO_3H, which had -COOH, -SO_3H, and -OH groups, and tested the production of EMF from fructose-based carbohydrates or HMF. The carbonaceous–SO_3H-functionalized catalyst proved to have relatively high catalytic activity for the conversion of fructose in one pot and direct conversion of HMF with 64.2% and 85.6%. The remarkable catalytic activity of Fe_3O_4@C-SO_3H is attributed to the fact that the binary solvent assembles DMSO and ethanol and the various functionalized groups. The prepared catalyst could be easily reused by exposure to the external magnetic field, and the recycled catalyst kept its catalytic activity without the loss of target product yield. The new catalytic system has exhibited relatively high catalytic activity for the production of EMF from a renewable biomass resource.

Yin et al. (2015) reported a magnetic catalyst Fe_3O_4@SiO_2-SH-Im-HSO_4, which is the modified silica-coated Fe_3O_4 magnetic material grafted on acid polyionic liquids. The prepared Fe_3O_4@SiO_2-SH-Im-HSO_4 proved superior activity for HMF etherification into EMF up to 89.6% yield at 100 °C. The Fe_3O_4@SiO_2-SH-Im-HSO_4 catalyst exhibited high stability in the reuse experiments without loss of its catalytic activity and can be easily collected from the reaction medium by the external magnet.

Grafting soluble acids on the inorganic insoluble support is a technique to prepare high-performance heterogeneous solid acidic catalysts. Liu and Zhang (2013) chose propyl sulfonic acid as the functional group and mesoporous silica as an inorganic insoluble support to construct a new catalyst. The prepared silica-SO_3H was utilized to convert HMF for the production of EMF in 83.8% yield under the optimized reaction conditions (100 °C, 10 h), which demonstrated the effective catalytic activity for the etherification of HMF.

Carbon-based materials have become green and sustainable catalysts due to their unique chemical properties (metal-free, stable, and recyclable), high mechanical resistance, and their propitiousness charge generation and transformations in catalytic processes in recent years. GO as an emergent new type of promising catalytic material, its unique properties (2D structure, high stability, and high surface areas) and the engineered catalytic performance have attracted much attention. With the Hummers oxidation of graphite, Wang et al. (2013) prepared graphene oxide (GO) and used it for the conversion of HMF into EMF production. A relatively high yield of EMF (92%) was shown when the GO catalyst was used in ethanol. When comparing the catalytic activity of GO with other available acid catalysts, GO showed superior catalytic activity in similar reaction conditions. Antunes et al. (2014) modified GO by partially reducing GO with sulfuric acid (S-RGO), which is a highly active, selective, and stable catalyst for the preparation of EMF converted from HMF. Amorphous carbon black (CB) and carbon nanotubes (CNTs) for EMF synthesized from HMF were explored at the same time. The catalytic activity of S-RGO proved that S-RGO was superior to S-CB, S-CNTs, and the classic commercial sulfonic acid resin Amberlyst-15, which can be attributed to the higher sulfonic acid group and stronger acidity.

Besides, sulfonation of natural biopolymers has emerged as another promising solid acid catalyst for the conversion of biomass into high-value-added. Liu et al. (2013b) developed the cellulose sulfuric acid for the etherification of HMF for the synthesis of EMF and observed a high yield (84.4%) under optimal reaction conditions. Thombal et al. (2016) studied the application of glucose-derived magnetic solid acid, glu-Fe_3O_4-SO_3H, as the HMF etherification catalyst. During the optimum catalyst loading, the isolated yield of EMF was 92%.

Etherification of HMF with ethanol over acidic Zr-montmorillonite (Zr-Mont) catalyst for a mixture production including EMF and 2-(diethoxymethyl)- 5-(ethoxy-methyl)furan reported by Shined and Rode (2017). The latter was selectively translated into EMF by conducting in water catalyzed by the same Zr-Mont. Under the reaction conditions, etherification of HMF with ethanol the target ethers EMF were produced in the highest yields of 91%. The catalyst Zr-Mont was prepared by simply mixing montmorillonite with aqueous $ZrOCl_2{\cdot}8H_2O$ and shows various properties consisting of Brønsted/Lewis acidity, high thermal and water stability. The introduction of Zr-Mont has the potential to convert multiple biomass-derived and petroleum-based substrates into high-value-added products.

In summary, solid acid catalysts are used in order to benefit from the ease of separation and economies of scale despite the reported high EMF yields with soluble acid catalysts. It is worth noting that HMF is not affordable as a feed for EMF production regardless of the high yields reported for HMF etherification to EMF by different groups. The techno-economic analysis carried out by Torres et al. (2010) estimated the HMF price of 2.16 $/kg based on the process proposed by Roman-Leshkov et al.

(2006). Due to this economical hurdle, other feedstocks such as fructose, glucose, polysaccharides, and biomass have been studied for EMF production as discussed in detail below.

8.2.3 Hexose Conversion to EMF

As mentioned earlier, various acid catalysts have been reported to convert HMF to EMF at high yields. Fructose dehydration to HMF is also possible with acid catalysts, thus, many research groups have studied the one-pot conversion of fructose to EMF to avoid extra costs for isolation and purification of HMF, and to take benefits from the low price of the feed. In a series of reactions, fructose dehydration to HMF is followed by HMF etherification with ethanol. EMF also participates in the rehydration reaction and produces EL.

8.2.3.1 Fructose Conversion to EMF

The most classic method to synthesize EMF is the utilization of fructose as the raw material. The involved reaction mechanism is shown in Scheme 8.2.

Homogeneous Mineral Salts

Zhou et al. (2014) investigated the catalytic activity of a series of metal chlorides such as $FeCl_3$, $CrCl_2$, $GeCl_4$, $IrCl_3$ and so on for the production of EMF. The presented toxicity, high price and instability drawbacks of other metal chlorides affect the further

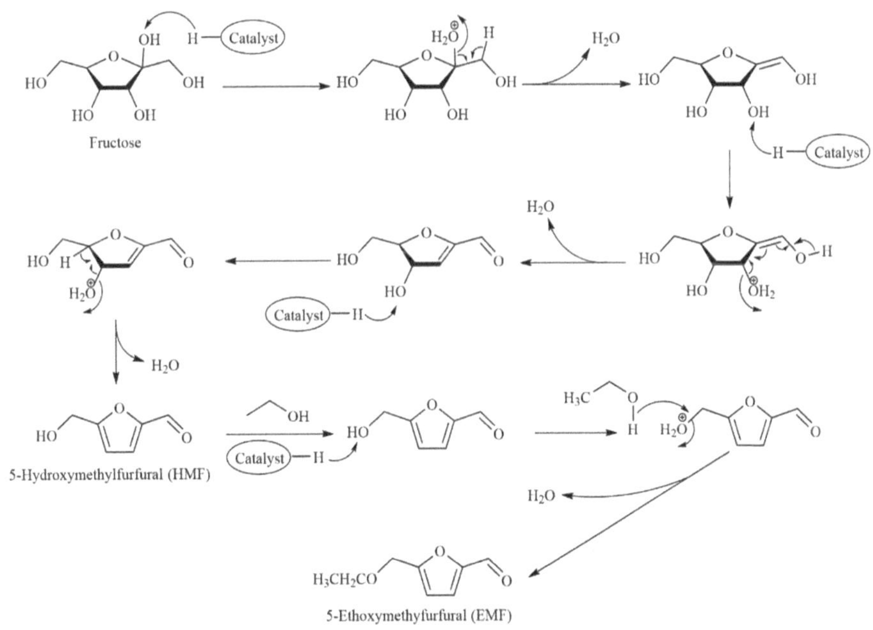

SCHEME 8.2 The reaction mechanism involved in the conversion of fructose to EMF.

explore. Catalyzed by $FeCl_3$, the maximum yield of EMF was 30.1% in a mixture solvent composed of [Bmim]Cl and ethanol.

Liu et al. (2012) examined various inorganic salts to promote EMF production from fructose. Among the multiple inorganic salts, NH_4Br, $CuCl_2·4H_2O$, and $NiCl_2·6H_2O$ can catalyze fructose converted to EMF with low yields. NH_4Cl demonstrated superior catalytic activity compared to others, 42% total yield of HMF and EMF were observed from fructose under optimum reaction conditions.

Considering Lewis acid is an efficient catalyst for the dehydration of carbohydrates, Jia et al. (2013) explored combinations of $AlCl_3$ with different co-catalysts such as $B(OH)_3$ and $BF_3·(Et)_2O$ or halide salts such as NaF, NaCl, and NaBr in order to prepare EMF from fructose in high productivity. Among various Lewis acids such as $SnCl_4·5H_2O$, $FeCl_3·6H_2O$, and $AlCl_3·6H_2O$, $AlCl_3·6H_2O$ showed the best Lewis acid catalytic performance. The efficacies of $BF_3·(Et)_2O/AlCl_3·6H_2O$ turned out to be significantly superior to other catalyst combinations, 55.0%, 45.4%, and 23.9% high yields of EMF derived from fructose, inulin, and sucrose were achieved, respectively, under the optimized conditions.

Heteropolyacid-Based Catalysts

Fructose acted as raw material, Yang et al. (2012) studied the one-pot dehydration-etherification in a mixture solvent composed of ethanol/THF (tetrahydrofuran) catalyzed by $H_3PW_{12}O_{40}$ (HPW). The introduction of microwave provided a high yield of 76% for EMF under the optimum reaction conditions. It was worth noting that THF as a co-solvent significantly enhanced fructose conversion, meanwhile, the EMF yield increased up to 76% from 65% by adding THF.

Wang et al. (2013) investigated various acid catalysts (HPW, phosphomolybdic acid (HPM), $AlCl_3$, H_3PO_4, and Amberlyst-15) in EMF synthesis from fructose in an ethanol/DMSO mixture. In the comparison of the catalytic effects, HPW and HPM exhibited superior catalytic performance over $AlCl_3$, H_3PO_4, and Amberlyst-15. Due to the stronger Brønsted acidity and the important heteropolyanion activity sites for the EMF selectivity, HPW catalyzed for the EMF high yield with 64% derived from fructose in DMSO and ethanol mixture within 130 min at 140 °C.

Liu et al. (2014) reported moderate 42.9% yields of EMF in the presence of 40 wt% MCM-41-HPW catalyst, which realized the one-pot conversion of fructose directly into EMF. The developed new catalyst MCM-41-HPW was obtained by using mesoporous MCM-41 supported on 12-tungstophosphoric acid and exhibited stable catalytic activity without the loss of acidic sites after using several cycles.

As mentioned before, $Fe_3O_4@SiO_2$ was used as a support for heterogeneous acid catalysts. Recently, Wang et al. (2013) reported $Fe_3O_4@SiO_2$-HPW as an acidic catalyst for the preparation of EMF converted from fructose by one-pot progress. The achieved EMF yield was 54.8%.

Liu et al. (2014) studied the application of 30 wt% K-10 clay-HPW for one-pot fructose conversion to EMF. With the optimized reaction conditions, they obtained 61.5% yield of EMF at 100 °C for 10 h. The achieved solid catalyst was highly stable and can recycled several times without obvious loss of catalytic activity.

With the prepared nanosphere PY-PW-1, Wang et al. (2017) concluded that fructose could also act as a substrate to be converted to EMF in a one-pot reaction process, obtaining a moderate yield of 55% with the enhanced temperature at 120 °C. Anymore, PY-PW-1 was stable enough for reused five times without evident catalytic activity loss. The published reports proved that mixture solvents, the introduction of THF or DMSO, can prevent the the formation of byproducts. Hereby, ethanol mixed with DMI (1,3-dimethyl-2-imidazolidinone) or GVL (γ-valerolactone) (7/3, v/v) was explored as co-solvent, while there was no increase of yields of EMF and downstream product EL. However, the addition of DMSO into the reaction medium enhanced the yields of EMF and EL with 58% and 28% compared to 55% and 20%, respectively. Notably, the activation energy (Ea) of PY-PW-1 and TEA-PW-1 were studied and calculated by kinetic research. The former was found to be two-fold higher than the latter. Except for the additive of DMSO, Xu et al. (2017) explored the ethanol/n-hexane solvent medium, and the results proved that the introduction of n-hexane can enhance the yield of EMF. The optimum volume ratio of ethanol to n-hexane was 6:4, a much higher EMF yield (66.3%) can be accumulated from fructose at 120 °C and 180 min. The yield ratio of EMF and EL is close to 4:1 the total yield of which can be up to 83.7%, increased by 5.7% when compared to the pure ethanol solvent.

With the same ethanol-DMSO mixture reaction system, Li et al. (2014) also developed a series of acid-base bifunctional hybrid nanosphere catalysts, including $H_4PW_{12}O_{40}$ and amino acids (e.g., lysine, arginine, or histidine). The study suggested the addition of base catalysts plays an important role in increasing EMF stability. The catalyst Lys/PW possessed optimum reaction activity, giving a high EMF yield of 76.6% with fructose as the raw material.

Ren et al. (2015) used Ag_1H_2PW as a solid acid catalyst for simultaneous fructose dehydration and HMF etherification. They reported a relatively high yield of EMF, 69.5%.

Sulfonic Acid Groups' Functionalized Catalyst

Amberlyst-15 as one of the classic solid strong acids has been applied in a range of reactions. The catalytic performance of the Amberlyst-15 was further explored by Zhu et al. (2011) for the production of HMF and then converted into 5-methoxymethylfurfural (MMF) in a low boiling point solvent with excellent activity. The mixture of fructose with THF, which enhanced the contact of active sites derived from Amberlyst-15 and reaction substrates promoted the catalytic progress and increased the selectivity of target molecules.

Mesoporous silica-supported sulfonic acid heterogeneous catalyst (silica-SO_3H) was also an effective solid acid catalyst used by Liu and Zhang (2013). The relatively high yield of EMF, 63.1% and 60.7%, respectively, were synthesized when this catalyst was reported by a one-pot reaction for the conversion of fructose and inulin. However, the reported silica-SO_3H showed no activity for the preparation of EMF from aldose-based carbohydrates such as glucose. It can effectively catalyze the glycosylation of glucose into ethyl D-glucopyranoside with a high yield of 91.7%. Morales et al. (2017) studied the conversion of fructose to EMF in the presence of arenesulfonic acid-modified SBA-15 mesostructured silica (Ar-SO_3H-SBA-15) in the ethanol/DMSO mixture. In order to optimize the reaction conditions (e.g., temperature, catalyst loading, and DMSO concentration), the response surface methodology was implemented. The maximum EMF yield was 63.4%.

By sulfonating Fe_3O_4@C, Yuan et al. (2015) reported that the prepared heterogeneous acidic catalyst was also active for the one-pot conversion of fructose into EMF with a relatively high yield up to 67.8%. Meanwhile, the magnetic Fe_3O_4@C-SO_3H could be easily recycled from the reaction medium by a permanent magnet and reused for several cycles without the catalysis activity loss. Similarly, Yao et al. (2016) prepared the same magnetic Fe_3O_4@C-SO_3H derived from wheat straw biomass for the comparable yield of EMF (64.2%). The only difference was that the EMF yield as high as 64.2% was reacted in the DMSO-ethanol binary solvent.

Inspired by the mentioned before, Wang et al. (2017) choose the promising carbon nanomaterials as the supporter to be functionalized, and the sulfonic acid-functionalized ordered mesoporous carbon (OMC) was obtained. With the novel catalyst OMC-SO_3H, a high EMF yield of 55.7% was afforded by the one-pot transformation of fructose.

Thombal et al. (2016) further adjusted the glu-Fe_3O_4-SO_3H loading rate to 50 wt% in order to synthesize EMF from fructose effectively compared to HMF, and the isolated EMF yield was reported to be 81%. The required catalyst for an optimum conversion observed in this result indicated that a higher amount of glu-Fe_3O_4-SO_3H is needed for efficient one-pot fructose conversion to EMF compared to HMF etherification to EMF.

Wang et al. (2013) proved that EMF yield was reduced when fructose instead of HMF acted as the starting material, which can be attributed to the existence of dehydrated intermediates and side reactions (e.g., polymerization of the intermediates, alcoholysis of HMF). Based on these, the author explored a mixture of solvents with a volume ratio of ethanol and DMSO in 3:7. The introduction of DMSO enhanced furan selectivity and obtained a high EMF yield of 71%.

THF has also been used as a co-solvent for one-pot fructose conversion to EMF. X. Liu et al. (2016) prepared sulfonic acid functionalized porous coordination polymer, MIL-101-SO_3H(x), and MIL-101-SO_3H(100) showed a better catalytic performance than the others. When using MIL-101-SO_3H(100) in the ethanol and THF mixture, the conversion of fructose to EMF and the EMF yield was reported to be 89.2% and 67.7%, respectively.

Li et al. (2014) synthesized a series of SO_3H-functionalized polymer solid acid catalysts for successive fructose dehydration and HMF etherification. With the optimum catalyst poly(VMPS)-PW, the desired EMF yield of 72.8% was achieved under the optimized reaction conditions at 110 °C within 10 h. Yin et al. (2015) showed that Fe_3O_4@SiO_2-SH-Im-HSO_4 catalyzed fructose dehydration and *in situ* HMF etherification at a relatively high yield of 60.4%.

The application of natural biopolymers including cellulose and chitosan act as catalyst support has attracted much attention. Liu et al. (2013b) published the cellulose sulfuric acid served as a promising heterogeneous acid catalyst for acid-catalyzed conversion. The renewable cellulose sulfuric acid catalyst utilized for the production of liquid fuel EMF from fructose has not been previously proved. The results' data demonstrated that EMF obtained at a high yield of 72.5% derived from fructose through the one-pot reaction pattern.

Ionic Liquid Catalysts

Kraus and Guney (2012) uniquely designed the tailored ionic liquid for EMF production from fructose, the imidazolium-based IL was a task-specific sulfonic acid-functionalized IL for the conversion of fructose in a one-pot procedure without the addition of solvent or acid catalysts. The introduction of 1-butyl-3-(3-sulfopropyl)-imidazolium chloride (4) and 1-methyl-3-(3-sulfopropyl)-imidazolium chloride (5) for EMF given the comparable isolated yields with 55% and 54% under the mild condition (100 °C, 80 min). It is noted that the novel biphasic medium comprised of hexanes and IL enhanced the EMF yields and showed a small amount of EL production.

The research of De et al. (2012) proved that Brønsted acidic IL $[DMA]^+[CH_3SO_3]^-$ was efficient for fructose conversion to EMF with a relatively high yield of 64% in ethanol at 120 °C for 16 h.

With the HPA-based IL hybrid catalyst $[MIMPS]_3PW_{12}O_{40}$, Liu et al. (2012) reported the highest fructose conversion to prepare EMF with 90.5% yield at 90 °C for 24 h. The $[MIMPS]_3PW_{12}O_{40}$ provided a promising strategy for large-scale production of EMF from renewable biomass.

Based on the superior physical and chemical properties and advantages of traditional ILs, Zuo et al. (2018) proposed an effective progress for the production of EMF from carbohydrates with a deep eutectic solvent (DES) mixture as solvent. 77.3% EMF yield was converted from fructose with Amberlyst-15 as the catalyst. Interestingly, the excellent catalytic performance of $CrCl_3$-modified Amberlyst-15 provided a high EMF yield of 46.7% with glucose as reaction feed.

Other Solid Acids

To further increase the recovery efficiency of solid acid catalysts, the researchers present the application of magnetic solid acid catalysts containing magnetic metal oxides for the synthesis of EMF. Transition-metal dichalcogenide is a novel 2D structure material primarily applied in hydrodeoxygenation reactions and consists of a central transition-metal atom layer surrounded by a layer of two chalcogen atoms (Veryasov et al. 2014; Lv et al. 2015). Quereshi et al. (2020) reported the application of a novel WS_2 solid catalyst in the EMF synthesis, which contained both Brønsted and Lewis acid sites caused by the defects of atoms, surface proton compensation, and the W-S-W delocalizing effect. With the microwave-assisted heating method at 160 °C and retaining 15 min, 62% yield of EMF was obtained with 100% fructose conversion in the presence of 30 mol% WS_2 catalysts.

Kumar and Srivastava (2019) prepared a $FeVO_4$ supported $-SO_3H$ functionalized polyaniline (S-PANI-$FeVO_4$) and investigated this recycled solid acid in the carbohydrates' dehydration. As the S-PANI-$FeVO_4$ contained both Brønsted and Lewis acid sites, EMF was successfully prepared with yields of 57.2% and 72.5% from sucrose and fructose at 90 °C in 24 h, respectively. Furthermore, molecular oxygen was employed as an environmental and cheap oxidant for the selective oxidation of aliphatic and aromatic alcohols into aldehydes in the presence of S-PANI-$FeVO_4$.

MXene is a hot researched 2D material with a graphene-like structure consisting of carbon-nitrides, nitrides, or transition metal carbides (Zhan et al. 2020). Bharath et al. (2021) riveted ultra-small Pd-Ru nanoparticles (NPs) on the surfaces of MXene nanosheets to form the Pd-Ru/MXene catalyst for EMF synthesis. Characterization results showed that Pd-Ru NPs with the hexagonal closed-pack structure were grown on the lamellar MXene surface, resulting in an increased specific surface area of 117 $m^2 \cdot g^{-1}$ with a higher acid-base sites density, then obtained 82% and 98% yield of EMF from fructose and HMF, respectively.

In summary, for EMF production from fructose, the one-pot method is conducted in the presence of various Lewis and Brønsted acid catalysts. In order to produce EMF from fructose at a desirable yield, high reaction temperature and retention time are required compared to the cases in which HMF is used as a feed. Ethanol/THF or ethanol/DMSO mixture is used as a reaction medium for EMF synthesis from fructose. Task-specific acidic ILs were also introduced to catalytic conversion of fructose to EMF. Very high yields (90.5%) of EMF from fructose were achieved in the presence of a solid acid catalyst $[MIMPS]_3PW_{12}O_{40}$ that was transformed from soluble phosphotungstic acid using IL. Running a one-pot fructose dehydration and HMF etherification may save HMF separation and isolation costs.

8.2.3.2 Glucose Conversion to EMF

Glucose is the most abundant and the cheapest available monosaccharide. Thus, its conversion into EMF is highly attractive. In order to produce EMF on a large scale, glucose should be considered as a feed since its price is considerably low compared to fructose and HMF. One-pot glucose conversion to EMF is desirable because it avoids the HMF isolation and purification steps that make the process more economically feasible. As a result, one-pot glucose conversion to EMF has been at the center of attention by many research groups.

Inspired by the high activity of HMF with the cheap, nontoxic, and abundant chemical $AlCl_3$ catalyst, Yang et al. (2012) explored the glucose conversion to EMF in the presence of Lewis acid $AlCl_3$. Although the total furans yield (including HMF and EMF) was relatively high in the ethanol-water solvent system, 57%, which is comparable to the fructose as the initial material with 58% total yield, the EMF selectivity was low and the isolated yield was 33% at 160 °C for 15 min. Following this, Liu et al. (2015) also used $AlCl_3$ as a homogenous acid catalyst to prepare EMF from glucose. They carried the reactions at 100 °C for 11 h; however, this modification did not enhance the EMF yield obviously and the reported yield was only 38.3% attributing to the aldoses properties of glucose. Not satisfied with the published effects, Xin et al. (2017) examined the transformation of glucose-EMF catalyzed by $AlCl_3 \cdot 6H_2O$ cooperated with a Brønsted solid acid catalyst (PTSA-POM). Multiple kinds of organic solvents were added to the ethanol-water system to study whether it can enhance the EMF yield. Data demonstrated the uselessness of the organic solvents introduced into the reaction medium and the optimum yield of EMF generated from glucose was 30.6% under the ethanol-water (9:1) system.

The production of EMF from glucose feedstock with high yield is urgently desired. Bai et al. (2018) employed the multifunctional MFI-Sn/Al zeolite as the catalyst for glucose conversion to the EMF product, which possesses dual meso-/microporosity and dual Lewis and Brønsted acidity. Sn and Al elements co-existence in the MFI-Sn/Al zeolite afford both Lewis acidic Sn- and Al-sites and Brønsted acidic Al-O(H)-Si sites for the successive isomerization-dehydration-etherification one-pot with three-step reaction. Over the multifunctional MFI-Sn/Al catalyst, a high EMF yield of 44% has been obtained from glucose conversion.

Inspired by the reported work that tin-containing zeolite Beta (Sn-BEA) was effective for the isomerization of glucose to fructose and hydrochloric acid was that to convert fructose to HMF, Lew et al. (2012) tried Lewis acid zeolite Sn-BEA catalyst combined with Brønsted acid catalyst Amberlyst-131 for one-pot glucose-to-fructose-to-HMF-to-EMF and produced EMF at 31% yields from glucose. Without Sn-BEA, the main products would be ethyl glucopyranoside and ethyl glucofuranoside (Lew et al. 2012).

Solid acid catalysts have been used in one-pot glucose conversion to EMF. The catalytic performance for glucose conversion to EMF was poor. Thombal and Jadhav (2016) also used glu-Fe_3O_4-SO_3H for glucose conversion to EMF in an ethanol and DMSO mixture. However, the production yield was not significant even after a long reaction duration. Similarly, in the presence of the MIL-101-SO_3H(100) solid acid catalyst, which possesses strong Brønsted acidity (X. Liu et al. 2016), the EMF production yield from glucose was less than 10% in a THF and ethanol mixture. Low yields of EMF could be attributed to the formation of stable ethyl glucopranoside and inefficient catalytic performance that could not catalyze glucose isomerization to fructose at high yields.

8.2.4 EMF Synthesis from Hexose-Based Carbohydrates

Production of EMF from disaccharides or polysaccharides is economically desirable due to the fact that the costs of saccharification, sugar isolation, and purification can be eliminated (as shown in Scheme 8.3). Sucrose is the most available disaccharide, which has the lowest price among the other disaccharides. Thus, EMF production from sucrose is considered valuable by many researchers around the world. In addition, inulin, an oligosaccharide, has received lots of attention because of its high fructose content.

8.2.4.1 Sulfonic Acid Groups' Functionalized Catalyst

Yuan et al. (2015) showed moderate EMF yields obtained from sucrose and inulin at 33.2% and 58.4%, which were catalyzed by the magnetic Fe_3O_4@C-SO_3H. In their research, the low yield of EMF converted from sucrose was attributed to the sucrose molecule composition, which contains one fructose and glucose unit, for the low yield of EMF obtained from glucose.

The acidic catalyst GO was applied for the production of EMF from disaccharide sucrose and polysaccharide inulin in an ethanol/DMSO mixture (Wang et al. 2013). For one fructose and one glucose unit, the sucrose and glucose molecules can always

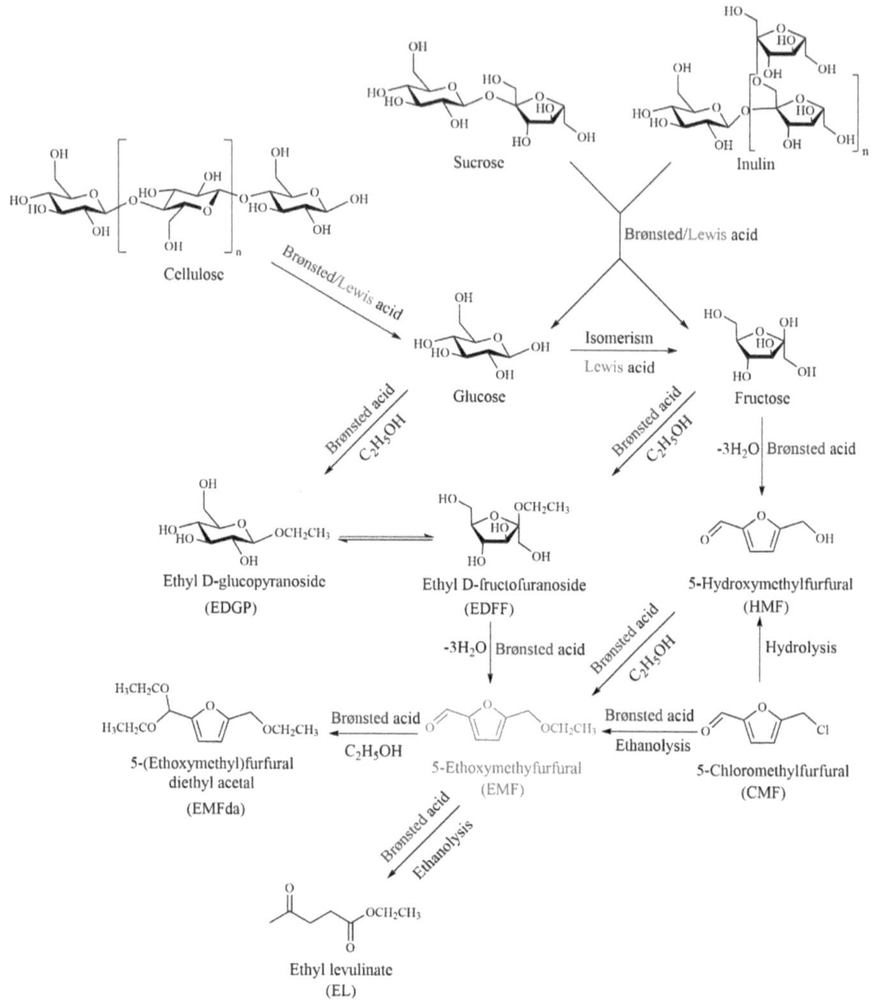

SCHEME 8.3 The reaction paths of synthesizing EMF from carbohydrates.

be converted to ethyl glucoside, EMF yield was 34% when sucrose was used as the substrate. Nevertheless, when inulin acts as the starting material for the production of EMF, a much higher EMF yield of 66% was obtained. The fact that one molecule of inulin contains one glucose unit and 1–59 fructose units explains the excellent yield of the target product.

The other solid acid catalyst examples found in the literature were Fe_3O_4@SiO_2-SH-Im-HSO_4 (Yin et al. 2015) and Fe_3O_4@SiO_2-SO_3H (Yuan et al. 2015). By dehydrating of carbohydrates catalyzed with acidic catalysts, HMF can be obtained. Based on the excellent catalytic activity of Fe_3O_4@SiO_2-SH-Im-HSO_4 for the yield of EMF from HMF, Yin et al. (2015) also explored the conversion of sucrose and inulin to prepare EMF in the presence of Fe_3O_4@SiO_2-SH-Im-HSO_4. The EMF yield derived from inulin (56.1%) was comparable with the substrate fructose. The majority of glucose always yielded ethyl glucoside, EMF production yield from sucrose was low with 34.4%. A similar trend was reported (Yuan et al. 2015) when Fe_3O_4@SiO_2-SO_3H was prepared and developed as a catalyst for the EMF synthesis. In this case, the EMF production yields were 40.2% from sucrose with 95.4% conversion and 63.3% when inulin was used as the initial substrate. Soon afterwards, Wang et al. (2017) explained the smooth promotion of sucrose and inulin to produce EMF with superior catalytic performance, 53.6% and 26.8%, respectively, under 140 °C for 24 h. Catalyst OMC-SO_3H showed excellent catalytic performance, indicating the promising for biomass conversion into value-added chemicals and liquid fuels.

Additionally, heterogeneous acidic catalysts have been used in one-pot EMF synthesis from di- and polysaccharides. For instance, Liu and Zhang (2013) explored the inulin, sucrose, and cellobiose conversion in ethanol catalyzed by silica-SO_3H catalyst. The inulin conversion yield to EMF high up to 60.7%, which was comparable to that of fructose for EMF production (63.1%). The EMF yields obtained from sucrose and cellobiose were exhibited at 34.9% and 0, respectively. The low EMF yields were due to the high reactivity of glucose units constituting the disaccharide with ethanol to form ethyl glucopyranoside.

X. Liu et al. (2016) also studied the one-pot conversion of sucrose, cellobiose, and inulin to EMF in an ethanol/THF mixture. As the acid catalyst, MIL-101-SO_3H(100), was used, EMF production yields from inulin and sucrose were 54.2% and 33.2%, respectively. However, this catalytic system performance for cellobiose conversion was very poor and HMF yield was less than 5%. Besides, the isolated yield of inulin conversion to EMF was reported (Thombal et al. 2016) to be 85% in the presence of glu-Fe_3O_4-SO_3H in ethanol and DMSO mixture.

8.2.4.2 Others Catalysts

Yang et al. (2012) introduced Lewis acid $AlCl_3{\cdot}6H_2O$ as a catalyst for the one-pot conversion of hexose-based carbohydrates including sucrose, maltose, cellobiose, starch, and cellulose to EMF in an ethanol/water mixture solvent system. Since the comparable catalytic performance converted from fructose and glucose for EMF, they reported a moderate EMF yield of 40% derived from sucrose. When maltose was used as a feed, a similar amount of furans was prepared compared with sucrose. However, the conversion efficiencies of cellobiose and starch to furan in this system were low and the EMF yields were only 30% and 26%, respectively. In the system of ethanol/water solvent at 160 °C, cellulose depolymerization and decrystallization were poor, and therefore, the total furan yield was less than 10%. The same as the Lewis acid catalyst, Yadav et al. (2014) investigated the cobalt (I , II, III) phthalocyanine (Co(x)Pc) catalyst for the isomerization of glucose to fructose, fructose dehydration, as well as the subsequent etherification. It is noted that Co(x)Pc was an effective catalyst for the direct conversion of carbohydrates to EMF in [EMIm]Cl ionic liquid and Co(III)Pc exhibited much higher catalytic activity.

With microwave heating, Yang et al. (2012) investigated EMF production from di- and polysaccharides (sucrose and inulin) in an ethanol/THF mixture with $H_3PW_{12}O_{40}$

acting as the catalyst. Similar to the other work, the glucose unit was converted to ethyl glucoside and only the available fructose became EMF, which resulted in the EMF production yield from sucrose was no more than 33%. Differently, the EMF synthesis from inulin retained 62% yield.

Except for the THF co-solvent, DMSO can also be used as a co-solvent for one-pot conversion of sucrose or inulin to EMF. Wang et al. (2013) applied a phosphotungstic acid (HPW) catalyst to prepare EMF from sucrose and inulin in an ethanol/DMSO mixture. The EMF production yield from sucrose was as low as 28% and a moderate yield of 54% from inulin. However, it was possible to convert inulin in a relatively high EMF yield, 66%, when a hybrid acid-base bifunctional catalyst Lys/PW was introduced for inulin and sucrose in a one-pot process to EMF in an ethanol/DMSO mixture (Li et al. 2014). Li et al. (2014) proved 58.5%, 42.4% and 36.5% yields of EMF from inulin, sorbose and sucrose, respectively. Because the catalytic system converted glucose into ethyl glucopyranoside rather than a high yield of EMF, the reported yield was very low for EMF using cellobiose as a feed, less than 5%.

The renewable, low toxicity/nontoxic compounds and cheap ingredients afforded the deep eutectic solvent (DES) mixture, giving 65.2% and 50.2% EMF high yields derived from inulin and sucrose catalytic by Amberlyst-15 and $CrCl_3$-Amberlyst-15, respectively (Zhou et al. 2014).

Recently, Li et al. (2016) used DeAl-H-beta as a catalyst for sucrose conversion to EMF and reported a moderate yield of 43%. They also introduced a one-pot two-step method that improved the EMF yield by up to 50%, while the total production yield was 69%. Besides, DeAl-H-beta and H-USY$^+$Amberlyst-15 catalyzed inulin in relatively high yields of 52% and 58%, respectively. However, the performances of both catalysts were poor for cellobiose conversion to EMF.

Recently, De et al. (2012) studied IL as a catalyst for one-pot cellulose and lignocellulosic biomass substrates, for example, sugarcane bagasse, conversion to EMF. In the presence of $[DMA]^+[CH_3SO_3]^-$, they reported 22% yield for EMF production from cellulose. Alam et al. (2012) showed EMF production converted directly from Foxtail and Red nut sedge weeds in the presence of ILs as a catalyst. The catalytic activity of $[DMA]^+[CH_3SO_3]^-$ (0.46 g) was better than $[NMP]^+[CH_3SO_3]^-$ (0.40 g) when converting foxtail with 1.0 g initial substrate to EMF. Similarly, $[DMA]^+[CH_3SO_3]^-$ and $[NMP]^+[CH_3SO_3]^-$ catalytic performances were almost comparable to EMF preparation from Red nut sedge biomass, the corresponding isolated yields of EMF were 0.28 and 0.27 g, respectively.

Grisel et al. (2014) studied alcoholysis of wheat straw in 94% (w/w) ethanol acidified with 30 mM H_2SO_4 at 200 °C. The reported yields for EMF and EL were 20% and 25%, respectively. Marine carbohydrate biomass was used to produce EMF as Kim et al. (2010) studied one-pot agar derived from red-algae conversion to EMF in the presence of [EMIM]Cl, $CrCl_2$, and Dowex resin mixture as a catalyst. They could isolate 3.9 g of EMF and EL (EMF to EL mole ration 5:2) from 10 g of agar.

8.3 SYNTHESIS OF MMF

Another important alkoxymethyl furfural is MMF synthesis by HMF reaction with methanol in the presence of an acid catalyst. MMF has different applications as a fuel additive and FDCA precursor (Eerhart et al. 2014). During the preparation of MMF, methyl levulinate (MLV) is produced as a reaction byproduct. Fructose as a substrate has received more attention to produce MMF compared to the other possible sources. Zhu et al. (2011) studied fructose conversion to MMF with a solid acid catalyst Amberlyst-15 in a methanol/THF mixture. They conducted the reaction in series at 120 °C and reported 47% and 14.4% yields for MMF and HMF, respectively. Besides, a soluble acid catalyst such as HCl or H_2SO_4 was reported for one-pot fructose conversion to MMF. Ren et al. (2015) conducted fructose dehydration in methanol in the presence of HCl. Running simultaneous fructose dehydration and HMF etherification at 80 °C resulted in a very low yield of MMF. However, by increasing the reaction temperature to 120 °C, the total production yield was improved to 40%. Bicker et al. (2005) studied fructose dehydration in hot compressed methanol in a continuous-flow reactor system. The MMF concentration reached its maximum at 30 s and then decreased due to the instability of MMF under the given reaction conditions. They reported 38.7 mM production of MMF while the fructose initial concentration was 55 mM.

In order to resolve the issues associated with homogenous acid catalysts in an attempt to produce MMF, Kraus and Guney (2012) introduced task-specific acidic ILs. They synthesized MMF from fructose in the sulfonic acid functionalized IL, namely 1-butyl-3-(3-sulfopropyl)-imidazolium chloride or 1-methyl-3-(3-sulfopropyl)-imidazolium chloride and methanol mixture. Hexane was also added to increase production yields. The results presented showed a superior performance of IL for MMF in a relatively high yield, 57%, to those reported earlier by Lai and Zhang (. 2011) and Zhu et al. (2011). However, through this method, the MLV yield was reported as only 9%. Garves (1988) conducted cellulose and biomass methanolysis in methanol in the presence of acid catalysts and various homogenous acid catalysts. For instance, HCl and HBr performances were not effective in producing MMF or MLV due to the partial solubility of cellulose in this reaction medium, while H_2SO_4 catalyzed cellulose more efficiently. However, MLV was the main product instead of MMF. Other acid catalysts such as CF_3SO_3H, p-Tolyl-SO_3 and SO_2 also catalyzed cellulose degradation, but severe reaction conditions were required compared to the case that H_2SO_4 was used as a catalyst. In addition, Garves reported that methanol participated in an undesired side reaction that resulted in dimethyl ether. In order to prevent dimethyl ether formation, methanol was partially replaced by water. Water suppressed not only the dimethyl ether production, but also reduced MMF and MLV yields. To address the issue, organic solvents such as CCl_4, toluene, and cyclohexane were introduced. In the presence of these solvents, dimethyl ether was not detected and the MLV yield improved while the MMF yield decreased. Moreover, different biomass conversions in the presence of methanol were investigated by Garves (1988). Similar to cellulose conversion, MLV was the main product and the MMF yield was low. Grisal et al. (2014) studied methanolysis of wheat straw in the presence of H_2SO_4. The MMF yield was negligible (< 4%) and the MLV production yield was also as low as 11.

In summary, di- and polysaccharides can be used for one-pot EMF production. Acid catalysts that have been introduced for glucose or fructose conversion to EMF were tested and their results were reported as promising in the case of inulin conversion. In general, EMF synthesis from sucrose did not result in high yields, due to the fact that glucose obtained from sucrose conversion is not converted to EMF at high yields. However, inulin, as a fructose polymer, showed better performance and produced EMF at high yields. One-pot conversion of cellulose or biomass to EMF resulted in low yields not only because of their complex structure, but also due to the low glucose conversion to EMF.

8.4 CONCLUSIONS AND PERSPECTIVES

Various catalytic materials have been demonstrated to have outstanding potential in producing EMF efficiently through single-step, two-step, and multi-step conversions of multi-substrates. EMF synthesis from hexose-fructose at the yield of 50%–70%, but high reaction temperature (90–130 °C) and long retention time (6–24 h) are needed compared to the HMF cases with yields of 70%–95%. Conversion of sucrose for EMF production did not support high yields for constitutional unit glucose was not converted to EMF in good yields in accordance with cellulose or biomass as feedstocks.

The problem of glucose conversion to ethyl glucopyranoside during the preparation of EMF has been a stumbling block. Therefore, an innovative process should be introduced to overcome this issue. The reaction conditions are also still to be advanced further and be addressed in future studies. Furthermore, a detailed kinetic study or reactor design has not been developed yet.

REFERENCES

Abhinav Kumar, Rajendra Srivastava. 2019. "$FeVO_4$ decorated-SO_3H functionalized polyaniline for direct conversion of sucrose to 2, 5-diformylfuran & 5-ethoxymethylfurfural and selective oxidation reaction." *Mol. Catal.* 465. Elsevier Science Ltd: 68–79. doi: 10.1016/j.mcat.2018.12.017.

A J J E Eerhart, W J J Huijgen, R J H Grisel, J C van der Waal, E de Jong, A de Sousa Dias, A P C Faaija, M K Patela. 2014. "Fuels and plastics from lignocellulosic biomass via the furan pathway: A technical analysis." *RSC Adv.* 4. Royal Soc Chemistry: 3536–3549. doi: 10.1039/C3RA43512A.

Akshay D Patel, Juan Carlos Serrano-Ruiz, James A Dumesic, Robert P Anex. 2010. "Techno-economic analysis of 5-nonanone production from levulinic acid." *Chem. Eng. J.* 160. Elsevier Science SA: 311–321. doi: 10.1016/j.cej.2010.03.034.

Aleksei Bredihhin, Sirvo Luiga, Lauri Vares. 2016. "Application of 5-ethoxymethylfurfural (EMF) for the production of cyclopentenones." *Synthesis-stuttgart* 48. Georg Thieme Verlag KG: 4181–4188. doi: 10.1055/s-0035-1562790.

Aleksei Bredihhin, Uno Mäeorg, Lauri Vares. 2013. "Evaluation of carbohydrates and lignocellulosic biomass from different wood species as raw material for the synthesis of 5-bromomethyfurfural." *Carbohydr. Res.* 375. Elsevier Science Ltd: 63–67. doi: 10.1016/j.carres.2013.04.002.

Ana I Torres, Prodromos Daoutidis, Michael Tsapatsis. 2010. "Continuous production of 5-hydroxymethylfurfural from fructose: a design case study." *Energy Environ. Sci.* 3. Royal Soc Chemistry: 1560–1572. doi: 10.1039/C0EE00082E.

Andreia A Rosatella, Svilen P Simeonov, Raquel F M Frade, Carlos A M Afonso. 2011. "5-Hydroxymethylfurfural (HMF) as a building block platform: biological properties, synthesis and synthetic applications." *Green Chem.* 13. Royal Soc Chemistry: 754–793. doi: 10.1039/C0GC00401D.

Anqiu Liu, Bing Liu, Yimei Wang, Rongshen Ren, Zehui Zhang. 2014. "Efficient one-pot synthesis of 5-ethoxymethylfurfural from fructose catalyzed by heteropolyacid supported on K-10 clay." *Fuel* 117. Elsevier Science Ltd: 68–73. doi: 10.1016/j.fuel.2013.09.072.

Anqiu Liu, Zehui Zhang, Zhongfeng Fang, Bing Liu, Kecheng Huang. 2014. "Synthesis of 5-ethoxymethylfurfural from 5-hydroxymethylfurfural and fructose in ethanol catalyzed by MCM-41 supported phosphotungstic acid." *J. Ind. Eng. Chem.* 20. Elsevier Science Ltd: 1977–1984. doi: 10.1016/j.jiec.2013.09.020.

Avelino Corma, Sara Iborra, Alexandra Velty. 2007. "Chemical routes for the transformation of biomass into chemicals." *Chem. Rev.*107. American Chemical Society: 2411–2502. doi: 10.1021/cr050989d.

Bart Vanderhaegen, Hedwig Neven, Stefan Coghe, Kevin J Verstrepen, Hubert Verachtert, Guy Derdelinckx. 2003. "Evolution of chemical and sensory properties during aging of top-fermented beer." *J. Agric. Food. Chem.* 23. American Chemical Society: 6782–6790. doi: 10.1021/jf034631z.

B F M Kuster. 1990. "5-Hydroxymethylfurfural (HMF). A review focussing on its manufacture." *Starch-Starke* 42. Wiley-VCH Verlag GmbH: 314–21. doi: 10.1002/star.19900420808.

Binder Joseph B, Raines Ronald T. 2009. "Simple chemical transformation of lignocellulosic biomass into furans for fuels and chemicals." *J. Am. Chem. Soc.* 131. American Chemical Society: 1979–1785. doi: 10.1021/ja808537j.

Bing Liu, Zehui Zhang. 2013. "One-pot conversion of carbohydrates into 5-ethoxymethylfurfural and ethyl D-glucopyranoside in ethanol catalyzed by a silica supported sulfonic acid catalyst." *RSC Adv.* 3. Royal Soc Chemistry: 12313–12319. doi: 10.1039/C3RA41043A.

Bing Liu, Zehui Zhang, Kejian Deng. 2012. "Efficient one-pot synthesis of 5-(ethoxymethyl)furfural from fructose catalyzed by a novel solid catalyst." *Ind. Chem. Eng. Res.* 51. American Chemical Society: 15331–15336. doi: 10.1021/ie3020445.

Bing Liu, Zehui Zhang, Kecheng Huang, Zhongfeng Fang. 2013a. "Efficient conversion of carbohydrates into 5-ethoxymethylfurfural in ethanol catalyzed by $AlCl_3$." *Fuel* 113. Elsevier Science Ltd: 625–631. doi: 10.1016/j.fuel.2013.06.015.

Bing Liu, Zehui Zhang, Kecheng Huang. 2013b. "Cellulose sulfuric acid as a bio-supported and recyclable solid acid catalyst for the synthesis of 5-hydroxymethylfurfural and 5-ethoxymethylfurfural from fructose." *Cellulose* 20. Springer: 2081–2089. doi: 10.1007/s10570-013-9944-0.

B Kamm, P R Gruber, M Kamm. 2007. "Biorefineries-industrial processes and products." Wiley-VCH Verlag GmbH.

Bora Kim, Jaewon Jeong, Seunghan Shin, Dohoon Lee, Sangyong Kim, Hyo-Jin Yoon, Jin Ku Cho. 2010. "Facile single-step conversion of macroalgal polymeric carbohydrates into biofuels." *ChemSusChem* 3. Elsevier Science Ltd: 1273–1275. doi: 10.1002/cssc.201000192.

Buntara Teddy, Sebastien Noel, Pim Huat Phua, Ignacio Melián-Cabrera, Johannes G de Vries, Hero J Heeres. 2011. "Caprolactam from renewable resources: catalytic conversion of 5-hydroxymethylfurfural into caprolactone." *Angew. Chem. Int. Ed.* 50. Wiley-VCH Verlag GmbH: 70837087. doi: 10.1002/ange.201102156.

Chidambaram Mandan, Bell Alexis T. 2010. "A two-step approach for the catalytic conversion of glucose to 2, 5-dimethylfuran in ionic liquids." *Green Chem.* 12. Royal Soc Chemistry: 1253–1262. doi: 10.1039/c004343e.

Christopher M Lew, Nafiseh Rajabbeigi, Michael Tsapatsis. 2012. "One-pot synthesis of 5-(ethoxymethyl)furfural from glucose using Sn-BEA and Amberlyst catalysts." *Ind. Eng. Chem. Res.* 51. American Chemical Society: 5364–5366. doi: 10.1021/ie2025536.

C J Moye, R J Goldsack. 1966. "Reaction of ketohexoses with acid in certain non-aqueous sugar solvents." *J. Appl. Chem.* 16. Wiley-VCH Verlag GmbH: 206–208. doi: 10.1002/jctb.5010160703.

Corma Avelino, de la Torre Olalla, Renz Michael. 2012. "Production of high quality diesel from cellulose and hemicellulose by the Sylvan process: catalysts and process variables." *Energy Environ. Sci.* 5. Royal Soc Chemistry: 6328–6344. doi: 10.1039/c2ee02778j.

Corma Canós Avelino, De la Torre Alfaro Olalla, Renz Michael. 2011. "High-quality diesel from hexose-and pentosederived biomass platform molecules." *ChemSusChem* 4. Wiley-VCH Verlag GmbH: 1574–1577. doi: 10.1002/cssc.201100296.

David W Brown, Arthur J Floyd, Richard G Kinsman, Yusuf Roshanhyphen Ali. 1982. "Dehydration reactions of fructose in non-aqueous media." *J. Chem. Technol. Biotechnol.* 32. Wiley-VCH Verlag GmbH: 920–924. doi: 10.1002/jctb.5030320730.

George A Kraus, Tezcan Guney. 2012. "A direct synthesis of 5-alkoxymethylfurfural ethers from fructose via sulfonic acid-functionalized ionic liquids." *Green Chem.* 14. Royal Soc Chemistry: 1593–1596. doi: 10.1039/C2GC35175G.

George W Huber, Juben N Chheda, Christopher J Barrett, James A Dumesic. 2005. "Production of liquid alkanes by aqueous-phase processing of biomass-derived carbohydrates." *Science* 308. Amer Assoc Advancement Science: 1446–1450. doi: 10.1126/science.111116.

Gerardus G Maria, F Dautzenberg. 2007. "Method for the synthesis of 5-alkoxymethylfurfural ethers and their use." European Patent EP 1834950 Sep 19.

G J M Gruter. 2010. "Hydroxymethylfurfural ethers from sugars or HMF and branched alcohols." U.S. Patent. US 2010/0218416.

G J M Gruter. 2010. "Hydroxymethylfurfural ethers from HMF and olefins." US Patent. US/0212217.

G J M Gruter. 2012. "Mixture of furfural and 5-(alkoxymethyl)furfural derivatives from sugars and alcohols." US Patent. US8133289 B2.

G J M Gruter, F Dautzenberg. 2010. "Method for the synthesis of 5-alkoxymethyl furfural ethers and their use." US Patent. US2009/131690.

G J M Gruter, L E Manzer, B V Furanix. 2008. "Hydroxymethylfurfural ethers from sugars and higher alcohols." US Patent. US 12/676,522. 2008.

G J M Gruter, F Dautzenberg. 2014. "Method for the synthesis of 5-hydroxymethyl furfural ethers and their use." U.S. Patent, 8877950.

Gleb Veryasov, Miha Grilc, Blaž Likozar, Adolf Jesih. 2014. "Hydrodeoxygenation of liquefied biomass on urchin-like MoS_2." *Catal. Commun.* 46. Elsevier Science Ltd: 18–186. doi: 10.1016/j.catcom.2013.12.011.

G Morales, M Paniagua, J A Melero, J Iglesias. 2017. "Efficient production of 5-ethoxymethylfurfural from fructose by sulfonic mesostructured silica using DMSO as co-solvent." *Catal. Today* 279. Elsevier Science Ltd: 305–316. doi: 10.1016/j.cattod.2016.02.016.

Govindan Bharath, Krishnamoorthy Rambabu, Abdul Hai, Pranay P Morajkar, Akshay V Salkar, Shadi W Hasan, Pau Loke Show, Fawzi Banat. 2021. "Highly selective etherification of fructose and 5-hydroxymethylfurfural over a novel Pd-Ru/MXene catalyst for sustainable liquid fuel production." *Int. J. Energy Res.*45(10). Wiley-VCH Verlag GmbH: 14680–14691. doi: 10.1002/er.6743.

G Raveendra, A Rajasekhar, M Srinivas, P S Sai Prasad, N Lingaiah. 2016. "Selective etherification of hydroxymethylfurfural to biofuel additives over Cs containing silicotungstic acid catalysts." *Appl. Catal. A: Gen.* 520. Elsevier Science Ltd: 105–113. doi: 10.1016/j.apcata.2016.04.017.

Guizhuan Xu, Binglin Chen, Zhangbin Zheng, Kai Li, Hongge Tao. 2017. "One-pot ethanolysis of carbohydrates to promising biofuels: 5-ethoxymethylfurfural and ethyl levulinate." *Asia-Pac. J. Chem. Eng.* 12. Wiley-VCH Verlag GmbH: 527–535. doi: 10.1002/apj.2095.

Hamid Omidvarborna, Ashok Kumar, Dong-Shik Kim. 2016. "Variation of diesel soot characteristics by different types and blends of biodiesel in a laboratory combustion chamber." *Sci. Total Environ.* 544. Elsevier Science Ltd: 450–459. doi: 10.1016/j.scitotenv.2015.11.076.

Haosheng Xin, Tingwei Zhang, Wenzhi Li, Mingxue Su, Song Li, Qun Shao, Longlong Ma. 2017. "Dehydration of glucose to 5-hydroxymethylfurfural and 5-ethoxymethylfurfural by combining Lewis and Brønsted acid." *RSC Adv.* 7. Royal Soc Chemistry: 41546–41551. doi: 10.1039/C7RA07684C.

Hongliang Wang, Tiansheng Deng, Yingxiong Wang, Xiaojing Cui, Yongqin Qi, Xindong Mu, Xianglin Hou, Yulei Zhu. 2013. "Graphene oxide as a facile acid catalyst for the one-pot conversion of carbohydrates into 5-ethoxymethylfurfural." *Green Chem.* 15. Royal Soc Chemistry: 2379–2383. doi: 10.1039/C3GC41109E.

Hongliang Wang, Tiansheng Deng, Yingxiong Wang, Yongqin Qi, Xianglin Hou, Yulei Zhu. 2013. "Efficient catalytic system for the conversion of fructose into 5-ethoxymethylfurfural." *Bioresour. Technol.* 136. Elsevier Science Ltd: 394–400. doi: 10.1016/j.biortech.2013.02.110.

Hong Zhu, Quan Cao, Chunhu Li, Xindong Mu. 2011. "Acidic resin-catalysed conversion of fructose into furan derivatives in low boiling point solvents." *Carbohydr. Res.* 346. Elsevier Science Ltd: 2016–2018. doi: 10.1016/j.carres.2011.05.026.

Huber George W, Iborra Sara, Corma Avelino. 2006. "Synthesis of transportation fuels from biomass: chemistry, catalysts, and engineering." *Chem. Rev.* 106. American Chemical Society: 4044–4098. doi: 10.1021/cr068360d.

Hugo Oliveira e Silva, Paula Guedes de Pinho, Beatriz P Machado, Tim Hogg, J C Marques, José S Câmara, F Albuquerque, Antonio C Silva Ferreira. 2008. "Impact of forced-aging process on Madeira wine flavor." *J. Agric. Food Chem.* 56. American Chemical Society: 11989–11996. doi: 10.1021/jf802147z.

Hu Li, Khokarale Santosh Govind, Ramakrishna Kotni, Saravanamurugan Shunmugavel, Anders Riisager, Song Yang. 2014. "Direct catalytic transformation of carbohydrates into 5-ethoxymethylfurfural with acid-base bifunctional hybrid nanospheres." *Energy Convers. Manag.* 88. Elsevier Science Ltd: 1245–1251. doi: 10.1016/j.enconman.2014.03.037.

Hu Li, Qiuyun Zhang, Song Yang. 2014. "Catalytic cascade dehydration-etherification of fructose into 5-ethoxymethylfurfural with SO_3H-functionalized polymers." *Int. J. Chem .Eng.* 1–7. doi: 10.1155/2014/481627.

Hu Li, Shunmugavel Saravanamurugan, Song Yang, Anders Riisager. 2016. "Direct transformation of carbohydrates to the biofuel 5-ethoxymethylfurfural by solid acid catalysts." *Green Chem.* 18. Royal Soc Chemistry: 726–734. doi: 10.1039/C5GC01043H.

I. Delidovich, K. Leonhard, R. Palkovits. 2014. Cellulose and hemicellulose valorisation: an integrated challenge of catalysis and reaction engineering. *Energy Environ. Sci.* 7. Royal Soc Chemistry: 2803–2830. doi: 10.1039/C4EE01067A.

Indrek Viil, Aleksei Bredihhin, Uno Mäeorg, Lauri Vares. 2014. "Preparation of potential biofuel 5-ethoxymethylfurfural and other 5-alkoxymethylfurfurals in the presence of oil shale ash." *RSC Adv.* 4. Royal Soc Chemistry: 5689–5693. doi: 10.1039/C3RA46570E.

Isabelle Cutzach, Pascal Chatonnet, Denis Dubourdieu. 1999. Study of the formation mechanisms of some volatile compounds during the aging of sweet fortified wines. *J. Agric. Food. Chem.* 47. American Chemical Society: 2837–2846. doi: 10.1021/jf981224s.

Jennifer D Lewis, Stijn Van de Vyver, Anthony J Crisci, William R Gunther, Vladimir K Michaelis, Robert G Griffin, Yuriy Román-Leshkov. 2014. "A continuous flow strategy for the coupled transfer hydrogenation and etherification of 5-(hydroxymethyl)furfural using lewis acid zeolites." *ChemSusChem* 7. Elsevier Science Ltd: 2255–2265. doi: 10.1002/cssc.201402100.

J F Yang, N Li, G Y Li, W T Wang, A Q Wang, X D Wang, Y Cong, T Zhang. 2013. "Solvent-free synthesis of C10 and C11 branched alkanes from furfural and methyl isobutyl detone." *ChemSusChem* 6. Wiley-VCH Verlag GmbH: 1149–1152. doi: 10.1002/cssc.201300318.

Jitian Liu, Yu Tang, Kaigui Wu, Caifeng Bi, Qiu Cui. 2012. "Conversion of fructose into 5-hydroxymethylfurfural(HMF) and its derivatives promoted by inorganic salt in alcohol." *Carbohydr. Res.* 350. Elsevier Science Ltd: 20–24. doi: 10.1016/j.carres.2011.12.006.

Joseph J Bozell, Gene R Petersen. 2010. "Technology development for the production of biobased products from biorefinery carbohydrates – the US Department of Energy's "top 10"revisited." *Green Chem.* 12. Royal Soc Chemistry: 539–554. doi: 10.1039/B922014C.

J S Câmara, M A Alves, J C Marques. 2006. "Changes in volatile composition of Madeira wines during their oxidative ageing." *Anal. Chim. Acta.* 563. Elsevier Science Ltd: 188–97. doi: 10.1016/j.aca.2005.10.031.

Junmei Wang, Zehui Zhang, Shiwei Jin, Xizhou Shen. 2017. "Efficient conversion of carbohydrates into 5-hydroxylmethylfurfan and 5-ethoxymethylfurfural over sufonic acid-functionalized mesoporous carbon catalyst." *Fuel* 192. Elsevier Science Ltd: 102–107. doi: 10.1016/j.fuel.2016.12.027.

Katia Barbera, Paola Lanzafame, Siglinda Perathoner, Gabriele Centi, Massimo Migliori, Alfredo Aloiseb, Girolamo Giordano. 2016. "HMF etherification using NH_4-exchanged zeolites." *New J. Chem.* 40. Royal Soc Chemistry: 4300–4306. doi: 10.1039/C5NJ03461B.

K Barbera, P Lanzafame, A Pistone, S Millesi, G Malandrino, A Gulino, S Perathoner, G Centi. 2015. "The role of oxide location in HMF etherification with ethanol over sulfated ZrO_2 supported on SBA-15." *J. Catal.* 323. Elsevier Science Ltd: 19–32. doi: 10.1016/j.jcat.2014.12.001.

Klaus Garve. 1988. "Acid catalyzed degradation of cellulose in alcohols." *J. Wood Chem. Technol.* 8. 121–134. doi: 10.1080/02773818808070674.

Kumar Karitkey Yadav, Sohail Ahmad, Shive M S Chauhan. 2014. "Elucidating the role of cobalt phthalocyanine in the dehydration of carbohydrates in ionic liquids." *J. Mol. Catal. A: Chem.* 394. Elsevier Science Ltd: 170–176. doi: 10.1016/j.molcata.2014.07.014.

Lewkowski Jarosław. 2001. "Synthesis, chemistry and applications of 5-hydroxymethyl-furfural and its derivatives." *ARKIVOC* (i). Arkat USA Inc: 17–54.

L Harivardhan Reddy, José L Arias, Julien Nicolas, Patrick Couvreur. 2012. "Magnetic nanoparticles: design and characterization, toxicity and biocompatibility, pharmaceutical and biomedical applications." *Chem. Rev.* 112. American Chemical Society: 5818–5178. doi: 10.1021/cr300068p.

Linke Lai, Yugen Zhang. 2011. "The production of 5-hydroxymethylfurfural from fructose in isopropyl alcohol: a green and efficient system." *ChemSusChem* 4. Elsevier Science Ltd: 1745–1748. doi: 10.1002/cssc.201100489.

Madhesan Balakrishnan, Eric R Sacia, Alexis T Bell. 2012. "Etherification and reductive etherification of 5-(hydroxymethyl) furfural: 5-(alkoxymethyl) furfurals and 2,5-bis (alkoxymethyl) furans as potential bio-diesel candidates." *Green Chem* 14. Royal Soc Chemistry: 1626–1634. doi: 10.1039/C2GC35102A.

M Bicker, D Kaiser, L Ott, H Vogel. 2005. "Dehydration of D-fructose to hydroxymethylfurfural in sub-and supercritical fluids." *J. Supercrit. Fluids* 36. Elsevier Science Ltd: 118–126. doi: 10.1016/j.supflu.2005.04.004.

Małgorzata E Zakrzewska, Ewa Bogel-Łukasik, Rafał Bogel-Łukasik. 2010. "Ionic liquid-mediated formation of 5-hydroxymethylfurfural, a promising biomass-derived building block." *Chem. Rev.* 111. American Chemical Society: 397–417. doi: 10.1021/cr100171a.

Maria J Climent, Avelino Corma, Sara Iborra. 2014. "Conversion of biomass platform molecules into fuel additives and liquid hydrocarbon fuels." *Green Chem.* 16. Royal Soc Chemistry: 516–547. doi: 10.1039/C3GC41492B.

Margarida M Antunes, Patrícia A Russo, Paul V Wiper, Jacinto M Veiga, Martyn Pillinger, Luís Mafra, Dmitry V Evtuguin, Nicola Pinna, Anabela A Valente. 2014. "Sulfonated graphene oxide as effective catalyst for conversion of 5-(hydroxymethyl)-2-furfural into biofuels." *ChemSusChem* 7. Elsevier Science Ltd: 804–812. doi: 10.1002/cssc.201301149.

Mark Mascal, Edward B Nikitin. 2008. "Direct, high-yield conversion of cellulose into biofuel." *Angew. Chem. Int. Ed.* 120. Wiley-VCH Verlag GmbH: 8042–8044. doi: 10.1002/ange.200801594.

Mark Mascal, Edward B Nikitin. 2009. "Towards the efficient, total glycan utilization of biomass." *ChemSusChem* 2. Wiley-VCH Verlag GmbH: 423–426. doi: 10.1002/cssc.200900071.

Md Imteyaz Alam, Sudipta De, Saikat Dutta, Basudeb Saha. 2012. "Solid-acid and ionic-liquid catalyzed one-pot transformation of biorenewable substrates into a platform chemical and a promising biofuel." *RSC Adv.* 2. Royal Soc Chemistry: 6890–6896. doi: 10.1039/C2RA20574B.

M E Janka, D M Lange, M C Morrow, B R Bowers, K R Parker, A Shaikh, L R Partin, C E Sumner. 2012. "Oxidation process to produce a crude and/or purified carboxylic acid product." US Patent. 2012/0302770A1.

Miao Zuo, Kai Le, Yunchao Feng, Caixia Xiong, Zheng Li, Xianhai Zeng, Xing Tang, Yong Sun, Lu Lin. 2018. "An effective pathway for converting carbohydrates to biofuel 5-ethoxymethylfurfural via 5-hydroxymethylfurfural with deep eutectic solvents (DESs)." *Industrial Crops & Products* 112. Elsevier Science Ltd: 18–23. doi: 10.1016/j.indcrop.2017.11.001.

M J Murphy, J D Taylor, R L McCormick. 2004. "Compendium of experimental cetane number data." Natl Renew Energy Lab.

M S Pérez-Coello, M A González-Viñas, E García-Romero, M C Díaz-Maroto, M D Cabezudo. 2003. "Influence of storage temperature on the volatile compounds of young white wines." *Food Control* 14. Elsevier Science Ltd: 301–306. doi: 10.1016/S0956-7135(02)00094-4.

Omidvarborna Hamid, Kumar Ashok, Kim Dong-Shik. 2015. "Recent studies on soot modeling for diesel combustion." *Renew. Sustain. Energy Rev.* 48. Elsevier Science Ltd: 635–647. doi: 10.1016/j.rser.2015.04.019.

Omidvarborna Hamid, Kumar Ashok, Kim Dong-Shik. 2014. "Characterization of particulate matter emitted from transit fueled with B20 in idle modes." *J. Environ. Chem. Eng.* 2. Elsevier Science Ltd: 2335–2342. doi: 10.1016/j.jece.2014.09.020.

Patrícia Neves, Margarida M Antunes, Patrícia A Russo, Joana P Abrantes, Sérgio Lima, Auguste Fernandes, Martyn Pillinger, Sílvia M Rocha, Maria F Ribeiro, Anabela A Valente. 2013. "Production of biomass-derived furanic ethers and levulinate esters using heterogeneous acid catalysts." *Green Chem.* 15. Royal Soc Chemistry: 3367–3376. doi: 10.1039/C3GC41908H.

Penghua Che, Fang Lu, Junjie Zhang, Yizheng Huang, Xin Nie, Jin Gao, Jie Xu. 2012. "Catalytic selective etherification of hydroxyl groups in 5-hydroxymethylfurfural over $H_4SiW_{12}O_{40}$/MCM-41 nanospheres for liquid fuel production." *Bioresour. Technol.* 119. Elsevier Science Ltd: 433–436. doi: 10.1016/j.biortech.2012.06.001.

P Lanzafame, D M Temi, S Perathoner, G Centi, A Macario, A Aloise, G Giordano. 2011. "Etherification of 5-hydroxymethyl-2-furfural (HMF) with ethanol to biodiesel components using mesoporous solid acidic catalysts." *Catal. Today* 175. Elsevier Science Ltd: 435–441. doi: 10.1016/j.cattod.2011.05.008.

P Lanzafame, K Barbera, S Perathoner, G Centi, A Aloise, M Migliori, A Macario, J B Nagy, G Giordano. 2015. "The role of acid sites induced by defects in the etherification of HMF on Silicalite-1 catalysts." *J. Catal.* 330. Elsevier Science Ltd: 558–568. doi: 10.1016/j.jcat.2015.07.028.

Raju S Thombal, V H Jadhav. 2016. "Application of glucose derived magnetic solid acid for etherification of 5-HMF to 5-EMF, dehydration of sorbitol to isosorbide, and esterification of fatty acids." *Tetrahedron Lett.* 57. Elsevier Science Ltd: 4398–4400. doi: 10.1016/j.tetlet.2016.08.061.

R J H Grisel, J C van der Waal, E de Jong, W J J Huijgen. 2014. "Acid catalysed alcoholysis of wheat straw: towards second generation furan-derivatives." *Catal. Today* 223. Elsevier Science Ltd: 3–10. doi: 10.1016/j.cattod.2013.07.008.

Robert-Jan van Putten, Jan C van der Waal, Ed de Jong, Carolus B Rasrendra, Hero J. Heeres, Johannes G de Vries. 2013. "Hydroxymethylfurfural, a versatile platform chemical made from renewable resources." *Chem. Rev.* 113. American Chemical Society: 1499–1597. doi: 10.1021/cr300182k.

Ruitao Lv, Joshua A Robinson, Raymond E Schaak, Du Sun, Yifan Sun, Thomas E Mallouk, Mauricio Terrones. 2015. "Transition metal dichalcogenides and beyond: synthesis, properties, and applications of single and few-layer nanosheets." *Acc. Chem. Res.* 48(1). American Chemical Society: 56–64. doi: 10.1021/ar5002846.

Ryan M West, Zhen Y Liu, Maximilian Peter, Christian A Gärtner, James A Dumesic. 2008. "Carbon-carbon bond formation for biomass-derived furfurals and ketones by aldol condensation in a biphasic system." *J. Mol. Catal. A: Chem.* 296. Elsevier Science Ltd: 18–27. doi: 10.1016/j.molcata.2008.09.001.

Shanshan Yin, Jie Sun, Bing Liu, Zehui Zhang. 2015. "Magnetic material grafted cross-linked imidazolium based polyionic liquids: an efficient acid catalyst for the synthesis of promising liquid fuel 5-ethoxymethylfurfural from carbohydrates." *J. Mater. Chem. A.* 3. Royal Soc Chemistry: 4992–4999. doi: 10.1039/C4TA06135G.

Shireen Quereshi, Ejaz Ahmad, Kamal K K Pant, Suman Dutta. 2020. "Insights into microwave-assisted synthesis of 5-ethoxymethylfurfural and ethyl levulinate using tungsten disulfide as a catalyst." *ACS Sustain. Chem. Eng.* 8(4). American Chemical Society: 1721–1729. doi: 10.1021/acssuschemeng.9b03231.

Shuguo Wang, Zehui Zhang, Bing Liu, Jinlin Lia. 2013. "Silica coated magnetic Fe_3O_4 nanoparticles supported phosphotungstic acid: a novel environmentally friendly catalyst for the synthesis of 5-ethoxymethylfurfural from 5-hydroxymethylfurfural and fructose." *Catal. Sci. Technol.* 3. Royal Soc Chemistry: 2104–2112. doi: 10.1039/C3CY00223C.

Shunmugavel Saravanamurugan, Anders Riisager. 2012. "Solid acid catalysed formation of ethyl levulinate and ethyl glucopyranoside from mono-and disaccharides." *Catal. Commun.*17. Elsevier Science Ltd: 71–75. doi: 10.1016/j.catcom.2011.10.001.

Siew Ping Teong, Guangshun Yi, Yugen Zhang. 2014. "Hydroxymethylfurfural production from bioresources: past, present and future." *Green Chem.* 16. Royal Soc Chemistry: 2015–2026. doi: 10.1039/C3GC42018C.

Stanisław K Tyrlik, Dorota Szerszeń, Marian Olejnik, Witold Danikiewicz. 1999. "Selective dehydration of glucose to hydroxymethylfurfural and a one-pot synthesis of a 4-acetylbutyrolactone from glucose and trioxane in solutions of aluminium salts." *Carbohydr. Res.* 315. Elsevier Science Ltd: 268–272. doi: 10.1016/S0008-6215(99)00037-3.

Sudipta De, Saikat Dutta, Basudeb Saha. 2012. "One-pot conversions of lignocellulosic and algal biomass into liquid fuels." *ChemSusChem* 5. Elsevier Science Ltd: 1826–1833. doi: 10.1002/cssc.201200031.

Suhas Shinde, Chandrashekhar Rode. 2017. "Cascade reductive etherification of bioderived aldehydes over Zr-based catalysts." *ChemSusChem* 10. Wiley-VCH Verlag GmbH: 4090–4101. doi: 10.1002/cssc.201701275.

Wenqin Shen, Geoffrey A Tompsett, Karl D Hammond, Rong Xing, Fulya Dogan, Clare P Grey, W Curtis Conner Jr., Scott M Auerbach, George W Huber. 2011. "Liquid phase aldol condensation reactions with $MgO-ZrO_2$ and shape-selective nitrogensubstituted NaY." *Appl. Catal. A: Gen.* 392. Elsevier Science Ltd: 57–68. doi: 10.1016/j.apcata.2010.10.023.

Xiaofang Liu, Hu Li, Hu Pan, Heng Zhang, Shan Huang, Kaili Yang, Wei Xue, Song Yang. 2016. "Efficient catalytic conversion of carbohydrates into 5-ethoxymcthylfurfural over MIL-101-based sulfated porous coordination polymers." *J. Energy Chem.* 25. Elsevier Science Ltd: 523–530. doi: 10.1016/j.jechem.2016.01.015.

Xiaoxue Zhan, Chen Si, Jian Zhou, Zhimei Sun. 2020. "MXene and MXene-based composites: synthesis, properties and environment-related applications." *Nanoscale Horiz.* 5(2). Royal Soc Chemistry: 235–258. doi: 10.1039/c9nh00571d.

Xiuquan Jia, Jiping Ma, Penghua Che, Fang Lu, Hong Miao, Jin Gao, Jie Xu. 2013. "Direct conversion of fructose-based carbohydrates to 5-ethoxymethylfurfural catalyzed by $AlCl_3{\cdot}6H_2O/BF_3(Et)_2O$ in ethanol." *J. Energy Chem.* 22. Elsevier Science Ltd: 93–97. doi: 10.1016/S2095-4956(13)60012-1.

Xuanmu Zhou, Zehui Zhang, Bing Liu, Quan Zhou, Shuguo Wang, Kejian Deng. 2014. "Catalytic conversion of fructose into furans using $FeCl_3$ as catalyst." *J. Ind. Eng. Chem.* 20. Elsevier Science Ltd: 644–649. doi: 10.1016/j.jiec.2013.05.028.

Yongshen Ren, Bing Liu, Zehui Zhang, Jintao Lin. 2015. "Silver-exchanged heteropolyacid catalyst (Ag_1H_2PW): an efficient heterogeneous catalyst for the synthesis of 5-ethoxymethylfurfural from 5-hydroxymethylfurfural and fructose." *J. Ind. Eng. Chem.* 21. Elsevier Science Ltd: 1127–1131. doi: 10.1016/j.jiec.2014.05.024.

Yuanyuan Bai, Lu Wei, Mengfei Yang, Huiyong Chen, Scott Holdren, Guanghui Zhu, Dat T Tran, Chunli Yao, Runcang Sun, Yanbo Pan, Dongxia Liu. 2018. "Three-step cascade over a single catalyst: synthesis of 5-(ethoxymethyl)furfural from glucose over a hierarchical lamellar multi-functional zeolite catalyst." *J. Mater. Chem. A.* 6. Royal Soc Chemistry: 7693–7705. doi: 10.1039/C8TA01242C.

Yuriy Roman-Leshkov, Juben Chheda, James A Dumesic. 2006. "Phase modifiers promote efficient production of hydroxymethylfurfural from fructose." *Science* 312. 1933–1937. doi: 10.1126/science.1126337.

Yuriy Román-Leshkov, Christopher J Barrett, Zhen Y Liu, James A Dumesic. 2007. "Production of dimethylfuran for liquid fuels from biomass-derived carbohydrates." *Nature* 447. Nature Portfolio: 982–985. doi: 10.1038/nature05923.

Yu Yang, Changwei Hu, Mahdi M Abu-Omar. 2012. "Conversion of glucose into furans in the presence of $AlCl_3$ in an ethanol-water solvent system." *Bioresour. Technol.* 116. Elsevier Science Ltd: 190–194. doi: 10.1016/j.biortech.2012.03.126.

Yu Yang, Mahdi M Abu-Omar, Changwei Hu. 2012. "Heteropolyacid catalyzed conversion of fructose, sucrose, and inulin to 5-ethoxymethylfurfural, a liquid biofuel candidate." *Appl. Energy* 99. Elsevier Science Ltd: 80–84. doi: 10.1016/j.apenergy.2012.04.049.

Y Yao, Z Gu, Y Wang, H J Wang, W Li. 2016. "Magnetically recoverable carbonaceous material: an efficient catalyst for the synthesis of 5-hydroxymethylfurfural and 5-ethoxymethylfurfural from carbohydrates." *Russ. J. Gen. Chem.* 86. Springer: 1698–1704. doi: 10.1134/S1070363216070276.

Zehui Zhang, Yimei Wang, Zhongfeng Fang, Bing Liu. 2014. "Synthesis of 5-ethoxymethylfurfural from fructose and inulin catalyzed by a magnetically recoverable acid catalyst." *ChemPlusChem* 79. Wiley-VCH Verlag GmbH: 233–240. doi: 10.1002/cplu.201300301.

Zhenhua Wang, Qianwang Chen. 2016. "Conversion of 5-hydroxymethylfurfural into 5-ethoxymethylfurfural and ethyl levulinate catalyzed by MOF-based heteropolyacid materials." *Green Chem.* 18. Royal Soc Chemistry: 5884–5889. doi: 10.1039/C6GC01206J.

Zhongwei Wang, Hu Li, Chengjiang Fang, Wenfeng Zhao, Tingting Yang, Song Yang. 2017. "Simply assembled acidic nanospheres for efficient production of 5-ethoxymethylfurfural from 5-hydromethylfurfural and fructose." *Energ. Technol.* 5. Wiley-VCH Verlag GmbH: 2046–2054. doi: 10.1002/ente.201700153.

Ziliang Yuan, Zehui Zhang, Judun Zheng, Jintao Lin. 2015. "Efficient synthesis of promising liquid fuels 5-ethoxymethylfurfural from carbohydrates." *Fuel* 150. Elsevier Science Ltd: 236–242. doi: 10.1016/j.fuel.2015.02.020.

9 Oxidative Esterification of 5-Hydroxymethylfurfural and Furfural

Dinesh Kumar Mishra, Cong Chien Truong, Sang Hyeok Ko, and Young-Woong Suh

9.1 INTRODUCTION

Over the last few decades, fossil fuels (crude oil, coal, and natural gas) have played a significant role in satisfying all our demands for many useful commodity chemicals and fuel products. However, the process of extracting these chemicals and fuel products from fossil fuels used as feedstocks is highly destructive to several aspects of the economy, energy saving, and the environment. Since there were several disadvantages of fossil fuels, it therefore became necessary to find alternative resources and develop technologies for economical and sustainability viewpoints. In these circumstances, biomass was recognized as a renewable resource that could indeed fulfil most of all of our demands. Many technologies have been developed using first-generation edible biomass (e.g., sugarcane, corn, potatoes, and rice) for the sustainable production of chemicals and fuels. One of the interesting examples is bio-based ethanol production from plant-derived corn starch (Inlow et al. 1988; Ashikari et al. 1989). In contrast, lignocellulosic biomass (or lignocellulose), which generally comes from non-edible plant species, is to be used as much as possible in sustainable production of bio-based chemicals and fuels, which is indeed highly desirable. In fact, the large-scale production of lignocellulose biomass is also possible in an eco-friendly manner. Approximately 181.5 billion tons of lignocellulose are produced each year in nature. A total of 8.2 billion tons of lignocellulose are used, which comes from a variety of many plant species (e.g., forest residues, agricultural residues, crops, and grasses) (Ashokkumar et al. 2022). The lignocellulose is made up mostly of three components, cellulose (40%), hemicellulose (35%), and lignin (25%), including other (5.0%), as shown in Scheme 9.1 (Ruppert et al. 2012). Cellulose is made up of D-glucose units joined to each other with 1,4-β-glycosidic linkage. Unlike cellulose, hemicellulose is a branched biopolymer, which is made up of pentoses (xylose and arabinose), hexoses (mannose, glucose, galactose, and rhamnose), and acetylated sugars (Gibson 2012). Apart from cellulose and hemicellulose, lignin is

DOI: 10.1201/9781003464839-9

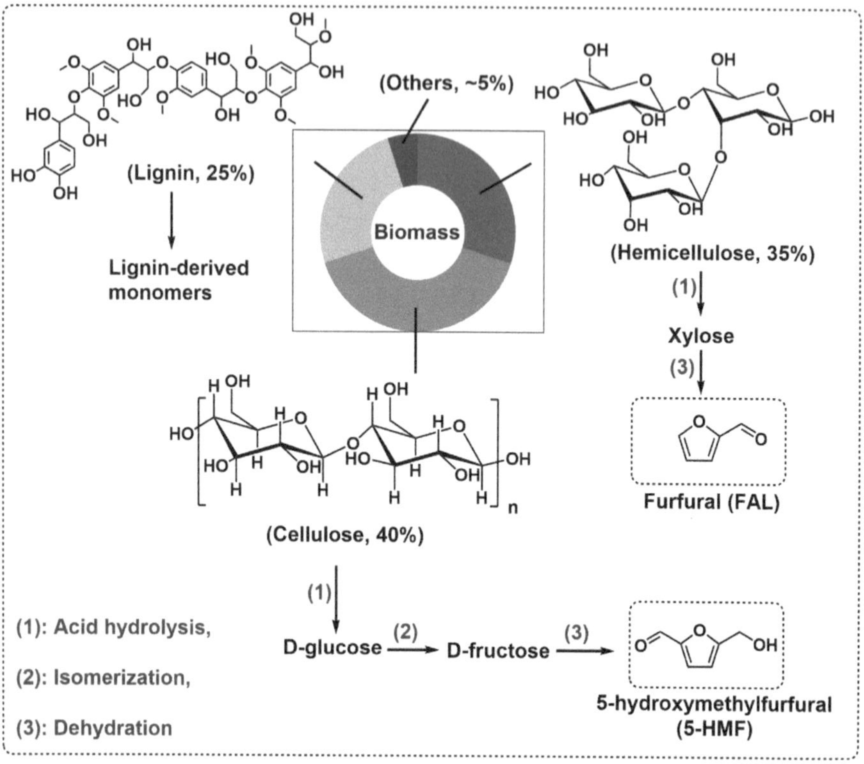

SCHEME 9.1 Composition of lignocellulosic biomass (originally drawn based on Ruppert et al. 2012).

a complex cross-linked polymer of three-dimensional monomeric units (p-coumaryl alcohol, coniferyl alcohol, and sinapyl alcohol) (Baruah et al. 2018; Ragauskas et al. 2014). Thus, using lignocellulose is very important to replace the conventional fossil fuels-based feedstocks and is indeed considered the greener approach.

There is no doubt that many efficient catalytic processes (e.g. hydrolysis, dehydration, oxidation, esterification, hydrogenation, and hydrodeoxygenation) have been developed for converting sustainable lignocellulosic biomass or its derived chemicals (Gallezot 2012; Hara et al. 2015; De et al. 2022; Mori 2023). Among these methodologies as abovementioned, oxidation of biomass-derived 5-hydroxymethyl furfural (5-HMF) is considered as one of the environmentally benign approaches to produce 2,5-furandicarboxylic acid (FDCA) (Kar et al. 2022; Jing et al. 2023). In this context, many researchers focused on aerobic oxidation of 5-HMF to produce FDCA using various supported metal-based catalysts (Gorbanev et al. 2009; Casanova, et al. 2009; Davis et al. 2011; Artz et al. 2015; Ait Rass et al. 2015; Mishra et al. 2017). 5-HMF is an important platform chemical with market size in 62 USD million in 2022. It estimated to increase 69 USD million by the end of 2029 at moderate compound annual growth rate (CAGR) rate of 2.5%. It could be noticed that FDCA, an

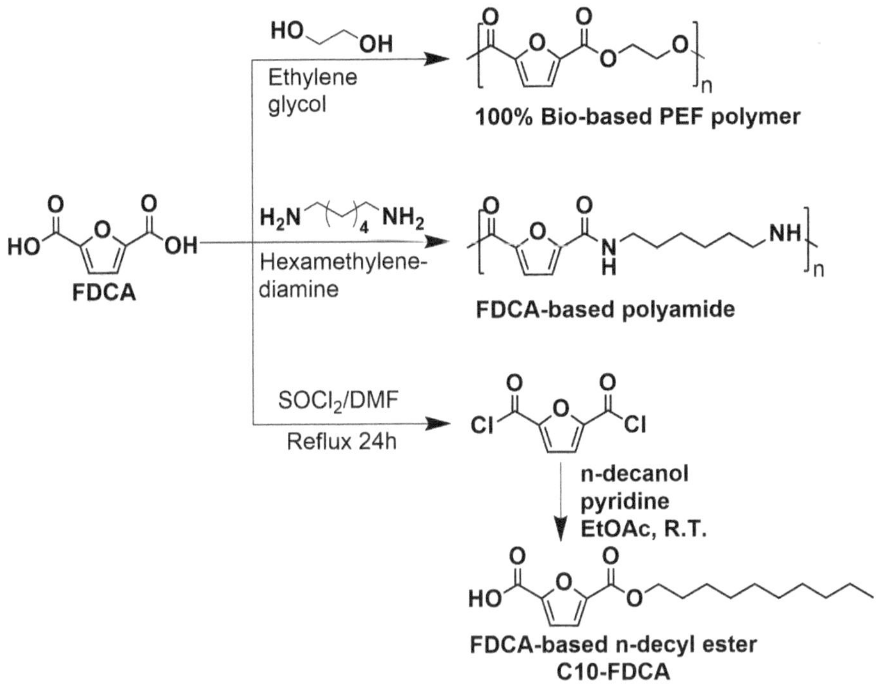

SCHEME 9.2 FDCA and its derived bio-based polymers, including surfactant.

oxidative product of 5-HMF, can further be used as a monomer to synthesize a variety of bio-based polymers, including surfactant and plasticizers (Scheme 9.2) (Cousin et al. 2018; https://polymerinnovationblog.com/polyethylene-furanoate-pef-100-biobased-polymer-to-compete-with-pet/; van Es et al. 2013; He et al. 2020). Owing to increased demand of environmentally friendly plastics, the global FDCA market size is anticipated to increase from 481.4 USD million (in 2022) to 980 USD million at a CAGR of 8.2% by the year 2031 (www.transparencymarketresearch.com/furandicarboxylic-acid-market.html). Despite the wide-range applications of FDCA products as aforementioned, FDCA is known to be less soluble in water and most of the organic solvents. Low solubility of FDCA is hampering the polymerization process for the synthesis of 100% bio-based PEF polymer. In contrast, oxidative esterification (one-pot coupling of oxidation and esterification) of 5-HMF has attracted great attention to produce the corresponding ester product, i.e., furan-2,5-dimethylcarboxylate (FDMC) (Zhao et al. 2022). FDMC can be used as a renewable alternative to FDCA monomer, which is highly soluble in most of the common solvents (Liu et al. 2022). Rather than FDMC, another ester, 2-methyl furoate, is used as flavor and fragrance components with huge industrial applications. Herein, we mainly focused on oxidative esterification of 5-HMF and furfural (FAL) used to produce their corresponding ester products (FDMC and 2-methyl furoate) with various supported noble and non-noble-based heterogeneous catalysts, which are described well. Furthermore, several new insights are presented into the supported catalysts as well as the reaction conditions (with and without base additives) used for oxidative esterification of 5-HMF and FAL. At the end of this chapter, the potential benefits and the future opportunities of esters and their derived bio-based polymers are also addressed.

9.2 SOURCE OF 5-HMF AND FAL

5-HMF is recognized as a major lignocellulosic-biomass-derived platform chemical. It has a low melting point (melts in between 28–34 °C). The color of 5-HMF is white while the commercially available samples of 5-HMF are often yellow. Many other physicochemical properties of 5-HMF are summarized in Table 9.1.

At first, 5-HMF is recognized as an intermediate during synthesis of levulinic acid (Grote and Tollens 1875). In fact, 5-HMF possesses multi-functionalities (aldehyde group, hydroxyl group, and furan ring) in its structure. Because of its unique

TABLE 9.1
Physicochemical Properties of 5-HMF and FAL

Name	5-HMF	Name	FAL
Molecular formula	$C_6H_6O_3$	Molecular formula	$C_5H_4O_2$
Molar mass	126.11	Molar mass	96.08
Density	1.243 g/mL at 25 °C (lit.)	Density	1.16 g/mL at 25 °C (lit.)
Melting point	28–34 °C	Melting point	−36 °C
Boiling point	114–116 °C/1 mmHg (lit.)	Boiling point	162 °C/1mmHg (lit.)
Flash point	175 °F (79 °C)	Flash point	137 °F (58 °C)
Solubility	Soluble in water, alcohol, ethyl acetate, acetone, carbon tetrachloride, and other conventional solvents	Solubility	95% ethanol (1mL/mL), clear solution, ether, 8.3% soluble in water
Vapor pressure	0.000891 mmHg at 25 °C	Vapour pressure	3.0 mmHg at 30 °C
Appearance	Liquid or crystalline powder and/or chunks	Appearance	Liquid
Color	Light yellow to yellow	Color	Very deep brown
pKa	12.82±0.10 (Predicted)	pKa	Not applicable
Storage condition	2–8 °C	Storage condition	Below 30 °C
Stability	Light sensitive and very hygroscopic	Stability	Avoid the contact with acid, base, and oxidizing agent
Sensitive	Air and light sensitive	Sensitive	Air and sunlight sensitive
Refractive Index	n20/D 1.562 (lit.)	Refractive Index	n20/D 1.527 (lit.)

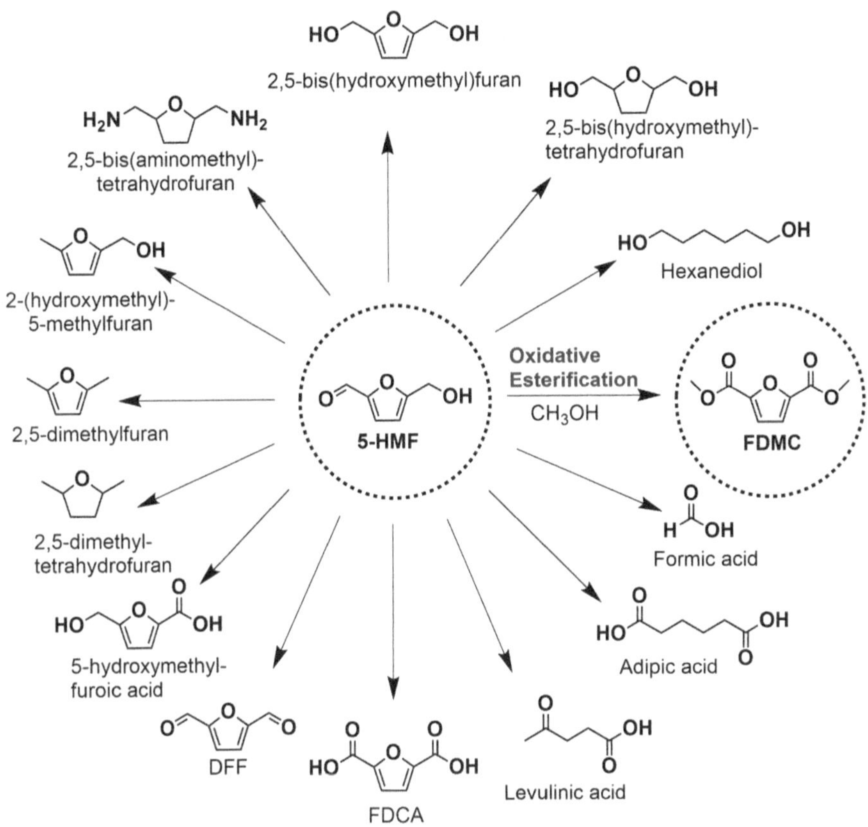

SCHEME 9.3 5-HMF and its derived valuable chemicals, including FDMC.

structure, 5-HMF is used as a precursor for the synthesis of various valuable chemicals, including FDMC product (Scheme 9.3).

In addition, 5-HMF is also listed as one of the "top 12 bio-based platform chemicals" suggested by US department of energy (DOE) (www.nrel.gov/docs/fy04osti/35523.pdf; Silva et al. 2022). 5-HMF is typically synthesized from renewable lignocellulosic biomass (LCB), which is abundantly available in nature (Lima et al. 2011; Zakrzewska et al. 2011). Lignocellulosic biomass is entirely made up of three major components, such as cellulose, hemicellulose, and lignin, including others as mentioned earlier (as shown in Scheme 9.1). It is also worth mentioning the fraction of cellulose, hemicellulose, and lignin may vary depending on the type of lignocellulosic biomass (Devi et al. 2020; Putro et al. 2016; Ruppert et al. 2012). As a result of various pretreatment techniques such as physical, chemical, thermophysical, thermochemical, and biological, lignocellulose biomass (LCB) is fractionated into cellulose, hemicellulose, and lignin (Scheme 9.4) (Putro et al. 2016; Zabed et al. 2019; Mishra et al. 2020). The cellulose is then broken into small fragments of monomeric D-glucose units (hexose) via acid-hydrolysis (Huang and Fu 2013; Suganuma

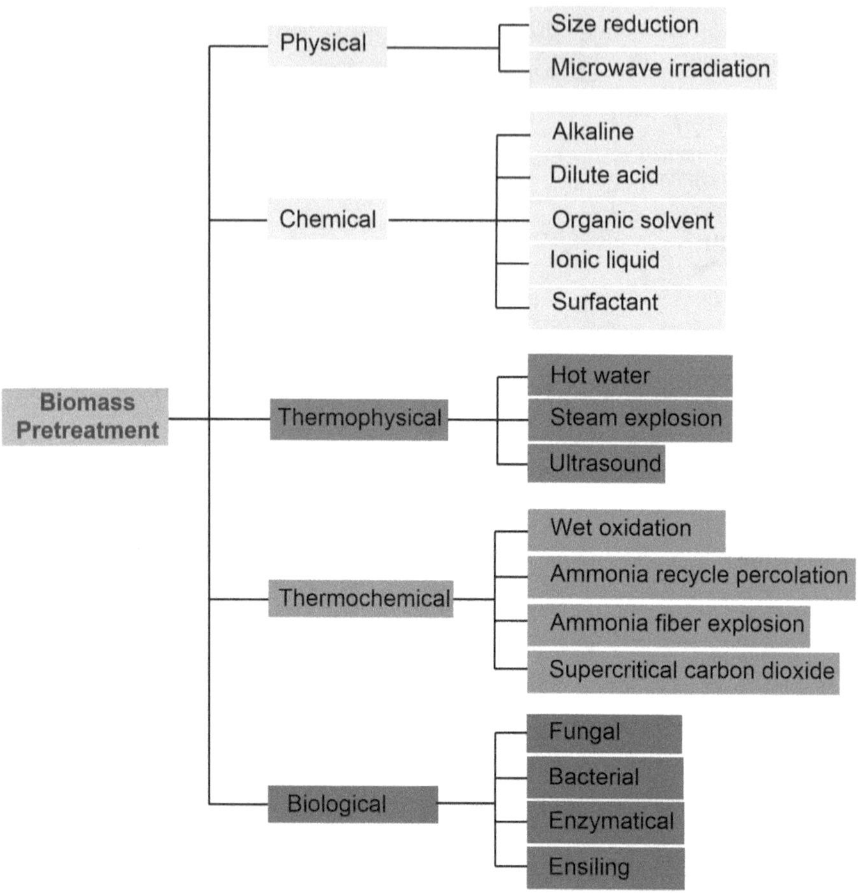

SCHEME 9.4 Pretreatment techniques of lignocellulosic biomass (originally drawn based on Putro et al. 2016; Zabed et al. 2019).

et al. 2008). The obtained D-glucose is isomerized to form D-fructose (Ju et al. 2019; Marianou et al. 2018). Finally, D-fructose undergoes dehydration to form 5-HMF in the presence of Brønsted acid (Scheme 9.1) (Wang et al. 2014; Nandiwale et al. 2014; Liu, et al. 2020; Shen et al. 2016; Swift et al. 2014).

On the other hand, FAL is a colorless motile liquid organic compound. It could be known that auto-oxidation of FAL occurs when it is exposed in sunlight in the presence of oxygen. So, the color of FAL is changed to dark/red color sometimes. Many other properties of FAL are as summarized in Table 9.1. After LCB pretreatment, hemicellulose is obtained (Alonso et al. 2013; Scapini et al. 2021; Kumar et al. 2009). Xylose (pentose), a hydrolyzed product of xylan (pentosan) present in hemicellulose, is dehydrated to form FAL in the presence of Brønsted acid (Scheme 9.1) (Mariscal et al. 2016; Peleteiro et al. 2015; Choudhary et al. 2012). FAL has also been

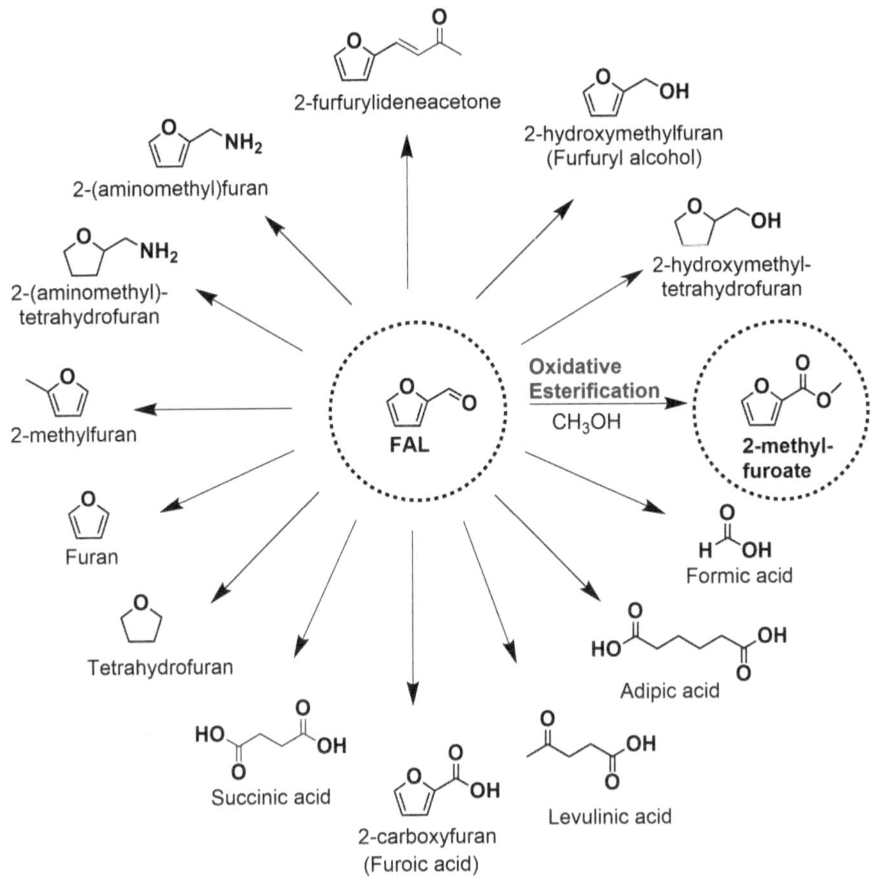

SCHEME 9.5 FAL and its derived valuable chemicals, including 2-methyl furoate.

used as a precursor for various chemicals, including 2-methyl furoate with wide range of applications (Scheme 9.5).

9.3 OXIDATIVE ESTERIFICATION OF 5-HMF TO FDMC

FDMC is a versatile chemical, which is used for the synthesis of plentiful advance materials and chemicals. Moreover, FDMC can also be used as such in polyester synthesis through the transesterification reaction (Taarning et al. 2008). In contrast to FDCA, FDMC can easily be separated and purified by vacuum distillation. Until today, there has been a growing focus on the synthesis of economically viable FDMC through oxidative esterification of 5-HMF derived from lignocellulosic biomass. In order to produce FDMC (with and without base additives), a variety of heterogeneous catalysts based on supported noble metals and non-metals were employed, which are described in the following sections.

9.3.1 Noble Metal Catalyzed Oxidative Esterification of 5-HMF

Among the vast majority of noble metal-based heterogeneous catalysts, which can execute the reaction under milder conditions, gold (Au) nanoparticles are proven to be excellent catalysts in many catalytic process including oxidative esterification of 5-HMF derived from lignocellulose (Bond et al. 1999; Turner et al. 2008; Liu et al. 2013; Weerathunga et al. 2021). For instance, Taarning et al. described oxidative esterification of 5-HMF, which was carried out in the presence of titania supported gold nanoparticles (Au/TiO_2) as the catalyst and sodium methoxide (CH_3ONa) as the base additive. The results showed that the FDMC product with high yield (98%) was accomplished only in 3 h at 130 °C under 4.0 bar O_2 (Taarning et al. 2008). Oxidative esterification of 5-HMF proceeded hardly even after longer reaction times in the absence of base additive. It could further be suggested that the FDMC product was formed via the formation of two 5-hydroxymethyl methylfuroate (HMMF) and methyl 5-formyl-2-furoate (MFF) intermediates (Scheme 9.6). The formation rate of HMMF (an intermediate) from 5-HMF was much faster than the formation rate of MMF (another intermediate) generated from HMMF. Finally, the MFF intermediate was converted to the FDMC product. It is also worth mentioning that the Au/TiO_2 catalyst was also easily recovered by the simple filtration method and reused.

In this continuation, Buonerba et al. reported a new class of polymer-based highly efficient catalyst, a syndiotactic poly(styrene)-cis-1,4-poly(butadiene) copolymer supported Au nanoparticles (Au/sPSB), for oxidative esterification of 5-HMF (Buonerba et al. 2018). The matrix of sPSB copolymer possess exceptional mechanical properties, good chemical, and thermal stability, physical cross-links, and complex polymorphism. Remarkably, these peculiar properties of polymer matrix and high dispersion of Au nanoparticles with particle size (10.1±3.1 nm) played a significant role to give an excellent yield of FDMC (>99%) from oxidative esterification of 5-HMF at 110 °C and 35 bar O_2. Moreover, cesium carbonate ($CsCO_3$) as a base additive and a mixture of solvents (DMA//methanol) were also used. Indeed, Au monometallic noble metal-based catalysts gave an excellent yield of FDMC from

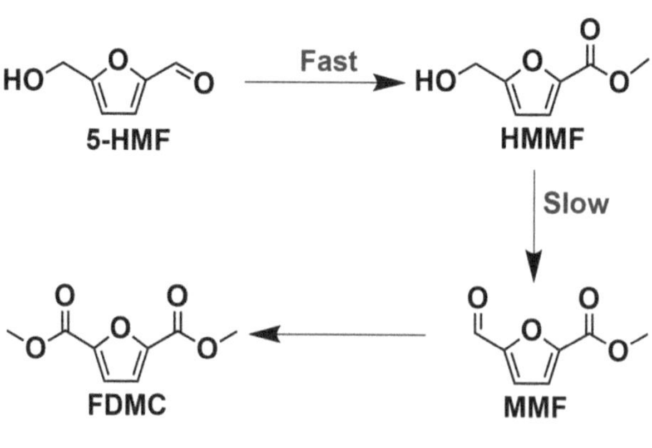

SCHEME 9.6 Reaction pathway of oxidative esterification of 5-HMF to FDMC.

oxidative esterification of 5-HMF, however, the harsh reaction conditions (high temperature and high pressure) were applied. Later, Kim and co-workers developed a bimetallic AuPd alloy nanoparticle supported on iron oxide (Fe_3O_4) for oxidative esterification of 5-HMF. Oxidative esterification of 5-HMF was performed using alloy AuPd/Fe_3O_4 as the catalyst and potassium carbonate (K_2CO_3) as an optimal base under the milder reaction conditions (at room temperature under atmospheric oxygen pressure). Interestingly, 92% yield of FDMC was obtained under the applied reaction conditions (at room temperature and 1 bar O_2) (Cho et al. 2019). The used mild reaction conditions diminished the degradation of 5-HMF or saponification, which has often been reported in the literature at harsh reaction conditions (Vuyyuru and Strasser 2012; Davis et al. 2011). In this continuation, Li and co-workers reported an interesting result on aerobic oxidation of 5-HMF to FDMC, which was carried out in the presence of bimetallic PdCoBi/C catalyst with ratio (Pd: Co: Bi 1: 1: 1) and K_2CO_3 used as the base. The high yield of FDMC (96%) was accomplished at low temperature (60 °C) and low pressure of oxygen (atmospheric pressure) (Li et al. 2018). Later, Salazar et al. performed the selective oxidative esterification of 5-HMF using a mixture of Co_xO_y-N@C+Ru@C used as a bimetallic catalyst in the presence of K_2CO_3. This Co_xO_y-N@C+Ru@C bimetallic catalyst afforded excellent yield of FDMC (99%) and 100% conversion of 5-HMF at 50 °C in 18 h (Salazar et al. 2019). This excellent yield of FDMC (99%) is ascribed due to the synergetic effect between Ru@C and Co_xO_y-N@C, which could generate superoxide radicals on the Co surface and could further migrate the active oxygen molecule to the Ru active sites. Under the identical reaction conditions, other metal-based catalysts (Pt, Ru, and Pd) gave only a trace amount of FDMC from oxidative esterification of 5-HMF. As mentioned above in most of the cases, bases such as CH_3ONa, Cs_2CO_3, and K_2CO_3 were used as additive materials. However, the use of base additives makes the whole process less green and more expensive. Indeed, the process needs extra efforts for the neutralization of products' mixture with acids after reaction. To overcome this challenge, the Corma research group described one-pot-base-free oxidative esterification of 5-HMF to FDMC with various metal oxides (e.g., Fe_2O_3, TiO_2, and ceria) and carbon supported Au nanoparticles. Among all, Au on nano-particulated ceria (Au/CeO_2) afforded 99% yield of FDMC at almost complete conversion of 5-HMF in 5 h, at 130 °C, and 10 bar O_2 (Casanova et al. 2009a). In fact, the availability of Lewis acid sites on the surface of Au nanoparticles stabilized by Ce atoms is found to be more cooperative for the oxidation of alcohol in the initial steps of the reaction (Abad et al. 2005). Taking into account the reaction pathway (Scheme 9.7), there were two intermediate compounds 3 and 4, the formation of these intermediates could be possible from the oxidation and acetalization of hemiacetal intermediate 2 obtained from reactant compound 1. It is noteworthy to mention that hemiacetal intermediate 2 was more prone to oxidation rather than acetalization, as the concentration of intermediate compound 3 was found higher than that of intermediate compound 4 during oxidative esterification. Compound 3 then gradually underwent oxidation to form monoester aldehyde 5, followed by hemi-acetalization into compound 7 and was further re-oxidized to form monoester acetal 9 (pathway 1). The whole process resulted in the diester 10. On the other hand, the second pathway (pathway 2) involved the formation of acetal alcohol

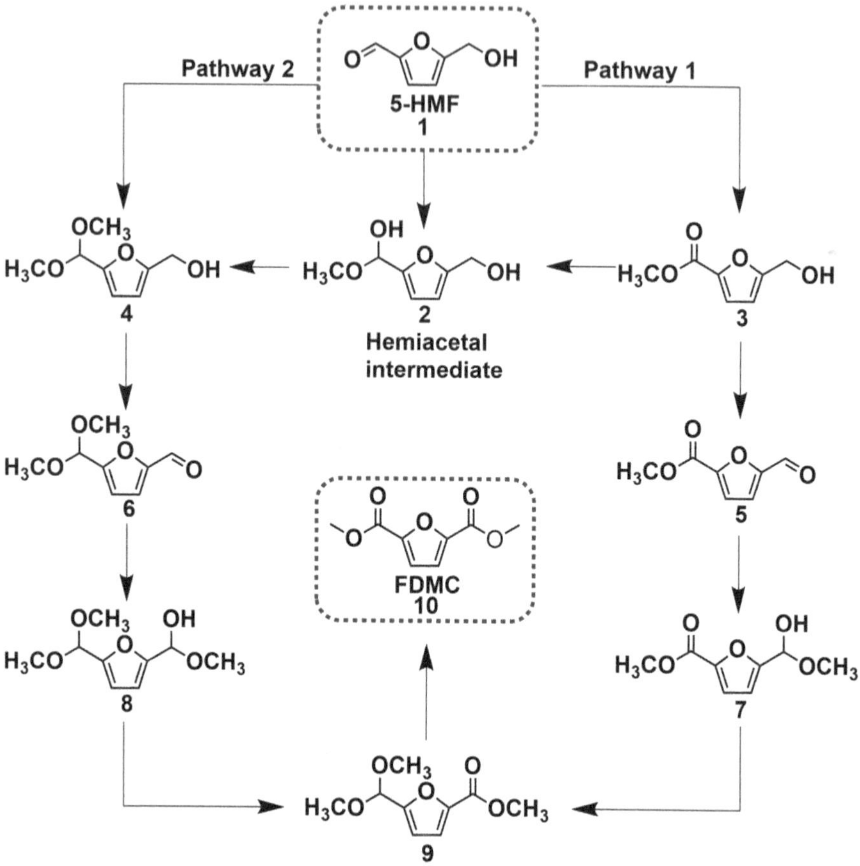

SCHEME 9.7 Reaction pathway of FDMC production from oxidative esterification of 5-HMF over Au/CeO_2 catalyst (originally drawn based on Casanova et al. 2009).

4 derived from acetalization of hemiacetal intermediate 2, the disguised primary and unstable compound. Furthermore, the aldehyde acetal 6 and the hemiacetal 8 were rapidly formed and detected at the trace level. Compound 9 was proven to be an obligatory stage for the two reaction pathways.

Moreover, it could be found that the Au/CeO_2 catalyst could easily recover and reuse with no significance loss of FDMC yield, but the selective of the FDMC product was maintained. On the other hand, the Au/TiO_2 catalyst could give 96% yield of FDMC even after a long time (24 h). Meanwhile, Au/Fe_3O_4 and Au/C catalysts exhibited only 6% and 7% yield of FDMC, respectively. In another example, Menegazzo et al. tested sulphated zirconia supported Au nanoparticles (Au/SZ with 2 wt% nominal sulphate content) in oxidative esterification of 5-HMF. The yield of FDMC was quite low as 32% at 100% conversion of 5-HMF. The poor selectivity of the FDMC product is attributed to the irregular plate-like Au nanoparticles (Menegazzo et al. 2015). Afterwards, Du et al. fabricated a series of bimetallic Au+Cu/γAl_2O_3 catalysts, which

TABLE 9.2
Noble Metal Catalyzed Oxidative Esterification of 5-HMF

Catalyst	Additive	*T*, °C	*t*, h	P_{O2}, bar	5-HMF Conv., %	FDMC Yield, %	Ref.
Au/TiO_2	CH_3ONa	130	3	4.0	99	98	(Taarning et al. 2008)
AuNPs-sPSB	Cs_2CO_3	110	5	35	99	>98	(Buonerba et al. 2018)
$AuPd\text{-}Fe_3O_4$	K_2CO_3	R.T.	24	1.0	100	92	(Cho et al. 2019)
PdCoBi@C	K_2CO_3	60	14	1.0	99	96	(Li et al. 2018)
Co_xO_y-N@C + Ru@C	K_2CO_3	50	16	1.0 (air)	100	99	(Salazar et al. 2019)
Au/CeO_2	No-base	130	5	10	>99	>99	(Casanova et al. 2009)
$Au\text{-}ZrO_2$	No-base	130	5	3.0	100	32	(Menegazzo et al. 2015)
$AuCu/Al_2O_3$	No-base	105	5	10	100	98	(Du et al. 2018)
Au-HAP	No-base	130	6	24 (air)	99.9	89.3	(Mishra et al. 2019; https://patentimages.storage.googleapis.com/57/c0/e0/55439b79e61134/US10947209.pdf)

were prepared by urea deposition-precipitation methods and followed by oxidation-reduction treatment. The excellent yield of FDMC (98%) was achieved from oxidative esterification of 5-HMF using the Au+Cu/γAl_2O_3 catalyst (atomic ratio of Au: Cu = 1: 1) at 105 °C in 5 h (Table 9.2). It could be suggested that the acidity of support and the synergism between Au and Cu metals played an important role towards high selectivity of the FDMC product (Du et al. 2018). Recently, Mishra et al. described another base-free protocol for oxidative esterification of 5-HMF using a hydroxyapatite supported Au (Au/HAP) nanocatalyst, which is the first base-free example in air condition. Interestingly, the Au/HAP nanocatalyst prepared by the standard homogeneous-deposition method showed 89.3% yield of FDMC (Mishra et al. 2019; https://patentimages.storage.googleapis.com/57/c0/e0/55439b79e61134/US10947209.pdf). In the presence of air condition, this remarkable yield of FDMC is due to the availability of strong basic sites on the surface of HAP (100 nm) used as the support material and its corresponding Au nanocatalyst with particle size of Au (2–4 nm). Furthermore, the Au/HAP nanocatalyst showed high stability even after five consecutive recycling runs.

9.3.2 Non-Noble Metal Catalyzed Oxidative Esterification of 5-HMF

There is no doubt that noble metal-based catalysts have shown the excellent yield of FDMC product from oxidative esterification of 5-HMF. However, the undeniable and high cost of noble-based catalysts circumscribed the scope for industrialization. For this purpose, synthesis of low-cost and eco-friendly non-noble metal-based catalysts is of great importance. Taking into consideration the low-cost and environmental impact related to the synthesis of non-noble-based catalysts, Deng and co-workers synthesized Co_xO_y-N@C as a highly active catalyst and applied it in oxidative esterification of 5-HMF to produce FDMC. The Co_xO_y-N@C was synthesized by wetness impregnation method and pyrolyzed in the temperature range from 600 to 800 °C (Deng et al. 2014). As reported, this catalyst was able to hold the chelating structures considered as active sites and these structures were uniformly distributed on carbon support. Above 800 °C, there was significant loss of active site densities due to breaking of the bond between metal and chelating ligands, resulting in decreased catalytic activity in oxidative esterification of 5-HMF. As reported, the Co_xO_y-N@C catalyst pyrolyzed at 800 °C showed the highest yield of FDMC (96%) at complete conversion of 5-HMF under the applied reaction conditions (100 °C, 10 bar O_2, K-OMS-2, and 0.2 eq. K_2CO_3). From the viewpoint of industrial-scale production, this Co_xO_y-N@C catalyst was successfully reused five times with no significance loss of original activity. In another interesting example, Feng et al. demonstrated a new class of ZIF-67-derived nano carbon composites (Co@C-N) used as an efficient catalyst in oxidative esterification of 5-HMF in the presence of Na_2CO_3 as the base. As shown in Table 9.3, this Co_xO_y-N@C(800) catalyst afforded 95% yield of FDMC, which is attributed to the electron transfer between nitrogen-doped carbon shells and Co nanoparticles (Feng et al. 2020). During oxidative esterification of 5-HMF, 5-HMF used as reactant, oxygen, and methanol molecules adsorbed and activated on the nitrogen-doped carbon (C-N) material. While, Co species acts as electron donor to enhance the electron density in C-N material. Moreover, this Co_xO_y-N@C(800) catalyst also exhibited good compatibility and stability during oxidative esterification of 5-HMF.

As mentioned earlier, base-free oxidative esterification of 5-HMF to produce FDMC is more preferable. In this continuation, Xu and co-workers described the base-metal approach for oxidative esterification of 5-HMF using the CoO_x-N/C catalyst, no additional base additive was used (Sun et al. 2016). Thanks to pyridinic nitrogen species in C-N material, which had Lewis base sites in its structure, the use of a base additive was avoided. This CoO_x-N/C catalyst in combination of αMnO_2 as the metal-base catalyst could afford 95% yield of FDMC under the applied reaction conditions (in 12 h at 100 °C and 6 bar O_2).

In another impressive work, Sun and co-workers developed a new strategy, the template protection-sacrifice (TPS) method, for the synthesis of hollow-yolk shell Co@CN nanocomposites following pyrolysis and etching (Sun et al. 2019). It is worthy to mention that SiO_2 played an important role in the catalyst synthesis process, which acted as a protecting shell in preventing agglomeration and fusion of Co/N nanoparticles during pyrolysis at high temperature 900 °C. The synthesized Co@CN catalyst with high surface area and pore volume showed a high yield of

FDMC (95%) from oxidative esterification of 5-HMF in 12 h (at 80 °C and atmospheric pressure of O_2) (Sun et al. 2019). In addition, the electron transfers from Co nanoparticles to Co/N nanoparticles was beneficial for the formation and release of active oxygen species. Moreover, the synthesized Co@CN catalyst was also easily recovered and reused five times in oxidative esterification of 5-HMF. In another new approach based on single atom catalysts (SACs), Zhou et al. reported the activity of various atoms (e.g., Fe, Co, Ni, and Cu) supported on nitrogen-doped carbon material as the whole catalyst in oxidative esterification of 5-HMF. In a short span of reaction time, in 9 h, this Co SAs-N@C catalyst showed 86% yield of FDMC at complete conversion of 5-HMF under such conditions (at 65 °C and 1 bar O_2) (Zhou et al. 2019). The synthesis approach could have the advantage of natural phenolic structure of lignin to spatially confine metal ions in metal-lignin complexes, opening a new route toward lignin valorization. Later, Lin and co-workers reported nitrogen-doped carbon supported Co/Zn (represented as Co@CN-Zn)-based catalyst for oxidative esterification of 5-HMF under mild conditions (at 50 °C) in 24 h. A high yield of FDMC (95%) complete conversion of 5-HMF was accomplished (Lin et al. 2020). Noticeably, Zn played very crucial roles not only in the synthesis process but also in the catalytic process of oxidative esterification of 5-HMF. At a pyrolysis temperature of 900 °C, evaporation of Zn resulted in the high surface area and the residual Zn in the whole catalyst was beneficial for increasing Lewis basic sites. Recently, Feng et al. reported another interesting result from oxidative esterification of 5-HMF using low-cost and nitrogen rich carbon nanosheets supported on the Co catalyst. At room temperature and atmospheric pressure of O_2, 92 % yield of FDMC was achieved (Feng et al. 2021). The high performance of the Co/CoN/CN700 catalyst in oxidative esterification of 5-HMF is due to electron transfer in between the metal and nitrogen-doped carbon nanosheets. Moreover, the Co/CoN/CN700 catalyst was reusable five times without loss of original activity.

In addition to the low-cost and earth-abundant Co-based catalysts, Tong et al. applied various catalysts, a series of non-noble metals (Cu, Cr, V, Mo, Mg, Ca, and Al) doped into MnO_2 support, for oxidative esterification of 5-HMF. Unfortunately, the conversion of 5-HMF was in between 39.4–77.6%. As shown in Table 9.3, the Cu-MnO_2 catalyst could give only 14% yield of FDMC under the applied reaction conditions (at 140 °C, 3 bar O_2, and 5 h), when the conversion of 5-HMF was 74.8% (Tong et al. 2017). In this continuation, Gupta et al. reported a new approach for oxidative esterification of 5-HMF using tert-butyl hydroperoxide (TBHP) as an oxidizing agent. For this purpose, the mesoporous alumina-nanospheres' embedded CuO nanoparticles' (CuO-mAl_2O_3) catalyst was prepared and used as the main catalyst. While, CuO-mAl_2O_3 catalysts exhibited 92% yield of FDMC at almost complete conversion of 5-HMF in 12 h and at 100 °C (Table 9.3) (Gupta et al. 2020). It is remarkable that both the CuO-mAl_2O_3 catalyst and TBHP used as the oxidizing agent played significance roles in oxidative esterification of 5-HMF. Initially, the reaction of TBHP and Cu(II) generates a *tert*-butoxyl radical and Cu(III) species in the mixture. The Cu(III) species on further reaction with an another molecule of TBHP get reduced to Cu(II) and simultaneously generates a *tert*-butylperoxyl radical. In the next step, *tert*-butoxyl radical gets converted into acetone and simultaneously generated a methyl

TABLE 9.3
Non-Noble Metal Catalyzed Oxidative Esterification of 5-HMF

Catalyst	Additive	*T*, °C	*t*, h	P_{O2}, bar	5-HMF Conv., %	FDMC Yield, %	Ref.
Co_xO_y-N@C	K_2CO_3	100	6	10	99	96	(Deng et al. 2014)
Co_xO_y-N@C(800)	Na_2CO_3	100	5	20	99	91	(Feng et al. 2020)
CoO_x-N/C + α-MnO_2	No-base	100	12	6	100	95.6	(Sun et al. 2016)
Co@NC	No-base	80	12	1	100	95	(Sun et al. 2019)
Co Sas-N@C	No-base	65	9	1	100	86.2	(Zhou et al. 2019)
Co@CN Zn	No-base	50	24	1	99	94	(Lin et al. 2020)
Co/CoN/CN-700	No-base	100	12	1	100	92	(Feng et al. 2021)
Cu/MnO_2	No-base	140	5	3	74.8	14	(Tong et al. 2017)
CuO/m-Al_2O_3	No-base	100	20	-	99	92	(Gupta et al.)
Co_7Cu_3-NC	No-base	80	4	2	100	95	(Liu et al. 2020)

radical. The substrate, 5-HMF or 2,5-dimethylfuran gets oxidized to FDCA in the presence of the oxidant TBHP. Subsequently, the reaction of 2,5-furandicarboxylic with *tert*-butoxyl or *tert*-butylperoxyl radicals generates acyloxyl radical intermediate. Finally, the successive coupling of the acyloxyl radical with methyl radicals results in the formation of the desired product FDMC. In the same year, another group of Liu et al. demonstrated nitrogen-doped carbon supported Co and Cu (Cu_xCo_y-NC) bimetallic catalyst for oxidative esterification of 5-HMF (Liu et al. 2020). The Cu-doping could effectively improve the amount of the active site (Co-N*x* species) with stronger electron intensity and the basicity of the catalyst through enhancing surface N content of the catalyst, thereby favoring the oxygen activation and hydrogen abstraction process during the oxidative esterification of 5-HMF. As shown in Table 9.3, the synthesized Cu_7Co_3-NC bimetallic catalyst could give a high yield of FDMC (95%) in 4 h at 80 °C and 2.0 bar O_2. Moreover, Cu_7Co_3-NC bimetallic catalyst was reused for six times with no significance loss of original activity.

9.4 OXIDATIVE ESTERIFICATION OF FAL TO 2-METHYL FUROATE

As one of the most promising molecules, 2-methyl furoate is used as a precursor to form the compounds that find use as flavor and fragrance components. The synthesis of 2-methyl furoate is an important chemical reaction in the fine chemical industry.

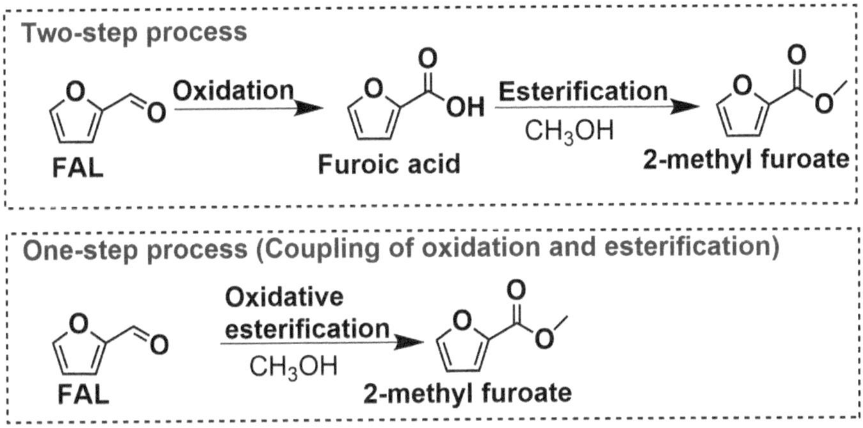

SCHEME 9.8 Formation of 2-methyl furoate from two-steps (oxidation and esterification) and one step (oxidative esterification) from FAL.

Traditionally, 2-methyl furoate was synthesized in two steps: (1) FAL was firstly converted into furoic acid in the presence of potassium permanganate ($KMnO_2$) as strong oxidant, preferably using acetone as solvent. (2) The synthesized furoic acid underwent esterification to form 2-methyl furoate in the presence methanol using concentrated H_2SO_4 (Scheme 9.8). However, these aggressive constituents, $KMnO_4$ and H_2SO_4, caused severe corrosion of the surface of used reaction equipment. It could further be noticed that the yield of 2-methyl furoate was not satisfactory and could not be improved. Recently, more attention is paid to perform one-pot oxidative esterification (one-pot coupling of oxidation and esterification) of FAL to produce 2-methyl furoate (Scheme 9.8), is most preferable and attractive sustainable approach. For this, various supported metals (noble and non-noble metals) used as catalysts were employed in oxidative esterification of FAL to produce 2-methyl furoate in the presence of base additives (with and without). Various supported noble and non-noble metal-based catalysts using oxidative esterification of FAL are discussed in the following sections.

9.4.1 Noble Metal Catalyzed Oxidative Esterification of FAL

Noble metal-based catalysts are known to be highly active to execute the reaction under milder conditions and stable in chemical reactions. Thus, various supported Au-based catalysts were tested in oxidative esterification of FAL and the results are represented in Table 9.4. In this context, Taarning et al. demonstrated oxidative esterification of FAL under very mild reaction conditions (at room temperature 22 °C and 1.0 bar O_2) with titania supported Au (Au/TiO_2) nanoparticles. In the presence of CH_3ONa used as a base additive, the excellent yield of 2-methyl furoate (100%) was obtained in 10–12 h (Taarning et al. 2008). A complete conversion of FAL (100%) was obtained under mild reaction conditions. Further, oxidative esterification of FAL to produce 2-methyl furoate has also been conducted by metal oxides (CeO_2 and

TABLE 9.4
Noble Metal Catalyzed Oxidative Esterification of FAL

Catalyst	Additive	*T*, °C	*t*, h	P_{O2}, bar	FAL Conv., %	2-methyl furoate Yield, %	Ref.
Au/TiO_2	CH_3ONa	22	10–12	1.0	100	100	(Taarning et al. 2008)
Au/FH	K_2CO_3	140	4	3.0	93.0	92.1	(Tong et al. 2015)
Au/UiO-66	K_2CO_3	140	4	3.0	100	100	(Ning et al. 2018)
Au/Z500	No base	120	3	6.0	100	98	(Signoretto et al. 2013)
Au/ZrO_2	No base	120	1.5	6.0	82	75.4	(Menegazzo et al. 2014)
Au/CeO_2	No base	120	1.5	6.0	66	~43	(Menegazzo et al. 2014)
Au/TiO_2	No base	120	1.5	6.0	20	18	(Menegazzo et al. 2014)
Au/CMK-3	No base	120	3.0	15	99.7	99.3	(Radhakrishnan et al. 2017)
Au/MgO	No base	110	2.0	6.0	98.0	97.0	(P. Ferraz et al. 2020)
$SA-NH_2-Au$	No base	140	0.5	20	100	99	(Song et al. 2022)

Fe_3O_4), hydroxyapatite (H) and FeOx modified hydroxyapatite (FH) supported Au-based catalysts, which were prepared by the deposition-precipitation method. Tong et al. reported the Au-FH catalyst among all other Au-based catalysts showed 91.8% yield of 2-methyl furoate at the 93% conversion of FAL in the presence of potassium carbonate (K_2CO_3) used as based additive under the reaction conditions (at 140 °C for 4 h under 3.0 bar O_2) (Tong et al. 2015). In another example, Ning et al. reported a series of metal organic frameworks supported Au nanoparticles prepared by impregnation-H_2-reduction (I-H) method (Ning et al. 2018). From oxidative esterification of FAL, the Au/UiO-66 catalyst showed excellent results with 100% yield of 2-methyl furoate and 100% conversion of FAL in 4 h at 140 °C and 3.0 bar O_2. There is no doubt that the excellent yield of 2-methyl furoate (100%) from oxidative esterification of FAL could be achieved, wherein base additives were employed. However, the use of base additives makes the whole process less green and expensive.

Base-free oxidative esterification of FAL is important for green and sustainability viewpoints, which can simplify downstream separation, reduce the number of reagent, and produce less waste. In this continuation, Signoretto and co-authors reported efficient, selective, and reusable zirconia supported Au nanoparticles (Au/Z500) for oxidative esterification of FAL in base-free conditions (Signoretto et al.

2013). For this, zirconia hydroxide prepared by the precipitation method (at constant pH = 8.6) was sulphated with ammonium sulphate (NH_4SO_4) and then the obtained sample was further calcined at 650 °C for 3 h (note: the synthesized sulphated zirconia support material is represented as Z). After loading of Au nanoparticles by the deposition-precipitation method, samples of Au/Z were finally calcined at different temperatures (150, 300, 500, 600, and 650 °C) in order to modulate the Au nanoparticles. The availability of sulphate ions (SO_4^{2-}) on the surface of zirconia support played an important role in modulation of highly dispersed Au nanoparticles. The Au/Z500 catalyst with modulated Au particle size (below 3.0 nm) showed the highest yield of 2-methyl furoate (98%) at complete conversion of FA (100%) under the identical conditions (at 120 °C and 6.0 bar O_2 in 3 h). In order to elucidate the role of nanosize of Au nanoparticles, Pinna et al. further studied the activity of the zirconia supported Au nanoparticles (Au/ZrO_2) catalyst in oxidative esterification of FAL. It could be suggested that the thermal treatment (at lower temperature 150 °C) gave the small nanosize of Au nanoparticles (1.68±0.38 nm), which is considered as an active phase for both conversion and product selectivity viewpoints. However, the thermal treatment at higher temperature resulted in the formation of a larger size of Au nanoparticles (> 4.0 nm), making the catalyst less active and selective in oxidative esterification of FAL (Pinna et al. 2013). Later, Menegazzo and co-workers reported various other metal oxides (zirconia, ceria, and titania) supported Au nanoparticles, which were evaluated in oxidative esterification of FAL and no base additive was used (Menegazzo et al. 2014). The activity of supported Au nanoparticles used as catalysts was in the following order: Au/ZrO_2 > Au/CeO_2 > Au/TiO_2. At 120 °C in 1.5 h, Au/ZrO_2 catalyst resulted 82% conversion of FAL and 92% selectivity of 2-methyl furoate. The highest activity of Au/ZrO_2 catalyst prepared by precipitation method (at a constant pH of 8.6) followed by calcination at 650 °C, is attributed to acid-base properties of ZrO_2 support and high dispersion of Au nanoparticles. Radhakrishnan et al. described oxidative esterification of FAL by various mesoporous CMK-3 supported Au nanoparticle (Au/CMK-3) catalysts. Among all, the Au (5.0%)/CMK-3 (with 2.5 nm size of Au nanoparticles) catalyst showed a remarkable selectivity of 2-methyl furoate (99.6%) with high conversion of FAL (99.7%) under the applied reaction conditions (at 120 °C and 15 bar O_2 in 3 h) (Radhakrishnan et al. 2017). Ferraz and co-authors demonstrated oxidative esterification of FAL by alkaline earth metal oxides (MgO, CaO, BaO, and SrO) supported Au nanoparticles (Ferraz et al. 2020). From TPD-CO_2 analysis results, the basic strength of metal oxides was in the following order: BaO > SrO > CaO > MgO. The presence of basic sites on the surface of metal oxides favored the formation of 2-methyl furoate, preventing the formation of acetal as a by-product. Thus, the Au/MgO catalyst with particle size (3.0 nm) showed 97% yield of 2-methyl furoate and 98% conversion of FAL in 3 h (at 110 °C and 6.0 bar O_2). Recently, Song et al. reported Au nanoparticles anchored in basic –NH_2 modified silica aerogels (SA-NH_2) used as an efficient catalyst for oxidative esterification of FAL. In typical synthesis of SA-NH_2-Au nanoparticles' catalyst, SA-NH_2 was firstly prepared by grafting of amino propyl (–$CH_2CH_2CH_2$-NH_2) on silica aerogels (SA, with average pore diameter of >30 nm), then synthesized SA-NH_2 was mixed with hydrogen tetrachloroaurate (III) trihydarte (used Au precursor), followed by addition of sodium borohydride ($NaBH_4$) to afford SA-NH_2-Au nanoparticles as the final catalyst. It could be revealed that the basic $-NH_2$ group played an important role in maintaining the alkaline environment to inhibit the side reaction of acetalization by acid, thereby 2-methyl furoate could be produced selectively. Thus, the SA-NH_2-Au catalyst with particle size 1.8 nm, could exhibit a quantitative yield of 2-methyl furoate (~100%) at complete conversion of FAL (at 140 °C in 0.5 h). As mentioned in all cases, Au-based catalysts showed excellent performance in oxidative esterification of FAL. However, the high price, scarcity, and critically low-abundance of Au metal as catalysts make the process expensive.

9.4.2 Non-Noble metal Catalyzed oxidative Esterification of FAL

The non-precious metal and base-free approach for the production of 2-methyl furoate from oxidative esterification of FAL is the most preferable and is of great interest, which is very important from the perspectives of green chemistry and sustainable development. The results obtained from non-noble metal catalyzed oxidative esterification of FAL are summarized in Table 9.5, Huo and co-workers reported non-precious and an efficient Co-based catalyst. Co-based catalyst (Co-N-C/MgO) was prepared by impregnation and pyrolysis of a cobalt (II) phenanthroline complex on MgO at 800 °C in N_2 atmosphere. Interestingly, Co-NC/MgO showed 98% selectivity at 93% conversion of FAL under the applied reaction conditions (at 100 °C and 5.0 bar O_2 in 12 h) (Huo et al. 2017). Recently, Yin et al. deals with a series of nitrogen-doped porous carbon (Co-NOPC) supported Co-based catalysts prepared by sacrificial-template method. Co-NOPC catalysts were further screened in oxidative esterification of FAL without base additive (Yin et al. 2021). Under mild reaction conditions (at 80 °C and 5.0 bar O_2), Co-NOPC-M showed the excellent results from oxidative esterification of FAL, both the conversion of FAL and the selectivity of 2-methyl furoate were 99.6% and 96.0%, respectively. The best catalytic performance was attributed to high dispersion of cobalt, ordered mesoporous structure, and larger surface area. In addition, the Co-NOPC-M catalyst showed superior stability even after six cycles with no significance loss of original activity. More recently, Desai and Yadav found montmorillonite K10 clay supported copper catalyst synthesized

TABLE 9.5
Non-Noble Metal Catalyzed Oxidative Esterification of FAL

Catalyst	Additive	T, °C	t, h	P_{O2}, bar	FAL Conv., %	2-methyl furoate Yield, %	Ref.
Co-NC/MgO	No-base	100	12	5.0	93.0	91.6	(Huo et al. 2017)
Co-NOPC-M	No-base	80	2	5.0	99.6	95.6	(Yin et al. 2021)
Cu-TK10	No-base	120	5	-	91.0	91.0	(Desai and Yadav 2022)

by the incipient wetness impregnation method using heteropolyacid, wherein proton (H^+) of dodeca-tungustophosphoric acid (TPA) was exchanged by Cu ion (Desai and Yadav 2022). From base-free approach oxidative esterification of FAL, this Cu-TK10 catalyst showed 100% selectivity of 2-methyl furoate and 91% conversion of FAL in the presence of *tert*-butyl hydroperoxide used as oxidant. The excellent selectivity of 2-methyl furoate (100%) is attributed to the strong Brønsted acidity of Cu-TK10 catalyst (total acidity: 2.18 mmol of NH_3 per a gram of sample), which is active and economically viable as well. More importantly, the Cu-TK10 catalyst was reusable with high activity and economically viable.

9.5 REACTION MECHANISM

Sun and co-workers proposed a plausible reaction mechanism for α-MnO_2/CoO_x-N/C catalyzed oxidative esterification of 5-HMF to FDMC (Sun et al. 2016). The synthesized CoO_x-N/C catalyst was combined with α-MnO_2 to catalyze the oxidative esterification of **1** to **5**, which occurred by following two different routes. As the reaction pathways illustrated in Scheme 9.9, the aldehyde group of **1** first proceeds with oxidative esterification to get compound **2** (route A). Compound **2** is relatively more stable than diformylfuran (DFF, **3**), which is obtained from the oxidation of alcohol group of **1** (route B). Intermediate compound **3** in the synthesis mechanism is able to further convert into diester compound **5** via the formation of another compound **4**, which represents also the involvement of route A. Side reactions could also occur to all aldehyde groups of the intermediates to form many acetals (highlighted with dotted line).

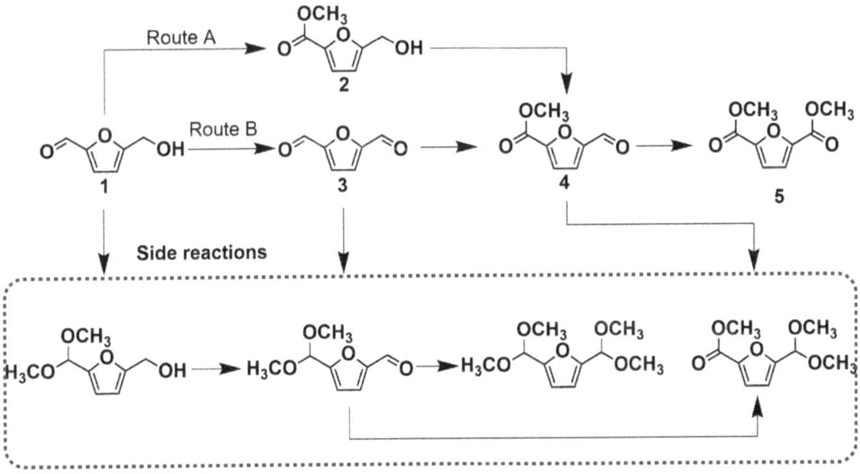

SCHEME 9.9 Two possible routes for oxidative esterification of 5-HMF (originally drawn based on Sun et al. 2016).

9.6 CONCLUSIONS AND PERSPECTIVES

The main aim of this chapter is therefore to describe environmentally friendly oxidative esterification 5-HMF and FAL to produce their corresponding esters (FDMC and 2-methyl furoate), and further their applications in synthesis of chemical intermediates and bio-based polymers. FDMC can be used as such in synthesis of 100% bio-based polymer. Aside from FDMC, 2-methyl furoate derived from oxidative esterification of FAL, is found to have huge industrial applications. Thus, we understood the potential of lignocellulosic biomass (especially second-generation of non-edible biomass), which is abundantly available in nature. Predominantly, lignocellulosic biomass is believed to be preferred renewable feedstock for sustainable production of bio-based chemicals, fuels, and polymers. A plethora of bio-based platform chemicals and biofuels could be produced directly either from lignocellulosic biomass or from its derived chemicals. In this context, various developed methodologies (e.g., hydrolysis, dehydration, oxidation, oxidative esterification, hydrogenation, and hydrodeoxygenation) were applied. Among all various methodologies, a well-known process is the oxidation process. Oxidation of biomass derived 5-HMF gives the FDCA product, which is not only used as a monomer but also used as a precursor for the production of a variety of sustainable chemicals. Indeed, FDCA and EG monomers combined with each other via condensation polymerization to form 100% bio-based PEF polymer. This bio-based PEF polymer showed improved mechanical and thermal properties, high glass transition temperature, and lower melting point along with superior barrier performance. Today, bio-based polymers, not only the PEF polymer but also others, are attracting more attention due to several improved properties and environmental footprints. Despite the versatility of FDCA as aforementioned, it is found to be less soluble in most of the organic solvents, including water. This is the reason that it hampers the whole process of polymerization. As an alternative to FDCA synthesis from oxidation of 5-HMF, synthesis of FDMC form oxidative esterification (one-pot coupling of oxidation and esterification) of 5-HMF is more attractive. In fact, FDMC, ester of the FCDA product, is highly soluble in most of organic solvents. As compared to FDCA, FDMC can also be separated and purified easily after reaction. Across the globe, people are increasingly turning to sustainable bio-based polymers for their manufacturing needs. In the near future, large-scale production of FDMC from oxidative esterification of biomass-derived 5-HMF is likely to play a vital role in the development of bio-based PEF polymer. Furthermore, the commercial availability of PEF polymers is expected to provide a positive response to new applications in the future. Aside from FDMC, 2-methyl furoate has also huge industrial applications.

REFERENCES

Abad, Alberto, Patricia Concepción, Avelino Corma, and Hermenegildo García. 2005. "A collaborative effect between gold and a support induces the selective oxidation of alcohols." *Angewandte Chemie International Edition* 44 (26). John Wiley & Sons, Ltd. 4066–4069. doi: 10.1002/anie.200500382.

Ait Rass, Hicham, Nadine Essayem, and Michèle Besson. 2015. "Selective Aerobic oxidation of 5-HMF into 2,5-furandicarboxylic acid with Pt catalysts supported on TiO_2- and ZrO_2-based supports." *ChemSusChem* 8 (7). John Wiley & Sons, Ltd. 1206–1217. doi: 10.1002/cssc.201403390.

Alonso, David Martin, Stephanie G. Wettstein, Max A. Mellmer, Elif I. Gurbuz, and James A. Dumesic. 2013. "Integrated conversion of hemicellulose and cellulose from lignocellulosic biomass." *Energy & Environmental Science* 6 (1). The Royal Society of Chemistry. 76–80. doi: 10.1039/C2EE23617F.

Artz, Jen and Regina Palkovits. 2015. "Base-free aqueous-phase oxidation of 5-hydroxymethylfurfural over ruthenium catalysts supported on covalent triazine frameworks." *ChemSusChem* 8 (22). John Wiley & Sons, Ltd. 3832–3838. doi: 10.1002/cssc.201501106.

Ashikari, Toshihiko, Shin-ichi Kunikasi, Nobuya Matsumoto, Teruo Amachi, and Hajime Yoshizumi. 1989. "Direct fermentation of raw corn to ethanol by yeast transformants containing a modified Rhizopus glucoamylase gene." *Applied Microbiology and Biotechnology* 32 (2). Springer Nature. 129–133. doi: 10.1007/BF00165875.

Ashokkumar, Veeramuthu, Radhakrishnan Venkatkarthick, Shanmugam Jayashree, Santi Chuetor, Selvakumar Dharmaraj, Gopalakrishnan Kumar, Wei-Hsin Chen, and Chawalit Ngamcharussrivichai. 2022. "Recent advances in lignocellulosic biomass for biofuels and value-added bioproducts- A critical review." *Bioresource Technology* 344. Elsevier B.V. 126195. doi: 10.1016/j.biortech.2021.126195.

Julie, Baruah, Bikash Kar Nath, Ritika Sharma, Sachin Kumar, Ramesh Chandra Deka, Deben Chandra Baruah, and Eeshan Kalita. 2018. "Recent trends in the pretreatment of lignocellulosic biomass for value-added products." *Frontiers in Energy Research* 6 (141). Frontiers Media S.A. doi: 10.3389/fenrg.2018.00141.

Bond, Geoffrey C., and David T. Thompson. 1999. "Catalysis by gold." *Catalysis Reviews* 41(3–4). Tayler & Francis. 319–388. doi: 10.1081/CR-100101171.

Buonerba, Antonio, Salvatore Impemba, Antonella Dentoni Litta, Carmine Capacchione, Stefano Milione, and Alfonso Grassi. 2018. "Aerobic oxidation and oxidative esterification of 5-hydroxymethylfurfural by gold nanoparticles supported on nanoporous polymer host mMatrix." *ChemSusChem* 11 (18). John Wiley & Sons, Ltd. 3139–3149. doi: 10.1002/cssc.201801560.

Casanova, O., S. Iborra, and A. Corma. 2009a. "Biomass into chemicals: One pot-base free oxidative esterification of 5-hydroxymethyl-2-furfural into 2,5-dimethylfuroate with gold on nanoparticulated ceria." *Journal of Catalysis* 265 (1). Elsevier Inc. 109–116. doi: 10.1016/j.jcat.2009.04.019.

Casanova, Onofre, Sara Iborra, and Avelino Corma. 2009. "Biomass into chemicals: Aerobic oxidation of 5-hydroxymethyl-2-furfural into 2,5-furandicarboxylic acid with gold nanoparticle catalysts." *ChemSusChem* 2 (12). John Wiley & Sons, Ltd. 1138–1144. doi: 10.1002/cssc.200900137.

Cho, Ahra, Sangmoon Byun, Jin Hee Cho, and Byeong Moon Kim. 2019. "AuPd-Fe_3O_4 nanoparticle-catalyzed synthesis of Furan-2,5-dimethylcarboxylate from 5-hydroxymethylfurfural under mild conditions." *ChemSusChem* 12 (10). John Wiley & Sons, Ltd. 2310–2317. doi: 10.1002/cssc.201900454.

Choudhary, Vinit, Stanley I. Sandler, and Dionisios G. Vlachos. 2012. "Conversion of xylose to furfural using Lewis and Brønsted acid catalysts in aqueous media." *ACS Catalysis* 2 (9). Americal Chemical Society. 2022–2028. doi: 10.1021/cs300265d.

Cousin, Thibault, Jocelyne Galy, Alain Rousseau, and Jérôme Dupuy. 2018. "Synthesis and properties of polyamides from 2,5-furandicarboxylic acid." *Journal of Applied Polymer Science* 135 (8). John Wiley & Sons, Ltd. 45901. doi: 10.1002/app.45901.

Davis, Sara E., Levi R. Houk, Erin C. Tamargo, Abhaya K. Datye, and Robert J. Davis. 2011. "Oxidation of 5-hydroxymethylfurfural over supported Pt, Pd and Au catalysts." *Catalysis Today* 160 (1). Elsevier B.V. 55–60. doi: 10.1016/j.cattod.2010.06.004.

De, Sudipta, Anand S. Burange, and Rafael Luque. 2022. "Conversion of biomass-derived feedstocks into value-added chemicals over single-atom catalysts." *Green Chemistry* 24 (6). Royal Society of Chemistry. 2267–2286. doi: 10.1039/D1GC04285H.

Deng, Jin, Hai-Jie Song, Min-Shu Cui, Yi-Ping Du, and Yao Fu. 2014. "Aerobic oxidation of hydroxymethylfurfural and furfural by using heterogeneous CoxOy–N@C catalysts." *ChemSusChem* 7 (12). John Wiley & Sons, Ltd. 3334–3340. doi: 10.1002/cssc.201402843.

Desai, Deepak S., and Ganapati D. Yadav. 2022. "Solvent-free oxidative esterification of furfural to 2-methyl furoate using novel copper-exchanged tungstophosphoric acid supported on montmorillonite K-10 catalyst." *Molecular Catalysis* 524. Elsevier B.V. 112256. doi: 10.1016/j.mcat.2022.112256.

Devi, Manolata Mayanglambam, Nidhi Aggarwal, and Shunmugavel Saravanamurugan. 2020. "Rice straw: A major renewable lignocellulosic biomass for value-added carbonaceous materials." *Current Green Chemistry* 7 (3). Bentham Science. 290–303. doi: 10.2174/2213346106666191127120259.

Du, Junmou, Huihuang Fang, Hongyan Qu, Jinping Zhang, Xinping Duan, and Youzhu Yuan. 2018. "Fabrication of supported Au-CuOx nanohybrids by reduction-oxidation strategy for efficient oxidative esterification of 5-hydroxymethyl-2-furfural into dimethyl furan-2,5-dicarboxylate." *Applied Catalysis A: General* 567. Elsevier B.V. 80–89. doi: 10.1016/j.apcata.2018.09.004.

Feng, Yunchao, Wenlong Jia, Guihua Yan, Xianhai Zeng, Jonathan Sperry, Binbin Xu, Yong Sun, Xing Tang, Tingzhou Lei, and Lu Lin. 2020. "Insights into the active sites and catalytic mechanism of oxidative esterification of 5-hydroxymethylfurfural by metal-organic frameworks-derived N-doped carbon." *Journal of Catalysis* 381 Elsevier B.V. 570–578. doi: 10.1016/j.jcat.2019.11.029.

Feng, Yunchao, Sishi Long, Guihua Yan, Wenlong Jia, Yong Sun, Xing Tang, Zehui Zhang, Xianhai Zeng, and Lu Lin. 2021. "Highly dispersed Co/N-rich carbon nanosheets for the oxidative esterification of biomass-derived alcohols: Insights into the catalytic performance and mechanism." *Journal of Catalysis* 397. Elsevier B.V. 148–155. doi: 10.1016/j.jcat.2021.03.031.

Gallezot, Pierre. 2012. "Conversion of biomass to selected chemical products." *Chemical Society Reviews* 41 (4). The Royal Society of Chemistry. 1538–1558. doi: 10.1039/C1CS15147A.

Gibson, Lorna J. 2012. "The hierarchical structure and mechanics of plant materials." *Journal of The Royal Society Interface* 9 (76). Royal Society. 2749–2766. doi: 10.1098/rsif.2012.0341.

Gorbanev, Yury Y., Søren K. Klitgaard, John M. Woodley, Claus H. Christensen, and Anders Riisager. 2009. "Gold-catalyzed aerobic oxidation of 5-hydroxymethylfurfural in water at ambient temperature." *ChemSusChem* 2 (7). John Wiley & Sons, Ltd. 672–675. doi: 10.1002/cssc.200900059.

Grote, A. Freiherrn V., and B. Tollens. 1875. "Untersuchungen über Kohlenhydrate. I. Ueber die bei Einwirkung von Schwefelsäure auf Zucker entstehende Säure (Levulinsäure)." *Justus Liebigs Annalen der Chemie* 175 (1–2). John Wiley & Sons, Ltd. 181–204. doi: 10.1002/jlac.18751750113.

Gupta, Shyam Sunder R., Ajayan Vinu, and Mannepalli Lakshmi Kantam. 2020. "Copper-catalyzed oxidative methyl-esterification of 5-hydroxymethylfurfural using TBHP as an oxidizing and methylating reagent: A new approach for the synthesis of furan-2,5-dimethylcarboxylate." *Journal of Catalysis* 389. Elsevier B.V. 259–269. doi: 10.1016/j.jcat.2020.05.032.

Hara, Michikazu, Kiyotaka Nakajima, and Keigo Kamata. 2015. "Recent progress in the development of solid catalysts for biomass conversion into high value-added chemicals." *Science and Technology of Advanced Materials* 16 (3). IOP Publishing. 034903. doi: 10.1088/1468-6996/16/3/034903.

He, Zhuang, Yanyu Lu, Changqu Lin, Honghua Jia, Hongli Wu, Fei Cao, and Pingkai Ouyang. 2020. "Designing anti-migration furan-based plasticizers and their plasticization properties in poly (vinyl chloride) blends." *Polymer Testing* 91. Elsevier Ltd. 106793. doi: 10.1016/j.polymertesting.2020.106793.

https://patentimages.storage.googleapis.com/57/c0/e0/55439b79e61134/US10947209.pdf.

https://polymerinnovationblog.com/polyethylene-furanoate-pef-100-biobased-polymer-to-compete-with-pet/.

www.nrel.gov/docs/fy04osti/35523.pdf.

www.transparencymarketresearch.com/furandicarboxylic-acid-market.html.

Huang, Yao-Bing, and Yao Fu. 2013. "Hydrolysis of cellulose to glucose by solid acid catalysts." *Green Chemistry* 15 (5). The Royal Society of Chemistry. 1095–1111. doi: 10.1039/C3GC40136G.

Huo, Na, Hong Ma, Xinhong Wang, Tianlong Wang, Gang Wang, Ting Wang, Leilei Hou, Jin Gao, and Jie Xu. 2017. "High-efficiency oxidative esterification of furfural to methylfuroate with a non-precious metal Co-N-C/MgO catalyst." *Chinese Journal of Catalysis* 38 (7). Elsevier B.V. 1148–1154. doi: 10.1016/S1872-2067 (17) 62841–62849.

Inlow, D., J. McRae, and A. Ben-Bassat. 1988. "Fermentation of corn starch to ethanol with genetically engineered yeast." *Biotechnoloy Bioengineering* 32 (2). John Wiley & Sons, Inc. 227–234. doi: 10.1002/bit.260320213.

Jing, Tianyun, Shaokang Yang, Yonghai Feng, Tingting Li, Yunpeng Zuo, and Dewei Rao. 2023. "Selective and effective oxidation of 5-hydroxymethylfurfural by tuning the intermediates adsorption on Co-Cu-CNx." *Nano Research* 16. Springer Nature 6670–6678. doi: 10.1007/s12274-023-5450-3.

Ju, Zhaoyang, Yuxuan Zhang, Tingting Zhao, Weihua Xiao, and Xiaoqian Yao. 2019. "Mechanism of glucose–fructose isomerization over aluminum-based catalysts in methanol media." *ACS Sustainable Chemistry & Engineering* 7 (17). American Chemical Society 14962–14972. doi: 10.1021/acssuschemeng.9b03241.

Kar, Sayan, Quan-Quan Zhou, Yehoshoa Ben-David, and David Milstein. 2022. "Catalytic furfural/5-hydroxymethyl furfural oxidation to furoic acid/Furan-2,5-dicarboxylic acid with H_2 production using alkaline water as the formal oxidant." *Journal of the American Chemical Society* 144 (3). American Chemical Society. 1288–1295. doi: 10.1021/jacs.1c10908.

Kumar, Parveen, Diane M. Barrett, Michael J. Delwiche, and Pieter Stroeve. 2009. 'Methods for pretreatment of lignocellulosic biomass for efficient hydrolysis and biofuel production." *Industrial & Engineering Chemistry Research* 48 (8). American Chemical Society. 3713–3729. doi: 10.1021/ie801542g.

Li, Feng, Xing-Long Li, Chuang Li, Jing Shi, and Yao Fu. 2018. "Aerobic oxidative esterification of 5-hydroxymethylfurfural to dimethyl furan-2,5-dicarboxylate by using homogeneous and heterogeneous PdCoBi/C catalysts under atmospheric oxygen." *Green Chemistry* 20 (13). The Royal Society of Chemistry. 3050–3058. doi: 10.1039/C8GC01393D.

Lima, Sérgio, Margarida M. Antunes, Martyn Pillinger, and Anabela A. Valente. 2011. "Ionic liquids as tools for the acid-catalyzed hydrolysis/dehydration of saccharides to furanic aldehydes." *ChemCatChem* 3 (11). John Wiley & Sons, Ltd. 1686–1706. doi: 10.1002/cctc.201100105.

Lin, Yamei, Guo-Ping Lu, Xin Zhao, Xun Cao, Lele Yang, Baojing Zhou, Qin Zhong, and Zhong Chen. 2020. "Porous cobalt@N-doped carbon derived from chitosan for oxidative esterification of 5-Hydroxymethylfurfural: The roles of zinc in the synthetic and catalytic process." *Molecular Catalysis* 482. Elsevier B.V. 110695. doi: 10.1016/j.mcat.2019.110695.

Liu, Huai, Ning Ding, Junnan Wei, Xing Tang, Xianhai Zeng, Yong Sun, Tingzhou Lei, Huayu Fang, Tianyuan Li, and Lu Lin. 2020. "Oxidative esterification of 5-hydroxymethylfurfural with an N-doped carbon-supported CoCu bimetallic catalyst." *ChemSusChem* 13 (16). John Wiley & Sons, Ltd. 4151–4158. doi: 10.1002/cssc.202000537.

Liu, Huai, Xing Tang, Xianhai Zeng, Yong Sun, Xixian Ke, Tianyuan Li, Jiaren Zhang, and Lu Lin. 2022. "Catalyst design strategy toward the efficient heterogeneously-catalyzed selective oxidation of 5-hydroxymethylfurfural." *Green Energy & Environment* 7 (5). Elsevier B.V. 900–932. doi: 10.1016/j.gee.2021.10.004.

Liu, Xiao, Xue Min, Hui Liu, Yuanqiao Cao, Yadong Liu, Miaomiao Han, Zhong-Ming Sun, and Shengxiang Ji. 2020. "Efficient conversion of cellulose to 5-hydroxymethylfurfural catalyzed by a cobalt-phosphonate catalyst." *Sustainable Energy & Fuels* 4 (11). The Royal Society of Chemistry 5795–5801. doi: 10.1039/D0SE01006E.

Liu, Xiao Yan, Aiqin Wang, Tao Zhang, and Chung-Yuan Mou. 2013. "Catalysis by gold: New insights into the support effect." *Nano Today* 8 (4). Elsevier B.V. 403–416. doi: 10.1016/j.nantod.2013.07.005.

Marianou, Asimina A., Chrysoula M. Michailof, Dimitrios K. Ipsakis, Stamatia A. Karakoulia, Konstantinos G. Kalogiannis, Haris Yiannoulakis, Konstantinos S. Triantafyllidis, and Angelos A. Lappas. 2018. "Isomerization of glucose into fructose over natural and synthetic MgO catalysts." *ACS Sustainable Chemistry & Engineering* 6 (12). American Chemical Society. 16459–16470. doi: 10.1021/acssuschemeng.8b03570.

Mariscal, R., P. Maireles-Torres, M. Ojeda, I. Sádaba, and M. López Granados. 2016. "Furfural: A renewable and versatile platform molecule for the synthesis of chemicals and fuels', *Energy & Environmental Science* 9 (4). The Royal Society of Chemistry. 1144–1189. doi: 10.1039/C5EE02666K.

Menegazzo, Federica, Michela Signoretto, Damiano Marchese, Francesco Pinna, and Maela Manzoli. 2015. "Structure–activity relationships of Au/ZrO_2 catalysts for 5-hydroxymethylfurfural oxidative esterification: Effects of zirconia sulphation on gold dispersion, position and shape." *Journal of Catalysis* 326. Elsevier B.V. 1–8. doi: 10.1016/j.jcat.2015.03.006.

Menegazzo, Federica, Michela Signoretto, Francesco Pinna, Maela Manzoli, Valentina Aina, Giuseppina Cerrato, and Flora Boccuzzi. 2014. "Oxidative esterification of renewable furfural on gold-based catalysts: Which is the best support?." *Journal of Catalysis* 309. Elsevier B.V. 241–247. doi: 10.1016/j.jcat.2013.10.005.

Mishra, Dinesh Kumar, Jin Ku Cho, Yongjin Yi, Hye Jin Lee, and Yong Jin Kim. 2019. "Hydroxyapatite supported gold nanocatalyst for base-free oxidative esterification of 5-hydroxymethyl-2-furfural to 2,5-furan dimethylcarboxylate with air as oxidant." *Journal of Industrial and Engineering Chemistry* 70. Elsevier B.V. 338–345. doi: 10.1016/j.jiec.2018.10.034.

Mishra, Dinesh Kumar, Santosh Kumar, and Ram Sambhar Shukla. 2020. "Chapter 12- Furfuryl alcohol–a promising platform chemical." In *Biomass, Biofuels, Biochemicals: Recent Advances in Development of Platform Chemicals*, 323–353 Elsevier Inc. doi: 10.1016/B978-0-444-64307-0.00012-3.

Mishra, Dinesh Kumar, Hye Jin Lee, Jinsung Kim, Hong-Shik Lee, Jin Ku Cho, Young-Woong Suh, Yongjin Yi, and Yong Jin Kim. 2017. "$MnCo_2O_4$ spinel supported ruthenium catalyst for air-oxidation of HMF to FDCA under aqueous phase and base-free conditions."

Green Chemistry 19 (7). The Royal Society of Chemistry. 1619–1623. doi: 10.1039/C7GC00027H.

Mori, Ryohei. 2023. "Replacing all petroleum-based chemical products with natural biomass-based chemical products: A tutorial review', *RSC Sustainability* 1 (2). The Royal Society of Chemistry. 179–212. doi: 10.1039/D2SU00014H.

Nandiwale, Kakasaheb Y., Nitish D. Galande, Pratika Thakur, Sanjay D. Sawant, Vishal P. Zambre, and Vijay V. Bokade. 2014. "One-pot synthesis of 5-hydroxymethylfurfural by cellulose hydrolysis over highly active bimodal micro/mesoporous H-ZSM-5 catalyst." *ACS Sustainable Chemistry & Engineering* 2(7). American Chemical Society. 1928–1932. doi: 10.1021/sc500270z.

Ning, Liangmin, Shengyun Liao, Xuguang Liu, Pengfei Guo, Zhenya Zhang, Haigang Zhang, and Xinli Tong. 2018. "A regulatable oxidative valorization of furfural with aliphatic alcohols catalyzed by functionalized metal-organic frameworks-supported Au nanoparticles." *Journal of Catalysis* 364. Elsevier B.V. 1–13. doi: 10.1016/j.jcat.2018.04.030.

P. Ferraz, Camila, Adriano H. Braga, Mohamed N. Ghazzal, Michał Zieliński, Mariusz Pietrowski, Ivaldo Itabaiana, Franck Dumeignil, Liane M. Rossi, and Robert Wojcieszak. 2020. "Efficient oxidative esterification of furfural using Au nanoparticles supported on group 2 alkaline earth metal oxides." *Catalysts* 10. MDPI. 4. doi: 10.3390/catal10040430.

Peleteiro, Susana, Andre M. da Costa Lopes, Gil Garrote, Juan Carlos Parajó, and Rafał Bogel-Łukasik. 2015. "Simple and efficient furfural production from xylose in media containing 1-Butyl-3-methylimidazolium hydrogen sulfate." *Industrial & Engineering Chemistry Research* 54 (33). American Chemical Society. 8368–8373. doi: 10.1021/acs.iecr.5b01771.

Pinna, F., A. Olivo, V. Trevisan, F. Menegazzo, M. Signoretto, M. Manzoli, and F. Boccuzzi. 2013. "The effects of gold nanosize for the exploitation of furfural by selective oxidation." *Catalysis Today* 203: Elsevier B.V. 196–201. doi: 10.1016/j.cattod.2012.01.033.

Putro, Jindrayani Nyoo, Felycia Edi Soetaredjo, Shi-Yow Lin, Yi-Hsu Ju, and Suryadi Ismadji. 2016. "Pretreatment and conversion of lignoccellulose biomass into valuable chemicals." *RSC Advances* 6 (52). The Royal Society of Chemistry. 46834–46852. doi: 10.1039/C6RA09851G.

Radhakrishnan, Ramakrishnan, Sivakumar Thiripuranthagan, Arulselvan Devarajan, Sakthivel Kumaravel, Elangovan Erusappan, and Kathiravan Kannan. 2017. "Oxidative esterification of furfural by Au nanoparticles supported CMK-3 mesoporous catalysts." *Applied Catalysis A: General* 545. Elsevier B.V. 33–43. doi: 10.1016/j.apcata.2017.07.031.

Ragauskas, A. J., G. T. Beckham, M. J. Biddy, R. Chandra, F. Chen, M. F. Davis, B. H. Davison, R. A. Dixon, P. Gilna, M. Keller, P. Langan, A. K. Naskar, J. N. Saddler, T. J. Tschaplinski, G. A. Tuskan, and C. E. Wyman. 2014. "Lignin valorization: Improving lignin processing in the biorefinery." *Science* 344 (6185). Science Publishing Group. 1246843. doi: 10.1126/science.1246843.

Ruppert, Agnieszka M., Kamil Weinberg, and Regina Palkovits. 2012. "Hydrogenolysis goes bio: From carbohydrates and sugar alcohols to platform chemicals." *Angewandte Chemie International Edition* 51 (11). John Wiley & Sons, Ltd. 2564–2601. doi: 10.1002/anie.201105125.

Salazar, Abel, Paul Hünemörder, Jabor Rabeah, Antje Quade, Rajenahally V. Jagadeesh, and Esteban Mejia. 2019. "Synergetic bimetallic oxidative esterification of 5-hydroxymethylfurfural under mild conditions." *ACS Sustainable Chemistry & Engineering* 7 (14). American Chemical Society. 12061–12068. doi: 10.1021/acssuschemeng.9b00914.

Scapini, Thamarys, Maicon S. N. dos Santos, Charline Bonatto, João H. C. Wancura, Jéssica Mulinari, Aline F. Camargo, Natalia Klanovicz, Giovani L. Zabot, Marcus V. Tres, Gislaine Fongaro, and Helen Treichel. 2021. "Hydrothermal pretreatment of lignocellulosic biomass for hemicellulose recovery." *Bioresource Technology* 342. Elsevier B.V. 126033. doi: 10.1016/j.biortech.2021.126033.

Shen, Yating, Yunlei Zhang, Yao Chen, Yongsheng Yan, Jianming Pan, Meng Liu, and Weidong Shi. 2016. "Combination of Brønsted and Lewis polymeric catalysts for efficient conversion of cellulose into 5-Hydroxymethylfurfural (HMF) in ionic liquids." *Energy Technology* 4 (5). John Wiley & Sons, Ltd. 600–609. doi: 10.1002/ente.201500415.

Signoretto, M., F. Menegazzo, L. Contessotto, F. Pinna, M. Manzoli, and F. Boccuzzi. 2013. "Au/ZrO_2: An efficient and reusable catalyst for the oxidative esterification of renewable furfural', *Applied Catalysis B: Environmental* 129. Elsevier B.V. 287–293. doi: 10.1016/j.apcatb.2012.09.035.

Silva, João Paulo Alves, Jéssica Siqueira Mancilha Nogueira, Celso Luiz de Aquino Santos, and Livia Melo Carneiro. 2022. "Chapter 9- 5-Hydroxymethylfurfural as a chemical platform for a lignocellulosic biomass biorefinery." In *Production of Top 12 Biochemicals Selected by USDOE from Renewable Resources: Status and Innovation,* 269–315, Elsevier Inc. doi: 10.1016/B978-0-12-823531-7.00004-4.

Song, Fan, Shuangshuang Cen, Cong Wan, and Lijun Wang. 2022. "Nano-Au anchored in organic base group-grafted silica aerogel: A durable and robust catalysts for green oxidative esterification of furfural." *ChemCatChem* 14 (18). John Wiley & Sons, Ltd. e202200704. doi: 10.1002/cctc.202200704.

Suganuma, Satoshi, Kiyotaka Nakajima, Masaaki Kitano, Daizo Yamaguchi, Hideki Kato, Shigenobu Hayashi, and Michikazu Hara. 2008. "Hydrolysis of cellulose by amorphous carbon bearing SO_3H, COOH, and OH groups." *Journal of the American Chemical Society* 130 (38). American Chemical Society 12787–12793. doi: 10.1021/ja803983h.

Sun, Kang-kang, Shu-jie Chen, Ze-lin Li, Guo-ping Lu, and Chun Cai. 2019. "Synthesis of a ZIF-derived hollow yolk–shell Co@CN catalyst for the oxidative esterification of 5-hydroxymethylfurfural." *Green Chemistry* 21 (7). The Royal Society of Chemistry. 1602–1608. doi: 10.1039/C8GC03868F.

Sun, Yuxia, Hong Ma, Xiuquan Jia, Jiping Ma, Yang Luo, Jin Gao, and Jie Xu. 2016. "A high-performance base-metal approach for the oxidative esterification of 5-hydroxymethylfurfural." *ChemCatChem* 8 (18). John Wiley & Sons, Ltd. 2907–2911. doi: 10.1002/cctc.201600484.

Swift, T. Dallas, Christina Bagia, Vinit Choudhary, George Peklaris, Vladimiros Nikolakis, and Dionisios G. Vlachos. 2014. "Kinetics of homogeneous Brønsted acid catalyzed fructose dehydration and 5-hydroxymethyl furfural rehydration: A combined experimental and computational study." *ACS Catalysis* 4 (1). American Chemical Society. 259–267. doi: 10.1021/cs4009495.

Taarning, Esben, Inger S. Nielsen, Kresten Egeblad, Robert Madsen, and Claus H. Christensen. 2008. "Chemicals from renewables: Aerobic oxidation of furfural and hydroxymethylfurfural over gold catalysts." *ChemSusChem* 1 (1–2). John Wiley & Sons, Ltd. 75–78. doi: 10.1002/cssc.200700033.

Tong, Xinli, Zonghui Liu, Linhao Yu, and Yongdan Li. 2015. "A tunable process: Catalytic transformation of renewable furfural with aliphatic alcohols in the presence of molecular oxygen." *Chemical Communications* 51 (17). The Royal Society of Chemistry. 3674–3677. doi: 10.1039/C4CC09562F.

Tong, Xinli, Linhao Yu, Hui Chen, Xuli Zhuang, Shengyun Liao, and Hongge Cui. 2017. "Highly efficient and selective oxidation of 5-hydroxymethylfurfural by molecular oxygen in the presence of Cu-MnO_2 catalyst." *Catalysis Communications* 90. Elsevier B.V. 91–94. doi: 10.1016/j.catcom.2016.11.024.

Turner, Mark, Vladimir B. Golovko, Owain P. H. Vaughan, Pavel Abdulkin, Angel Berenguer-Murcia, Mintcho S. Tikhov, Brian F. G. Johnson, and Richard M. Lambert. 2008. "Selective oxidation with dioxygen by gold nanoparticle catalysts derived from 55-atom clusters." *Nature* 454 (7202). Springer Nature. 981–983. doi: 10.1038/nature07194.

van Es, Daan S., Sinisa Marinkovic, Xavier Oduber, and Boris Estrine. 2013. "Use of furandicarboxylic acid and its decyl ester as additives in the Fischer's glycosylation of decanol by d-glucose: Physicochemical properties of the surfactant compositions obtained." *Journal of Surfactants and Detergents*, 16 (2). John Wiley & Sons Inc. 147–154. doi: 10.1007/s11743-012-1382-8.

Vuyyuru, Koteswara Rao, and Peter Strasser. 2012. "Oxidation of biomass derived 5-hydroxymethylfurfural using heterogeneous and electrochemical catalysis." *Catalysis Today* 195 (1). Elsevier B.V. 144–154. doi: 10.1016/j.cattod.2012.05.008.

Wang, Tianfu, Michael W. Nolte, and Brent H. Shanks. 2014. "Catalytic dehydration of C6 carbohydrates for the production of hydroxymethylfurfural (HMF) as a versatile platform chemical." *Green Chemistry* 16 (2). The Royal Society of Chemistry. 548–572. doi: 10.1039/C3GC41365A.

Weerathunga, Helapiyumi, Sarina Sarina, Huai-Yong Zhu, and Eric R. Waclawik. 2021. "Oxidative esterification of 5-hydroxymethylfurfural into dimethyl 2,5-furandicarboxylate using Gamma alumina-supported gold nanoparticles." *ACS Omega* 6 (7). American Chemical Society. 4740–4748. doi: 10.1021/acsomega.0c05541.

Yin, Defeng, Yanxia Zheng, Lixi Yang, Shuyue Li, Daqing Zhu, Yafei Guo, Cuncun Zuo, Yuchao Li, Haofei Huang, and Ming Wang. 2021. "Oxidative esterification of renewable furfural on cobalt dispersed on ordered porous nitrogen-doped carbon." *RSC Advances*, 11 (6). The Royal Society of Chemistry. 3280–3287. doi: 10.1039/D0RA09945G.

Zabed, Hossain M., Suely Akter, Junhua Yun, Guoyan Zhang, Faisal N. Awad, Xianghui Qi, and J. N. Sahu. 2019. "Recent advances in biological pretreatment of microalgae and lignocellulosic biomass for biofuel production." *Renewable and Sustainable Energy Reviews* 105: Elsevier B.V. 105–128. doi: 10.1016/j.rser.2019.01.048.

Zakrzewska, Małgorzata E., Ewa Bogel-Łukasik, and Rafał Bogel-Łukasik. 2011. "Ionic liquid-mediated formation of 5-hydroxymethylfurfural—A promising biomass-derived building block." *Chemical Reviews* 111 (2). American Chemical Society. 397–417. doi: 10.1021/cr100171a.

Zhao, Deyang, Ting Su, Christophe Len, Rafael Luque, and Zhenglong Yang. 2022. "Recent advances in the oxidative esterification of 5-hydroxymethylfurfural to furan-2,5-dimethylcarboxylate." *Green Chemistry* 24 (18). The Royal Society of Chemistry. 6782–6789. doi: 10.1039/D2GC01772E.

Zhou, Hua, Song Hong, Hao Zhang, Yongting Chen, Huanghui Xu, Xueke Wang, Zheng Jiang, Shengli Chen, and Yun Liu. 2019. "Toward biomass-based single-atom catalysts and plastics: Highly active single-atom Co on N-doped carbon for oxidative esterification of primary alcohols." *Applied Catalysis B: Environmental*, 256: Elsevier B.V. 117767. doi: 10.1016/j.apcatb.2019.117767.

10 Aldol Condensation of Furfural with Carbonyl Compounds

Cong Chien Truong, Dinesh Kumar Mishra, and Young-Woong Suh

10.1 INTRODUCTION

With the fast depletion of fossil fuels and mounting concerns on global climate changes, it is imperative to find out eco-friendly and sustainable alternatives for the mass production of petroleum-free fuels and chemicals. In this context, lignocellulose biomass (LCB), which comes from plants, food crops, and agricultural/forestry residues, stands out as a renewable and accessible candidate. LCB is mainly composed of cellulose (40–50%), hemicellulose (20–30%), and lignin (10–25%), in which cellulose and hemicellulose are predominantly adopted as starting feedstocks for valorization because they are easily depolymerized and processed compared to lignin (Wang et al. 2021; Wang et al. 2023). In addition to valuable products in energy sector (syngas, biohydrogen, and liquid biofuels), a myriad of top value-added platform molecules can be achieved through the catalytic transformation of LCB (Zhou et al. 2011; Luterbacher, Martin Alonso, and Dumesic 2014). Specifically, the acid-mediated depolymerization-dehydration cascade of cellulose/hemicellulose gives rise to furfural (FF), a C5 privileged building block for the manufacture of bio-based chemicals and fuel components (Mariscal et al. 2016). Although FF can be transformed into several fuel additives, such as 2-methylfuran (2-MF), 2-methyltetrahydrofuran (MTHF), ethyl furfuryl ether (EFE), ɣ-valerolactone (GVL) and alkyl levulinates through an assortment of catalytic reactions (Lange et al. 2012; Bohre et al. 2015), the upgrade of FF into furanic intermediates with extended carbon numbers/branches is a prerequisite for the delivery of petroleum-free task-specific alkanes (i.e. gasoline, transportation liquid fuels, and lubricants). Indeed, the direct conversion of FF-derived compounds (e.g. 2-MF, MTHF, EFE, GVL) into hydrocarbons generally provides short-chain molecules with low values of density, heat of combustion, and viscosity.

The aldol condensation is an important coupling reaction to establish new C–C bonds between two carbonyl compounds. Compared to other C–C bond coupling strategies for furanic derivatives, the condensation between FF and carbonyl compounds bearing α-hydrogen atom (α-H) tops the list of interest because the reaction can conveniently take place with a wide range of reactants in mild conditions over a diversity of catalysts (acid, base, and acid-base type), ultimately furnishing

DOI: 10.1201/9781003464839-10

furanic condensation products with extension of carbon chains and low oxygen content (Li et al. 2017). More importantly, these resulting oxygenates can be transformed into liquid fuel-/lubricant-range alkanes by hydrodeoxygenation. Although the condensation between FF and ketones was successfully implemented in the presence of homogeneous acid/base catalysts, the processes were teeming with several hurdles of equipment corrosion, product neutralization, waste disposal management, and catalyst recovery. Therefore, the development of efficient and reusable catalysts is highly required for this transformation. Stimulated by prominent advantages of heterogeneous catalysts in the domain of biomass valorization and aldol condensation, this chapter mainly focuses on the eco-friendly manufacture of fuel precursors (C8–C15) from the condensation of FF and typical biomass-derived carbonyl compounds (i.e. acetone, levulinic acid, and angelica lactone) over different types of solid acid, base, and acid-base catalyst. Particularly, several issues, such as the catalyst structure-reactivity relationship, recyclability, reaction mode, and catalytic pathways are elaborated.

10.2 ALDOL CONDENSATION OF FF WITH ACETONE

Biomass-derived acetone, which can be produced through the well-known acetone-butanol-ethanol (ABE) biological fermentation of carbohydrates (Ketabchi et al. 2020), is a typical example of carbonyl substrate containing active α-H atoms for the aldol condensation. Generally, the aldol reaction between FF and acetone (Scheme 10.1) can take place in the presence of either acid, base or acid-base bifunctional catalysts, delivering α,β-unsaturated ketone, i.e. furfurylidene acetone (FAc, C8 adduct) as the major condensation product. Depending on the characteristics of catalyst, acetone/FF molarity and reaction conditions, FAc can condensate with another molecule of FF to generate a furanic dimer product, i.e. difurfurylidene acetone (F_2Ac, C13 adduct). In this section, several types of heterogeneous acid, base, and acid-base bifunctional catalyst for the synthesis of FAc and/or F_2Ac are presented.

10.2.1 Acidic Catalysts

Technically, the production of FF normally takes place through the hydrolysis-dehydration cascade of sugars in acidic media. As such, the aldol condensation of FF

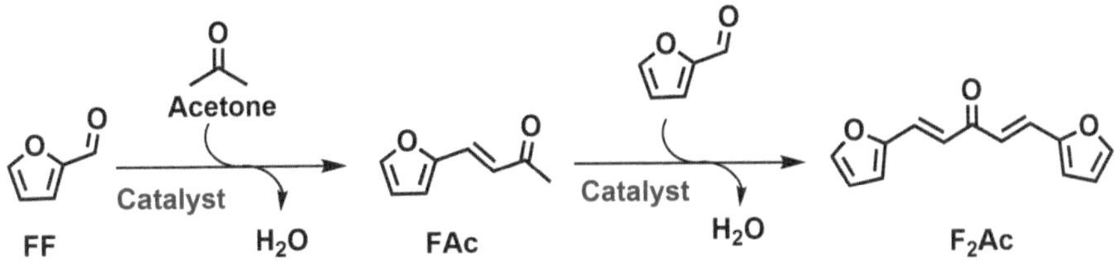

SCHEME 10.1 The catalytic aldol condensation of FF with acetone.

with acetone over acidic catalysts turns out to be convenient because initial steps of neutralization and/or purification of real upstream FF supply can be circumvented.

Jing and coworkers reported that niobium oxide (Nb_2O_5), a water-tolerant Lewis solid acid, could facilitate the condensation of FF with acetone in an effective manner (Jing et al. 2019). It is found that FAC (70.8%) and F_2AC (9.2%) were concomitantly yielded from an 88.9% conversion of FF under experimental conditions (130 °C, 4 h, acetone/FF molarity = 4). Extensive DRIFTS experiments of acetone adsorption over Nb_2O_5 verified that this solid acid could efficiently activate the C=O bond of acetone, inducing the active enolate intermediate for the nucleophilic addition into FF. Apart from acetone, the Nb_2O_5-mediated condensation between FF and other biomass-derived carbonyl compounds such as 4-heptanone, methyl isobutyl ketone and 2,4-pentanedione were also attempted, delivering the aldol products in satisfactory yields (70.8–81.7%). In the aspect of reusability, the spent Nb_2O_5 catalyst after treating at 400 °C showed no phase change and deterioration of reactivity relative to the pristine sample.

In another study, the impact of crystal structures (monoclinic vs. hexagonal) as well as thermal post-treatment of ceria phosphate materials (CeP) on their aldolization activity was extensively investigated (Li et al. 2019). To this end, CeP materials attained at hydrothermal temperature of 120 °C (hexagonal) and 180 °C (monoclinic), respectively, were subjected to the calcination at different temperatures (500, 700, and 900 °C) and examined for the aldol reaction. Among them, CeP-180-900, which was granted with good textural stability and high proportion of Ce^{4+} Lewis acid centers, displayed the highest catalytic performance. The FF conversion impressively reached up to 96% along with 68.4% yield of products (FAc: 54.4%, F_2Ac: 14%) when the aldolization was executed at 160 °C, 3 h with acetone/FF molar ratio 10:1. In-depth recyclability survey of CeP-180-900 revealed that the spent catalyst subjected to air calcination (550 °C, 5 h) could be reused for five cycles with acceptable results.

Zeolites, a pronounced class of porous aluminosilicate materials with unique characteristics (shape selectivity, tunable acid-base properties, and high stability), have been widely applied as both heterogeneous catalysts and solid supports in the biomass valorization (Perego et al. 2017). Driven by the success of zeolites in several acid-mediated transformations (e.g. hydrolysis, isomerization, dehydration, esterification, ketonization, and others) (Taarning et al. 2011), different types of acidic zeolites, such as H-ZSM-5(28), H-ZSM-5(50), H-BEA(25), H-BEA(38), H-MOR(20), HSDUSY(5), and HSDUSY(80), were applied to promote the FF-acetone condensation (Kikhtyanin, Kelbichová, et al. 2014). It is confirmed that the catalytic reactivity of tested zeolites were considerably influenced by their structural and textural properties. Of which, H-BEA(38) appeared to be most proficient, triggering the conversion of FF at 38.5% along with FAc and F_2Ac in 79.5% and 3.7% selectivity, respectively, under established conditions (100 °C, 2 h, acetone/FF molarity = 10). Herein, the unmatched catalysis of H-BEA(38) was rationalized by two key factors of: (i) wide-pore three-dimensional crystalline framework with small crystal sizes; and (ii) high concentration of Brønsted acid sites. Although H-BEA(38) became readily deactivated due to the accumulation of coke inside micropores, the recovered catalyst could efficiently restore the original activity with the aid of air calcination.

Prompted by these findings, a number of layered MWW-type zeolites (i.e. MCM-22, MCM-36, MCM-49, and MCM-56) to catalyze the assembly of FAc were also considered (Kikhtyanin, Chlubna, et al. 2014). It is noticed that the structural and textural properties of MWW catalysts exerted a strong impact on the product distributions. Regarding the FF conversion and FAc selectivity after 2 h at 100 °C, the performance of testing samples and benchmark H-BEA(25) followed the sequence of H-BEA(25) < MCM-36 < MCM-56 < MCM-49 < MCM-22. In this circumstance, the superiority of three-dimensional supercaged materials (MCM-22 and MCM-49) over delaminated MCM-56 and pillared MCM-36 in respect to efficiency was explicated by the prevalence of Brønsted acid centers and their population in large supercages. Although all re-used MWW catalysts (MCM-22, MCM-36, and MCM-49) encountered a drastic deterioration of reactivity due to the coke disposition, only MCM-36 suffered rapid and severe coking.

Stimulated by the success of Lewis acidic zeolites in several examples of aldol condensation (Lewis, Van de Vyver, and Roman-Leshkov 2015; Tolborg et al. 2016), Su and coworkers synthesized two tin-containing zeolites with 10-membered ring MFI (Sn-MFI) and 12-membered ring BEA (Sn-Beta) framework, to catalyze the title aldol reaction (Su et al. 2017). Interestingly, it is recognized that the selectivity of condensation products could be controlled by the structural characteristics of zeolites. Specifically, FAc (40%) and F_2Ac (20.5%) were simultaneously yielded at 100% conversion of FF over Sn-Beta, whereas FAc of 61% yield was solely produced over Sn-MFI under identical conditions (160 °C, 6 h, acetone/FF molarity = 10). This stark contrast arose from the fact that the Sn-MFI catalyst was endowed with smaller pore sizes and competence to lower the activation energy of aldolization compared to Sn-Beta. Although the introduction water was deemed inimical to the reactivity of both Sn-Beta and Sn-MFI in the aldolization, it is worth mentioning that the product selectivity (FAc vs. F_2Ac) in the case of Sn-Beta catalyst could be considerably regulated by the content of water. As an example, FAc of 21% yield and 42% selectivity was exclusively constituted over Sn-Beta catalyst when 10 wt% water was introduced into the reaction system. This appealing occurrence was interpreted by the reason that the competitive adsorption of water on active sites of Sn-Beta catalyst negated the diffusion of reactants, completely impeding the formation of F_2Ac from further condensation of FAc with FF. Besides, recyclability tests over spent catalysts (Sn-MFI and Sn-Beta) pointed out that these Lewis acidic zeolites after thermal treatment (550 °C in air) could bear striking similarities in both FF conversion and product selectivity relative to the fresh samples. On the other hand, iron-grafted mesoporous zeolite (Fe-MCM-41, 20 wt% Fe) developed by Su and coworkers furnished an 86.9% conversion of FF with 67.5% yield of condensation products (FAc: 60%, F_2Ac: 7.5%), exceeding those of microporous Sn-MFI catalyst under identical conditions of 160 °C and 24 h (Su et al. 2020). Furthermore, the thermal treatment at 500 °C guaranteed an acceptable catalytic regeneration of Fe-MCM-41 during five consecutive runs. By the same token, Kikhtyanin et al. adopted two Lewis acidic metal organic frameworks (MOFs), i.e. Cu-BTC (copper benzene-1,3,5-tricarboxylate) and Fe-BTC (iron benzene-1,3,5-tricarboxylate) to catalyze the title condensation (Kikhtyanin, Kubička, and Čejka 2015). Regardless of their similarity in physico-chemical properties, Fe-BTC surpassed Cu-BTC in the context of FF conversion (26.2% vs. 2.7%) and FAc yield (18.6% vs. 1.9%) under testing conditions (100 °C, 2 h, acetone/FF molarity = 10). Attractively, a set experiments with dehydrated and hydrated MOF samples illustrated that the existence of weak Brønsted acid sites as a consequence of structural defects or coordination of Fe^{3+} cation with surrounding water molecules, in conjunction with Lewis acid sites, was responsible for the supremacy of Fe-BTC.

The dispersion of metal oxides on solid support is considered as an attractive strategy to improve the catalytic performance and stability of these active species during the reaction or thermal treatment at high temperatures (Tichit and Álvarez 2022; Gao et al. 2023; Biswas, Pal, and Pal 2020). Taking this into account, Yuan et al. reported the construction and trial of NbO_x clusters entrapped into MFI zeolite (2.6 wt% Nb-MFI) towards the FF-acetone aldolization (Yuan et al. 2020). The characterization results disclosed that oligomeric NbO_x clusters were homogeneously distributed within MFI matrix, accounting for the existence of Brønsted acid and strong Lewis acid centers. With a pronounced enhancement of both acid strength and density of catalytic Lewis acid sites, this hybrid catalyst was able to deliver FAc in 51.8% yield and 100% selectivity at 140 °C and 1 h. In the matter of durability, the Nb-MFI catalyst reactivated by air calcination showcased marginal loss of activity after several runs. Identically, a recent example of Nb_2O_5 nanoparticles embedded in mesoporous silica (Nb_2O_5/m-SiO_2) for the assembly of FAc was described (Skrodczky et al. 2023). By the assistance of microwave irradiation and acetophenone as a solvent, the one-pot construction of Nb_2O_5/m-SiO_2 was accomplished in a simple and rapid manner without the usage of any templates. Compared to bare Nb_2O_5, the beneficial combination of Nb_2O_5 nanoparticles with m-SiO_2 led to an increase in both surface area and density of Lewis-Brønsted acid sites of Nb_2O_5/m-SiO_2. Particularly, textural properties, acidity strength, and catalytic performance of Nb_2O_5/m-SiO_2 series could be easily tailored by the molarity of Nb/Si. Regarding these features, Nb_2O_5/m-SiO_2 with a Nb content of 42% enabled the highest yield (62%) and selectivity (82.7%) of FAc after 5 h reaction at 140 °C. Concerning the durability, the spent catalyst after re-calcination (400 °C in air) could preserve the initial activity with negligible changes of product selectivity throughout multiple runs. Analogously, the ZrO_2/m-SiO_2 (25 wt% Zr) catalyst rendered an excellent FF conversion (94.6%) and FAc yield (73.4%) under optimized conditions (140 °C, 2 h, acetone/FF molarity = 10) (Balaga et al. 2022). This impressive outcome was elucidated by the existence of isolated monomeric Zr species on m-SiO_2, respectably boosting the density of surface Lewis acid sites (Zr^{4+}) to catalyze the aldolization. Strikingly, it is found that the spent ZrO_2/m-SiO_2 catalyst was invulnerable to the leaching of active sites, enabling five recycles without any decline of FAc selectivity. Subsequently, a variety of mesoporous materials (Al-MCM-41, Al-TUD-1, and FDU-12) impregnated with cooper oxide (20 wt% Cu) was constructed and appraised for the aldol reaction of FF with acetone (Gandhi et al. 2023). Structural characterizations and acidity measurement indicated that the surface area and acidic strength of resulting samples, which pivotally underlay their reactivity towards the aldolization, followed the order of CuO/FDU-12 < CuO/Al-TUD-1 < CuO/Al-MCM-41. As such, 20 wt% CuO/Al-MCM-41 offered the highest FAc outcome with

90.2% yield and 92% selectivity under harsh conditions (175 °C, 24 h, acetone/FF molarity = 40). More impressively, the lifespan of this Lewis-Brønsted acidic catalyst could be prolonged after three consecutive runs without any needs of thermal regeneration.

In place of Lewis acids, Talanova and coworkers designed the tetraphenylmethane-cored porous aromatic frameworks (PAFs) bearing sulfonic acid groups $-SO_3H$ as a Brønsted acid catalyst for the FF-acetone condensation (Talanova, Yarchak, and Karakhanov 2019). The synthetic material (PAF–SO_3H), which was constructed by a two-step protocol of: (i) Suzuki cross coupling between tetrakis(4-bromophenyl) methane and 1,4-benzenediboronic acid; (ii) sulfonation with chlorosulfonic acid, consisted of mesoporous aromatic skeletons with a high specific area (192 $m^2\ g^{-1}$) and a high acidity strength (1.45 mmol g^{-1} H^+). At given conditions of 130 °C and 4 h in 1.0 MPa Argon, the catalysis of (PAF–SO_3H) in the FF-acetone aldol reaction gave a low outcome with 30% FF conversion and 25% yield of FAC. Setting the low efficiency aside, the spent PAF–SO_3H catalyst simply recovered by filtration could be recycled with a minor loss of activity. All the reactivity of acidic catalysts for the FF-acetone condensation are summarized in Table 10.1.

TABLE 10.1
The Aldol Condensation of FF with Acetone over Different Acidic Catalysts

Catalyst	Reaction Condition	Conv. (%)	Yield (%)	Ref.
Nb_2O_5	FF (5 mmol), acetone (20 mmol), cat. (0.1 g), 130 °C, 4 h.	88.9	70.8[a], 9.2[b]	(Jing et al. 2019)
CeP-180–900	FF (1.25 mmol), acetone (12.5 mmol), cat. (0.1 g), 160 °C, 4 h.	96	54.3[a], 14.1[b]	(Li et al. 2019)
H-BEA(25)	FF (68 mmol), acetone (680 mmol), cat. (2 g), 100 °C, 2 h.	38.5	30.6[a], 1.4[b]	(Kikhtyanin, Kelbichová, et al. 2014)
MCM-22	FF (34 mmol), acetone (340 mmol), cat. (0.5 g), 100 °C, 2 h.	60.4	49.6[a], 4.0[b]	(Kikhtyanin, Chlubna, et al. 2014)
Sn-Beta	FF (1.25 mmol), acetone (12.5 mmol), cat. (0.1 g), 160 °C, 2 h.	93	40[a], 20.5 [b]	(Su et al. 2017)
Sn-MFI	FF (1.25 mmol), acetone (12.5 mmol), cat. (0.1 g), 160 °C, 6 h.	80	61[a]	
20 wt% Fe-MCM-41	FF (1.25 mmol), acetone (25 mmol), cat (0.1 g), 160 °C, 24 h.	86.9	60[a], 7.5[b]	(Su et al. 2020)

(continued)

TABLE 10.1 (Continued)
The Aldol Condensation of FF with Acetone over Different Acidic Catalysts

Catalyst	Reaction Condition	Conv. (%)	Yield (%)	Ref.
Fe-BTC	FF (34 mmol), acetone (340 mmol), cat. (1 g), 100 °C, 2 h.	26.2	18.6[a], 5.2[b]	(Kikhtyanin, Kubička, and Čejka 2015)
2.6 wt% Nb-MFI	FF (1.25 mmol), acetone (12.5 mmol), cat. (0.1 g), 140 °C, 2 h.	51.8	51.7[a]	(Yuan et al. 2020)
Nb_2O_5/m-SiO_2	FF (1.6 M in acetone) cat. (cat./FF mass ratio= 0.2), 140 °C, 5 h.	75	62[a], 7.4[b]	(Skrodczky et al. 2023)
ZrO_2/m-SiO_2 (25 wt% Zr)	FF (2 mmol) acetone (20 mmol), cat. (0.08 g), 140 °C, 2 h.	93.6	73.4[a], 15.9[b]	(Balaga et al. 2022)
20 wt% CuO/ Al-MCM-41	FF (66 mmol), acetone (2640 mmol), cat. (1 g), 175 °C, 24 h.	98	90.2[a]	(Gandhi et al. 2023)
PAF–SO_3H	FF (1.2 mmol), acetone (12 mmol), cat. (0.02 g), Ar (1 MPa), 130 °C, 4 h.	30	25[a]	(Talanova, Yarchak, and Karakhanov 2019)

Notes: [a] FAc.
[b] F_2Ac.

Based on the reaction profile, the plausible mechanism of FF-acetone condensation over acidic catalysts is depicted in Scheme 10.2 (Jing et al. 2019; Balaga et al. 2022). At first, both FF and acetone are adsorbed on the catalyst surface through the coordination of both C=O groups with acid sites. Later, the enolization of acetone taking place through the acid-assisted abstraction of one α-H atom results in the enol tautomer. Promptly, this enol species will undertake the nucleophilic attack to the C=O bond of FF, giving the β-hydroxy ketone intermediate (FAc-OH). Under the assistance of acidic catalyst at high temperature, this intermediate is spontaneously dehydrated into FAc product. Following the similar pathway, the assembly of F_2Ac can be established through the condensation of FAc with another FF molecule. In this proposed mechanism, other acid-mediated side reactions such as self-condensation of acetone into mesityl oxide, dimerization of FAc into $(FAc)_2$, or humin formation cannot be ruled out.

10.2.2 Basic Catalysts

Relative to acid catalysis, the aldol condensation of FF with acetone is commonly conducted in the presence of basic catalysts because the reaction can proceed at

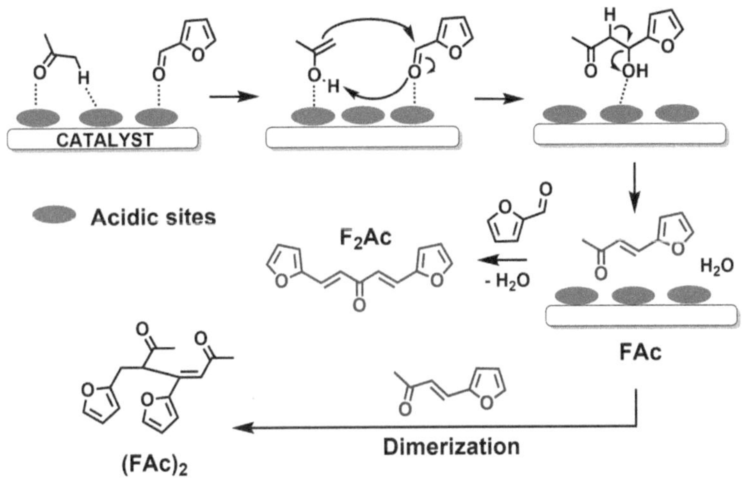

SCHEME 10.2 The proposed mechanism for the aldol condensation of FF with acetone in the presence of acidic catalyst.

milder conditions with fast reaction velocity and no side-reactions (i.e. oligomerization of FF or dimerization of FAc). Several studies showed that various Mg-Zr mixed oxides fabricated by numerous methods exhibited particular reactivity towards the FF-acetone condensation. For instance, 0.92Mg–Zr composite (Mg/Mg+Zr ratio = 0.92) constructed by sol-gel technique induced a mixture of products after 24 h reaction at 50 °C, where F_2Ac in 60% yield was favorably established as the main product (Faba, Díaz, and Ordóñez 2012). Meanwhile, co-precipitated 0.5Mg–Zr oxide (Mg/Mg+Zr ratio = 0.5), which encompassed two different phases of cubic-MgO and cubic-$Mg_{0.5}Zr_{0.5}O_{1.5}$, provided condensation products at the yield of 33.1% (FAc: 25.9%, F_2Ac:7.2%) when the reaction was executed in aqueous conditions (60 °C, 3 h, acetone/FF molarity = 11.3) (Sádaba et al. 2011). Analogously, the conversion of FF and FAc yield reached at 70% and 28%, respectively, over alcogel-derived 0.2Mg–Zr oxide (Mg/Mg+Zr ratio = 0.2) after 3 h reaction in water at 60 °C (Sadaba et al. 2012). Nevertheless, these Mg–Zr mixed oxides were prone to serious deactivation due to the surface disposition of carbonaceous materials, the leaching of active components (Mg^{2+}, Zr^{4+}) and/or hydration of MgO phase. Hence, other Mg-based mixed oxide catalysts have been alternatively developed to improve the reactivity and stability. As an example with 0.75Mg–Al oxide composite (Mg/Mg+Al ratio = 0.75) derived from calcined Mg–Al layered double hydroxides (LDHs), 98% conversion of FF coupled with 92.4% selectivity of condensation products (FAc: 71.2%, F_2Ac: 21.2 %) were achieved at 100 °C, 2 h and acetone/FF ratio 10:1 (Hora et al. 2014). On the contrary, the yield and selectivity of FAc were garnered at 70.8% and 88.1%, respectively, within 3 h at 100 °C in biphasic phase (water-toluene) over LDHs-derived 0.67Mg–Al catalyst (Mg/Mg+Al ratio =0.67) (Parejas et al. 2019). Relevant to the durability, negligible catalytic degeneration was observed for the recovered 0.67Mg–Al composite after three recycles. Alternatively, pyroaurite-based 0.91Mg–Fe oxide

(Mg/Mg+Fe ratio =0.91), which comprised a high concentration of basic sites and an abundant population of strong basic centers (O^{2-}), induced a 88% conversion of FF into condensation products with 71.4% selectivity (FAc: 37.4%, F_2Ac: 34%) at 50 °C (Kocík et al. 2018). Apart from Mg-based mixed oxides, other metal composites were also adopted for the title aldol condensation. For instance, mesoporous 0.6Co–Al (Co/Co+Al ratio = 0.6) spinel catalyst derived from the decomposition of corresponding Co–Al hydrotalcite precursor at 300 °C smoothly converted 96.4% FF into condensation products with 99% total selectivity (FAc: 67.4%, F_2Ac: 31.6%) after 5 h reaction at 140 °C (Xu et al. 2011). Later, Smoláková et al. demonstrated that the catalytic performance of xZn–Al oxides (Zn/Zn+Al ratio = 0.57, 0.66, 0.76, 0.81, 0.83) after thermal treatment at 400 °C was strongly correlated with the specific area and the size of ZnO crystallite (Smoláková et al. 2018). Accordingly, 0.66Zn–Al, which was highlighted with a high surface area and small ZnO crystallite, surpassed the counterparts in terms of catalytic efficiency, empowering an 81% FF conversion coupled with 66% selectivity of FAc and 26% selectivity of F_2Ac under mild conditions (50 °C, 6 h, acetone/FF molarity = 10). Indeed, the critical impact of ZnO species in various Zn–Al mixed oxide catalysts on the FF conversion was later elucidated, in which the lowest conversion of FF corresponded with the catalyst containing a large amount of well-crystalline ZnO phase (Dubnová et al. 2021). With reference to bulk La_2O_3 and bare Zn–Al mixed oxide, high-surface Zn–Al oxide doped with hexagonal lanthanum dioxycarbonate ($La_2O_2CO_3$) showed a remarkable improvement in the catalytic performance (Kong et al. 2021). 91% FF conversion together with 100% selectivity of condensation products (FAc: 61.7%, F_2Ac: 38.3%) was accomplished when the aldolization was facilitated by ZnAlLa catalyst (denoted as $La_2O_2CO_3$/ZnO–Al_2O_3, Zn:Al:La molar ratio = 2:1:2) at 80 °C, 4 h, and acetone/FF molar ratio 1:1. The prominence of this designed catalyst was rationalized by a plenitude of moderate basic sites (i.e. La^{3+}–O^{2-} pairs) on the catalyst surface. Conversely, Al_2O_3 doped with rod-shape $La_2O_2CO_3$ (denoted as $La_2O_2CO_3/Al_2O_3$) exhibited a different product distribution in water media, furnishing FAc and F_2Ac in 30.6% and 69.2% selectivity, respectively, when the conversion of FF reached at 92.3% under identical conditions. In conjunction with the surface disposition of organic compounds, the morphology change of $La_2O_2CO_3$ was blamed for the deactivation of lanthanum-based catalysts after repeated runs.

In another direction, a simple organic solvent-free synthetic protocol (mechanical grinding-calcination) was introduced to fabricate sodium Y zeolite supported MgO (MgO/NaY) as a nanohybrid basic catalyst for the title aldolization (Huang et al. 2012). According to the study, the catalytic performance of MgO/NaY series was correlated with the content of MgO, in which 20 wt% MgO/NaY emerged as the most competent catalyst. Particularly, FF was completely converted into FAc and F_2Ac at respective selectivity of 42.2% and 57.1% when the condensation was carried out in a biphasic solvent (water-ethanol) at 85 °C for 8 h. However, the recovered MgO/NaY catalyst needed to be thermally reactivated (600 °C, 4 h) to restore the initial activity, which was expounded by the hydrothermal transformation of MgO species into $Mg(OH)_2$ during the reaction. Afterwards, MgO decorated on nitrogen-doped carbon template (20 wt% MgO/CN), which was acquired from the thermal condensation of glucosamine hydrochloride and MgO, showed a preferential establishment of FAc from the condensation of FF with acetone (Li et al. 2014). The introduction of N-containing carbon matrix endowed with both hydrophilicity and porous structure was proposed to enhance the water dispersity and stability of MgO/CN composite, leading to the unparalleled competence of this hybrid catalyst. With such distinct features, MgO/CN empowered a full conversion of FF into FAc and F_2Ac with corresponding yield of 87% and 12% after reaction at 80 °C for 3 h in water media.

In an endeavor to improve the reactivity and stability of Mg–Zr mixed oxides, Faba and coworkers determined to adopt high surface-area graphites (HSAGs) to immobilize these active species (Faba, Diaz, and Ordonez 2013). Herein, the employment of HSAG with a high surface area of 100 m^2g^{-1} (HSAG100) and co-precipitation method for immobilizing Mg–Zr mixed oxides not only enabled a regular distribution of basic centers on the carbon surface, enhanced the extent of these active sites exposed to reactants, but also hindered the formation of coke during the reaction. Regarding the activity of Mg–Zr/HSAG100 at 50 °C and 24 h, FF was quantitatively converted into FAc and F_2Ac at 37% and 51% selectivity, respectively. Contrary to MgO hybrid catalysts, the 30 wt% CaO/ZSM-5 calcined at 700 °C proficiently induced the complete conversion of FF in water, affording 98.6% yield of condensation products (FAc: 38.4%, F_2Ac: 60.2%) after 2 h reaction at 80 °C (Fang et al. 2020). Afflicted by the leaching of Ca metal, it is no surprise that 30 wt% CaO/ZSM-5 became severely deactivated after the completion of aldolization.

Following the same concept, the incorporation of stronger basic species, i.e. potassium oxide clusters (K_2O) into carbonaceous materials was also adopted (Zhan et al. 2019). The one-pot fabrication of K_2O cluster-promoted N-doped microporous carbon (K_2O/C) was simply carried out by heating the mixture of dicyandiamide, potassium nitrate, and polyvinyl alcohol (PVA) at 1000 °C in N_2. The 10 wt% K_2O/C-PVA catalyst presented the exceptional performance with 85.4% yield and 89.6% selectivity to FAc at experimental conditions of 100 °C and 2 h, where the preponderance of K_2O clusters and their strong interaction with nitrogen and carbon atoms of template were deemed major factors. Instead of F_2Ac, the dominant assembly of FAc was expounded by the population of active species inside the micropores of N-doped carbon template. In other words, these micropores (< 1.4 nm) served as physical barriers to large-size guest molecules, substantially retarding the assembly of F_2Ac from the ensuing condensation of FAc with FF. Importantly, the reusability survey disclosed that the spent K_2O/C-PVA catalyst after re-calcination (1000 °C in N_2) was able to display the original activity with no diminution in FAc selectivity. Motivated by the construction of superbasic mesoporous composites (e.g. K_2O/MgO–SBA-15 and KF-modified Al–La/SBA-15) (Wu et al. 2006; Liu et al. 2017), the catalysis of potassium-modified bifunctional MgAl-SBA-15 composite (K_2O/12MgAl–SBA-15) for the aldol condensation was examined as well (Arumugam et al. 2023). It is found that the loading of K_2O species not only increased the density of strong basic sites but also considerably covered the Brønsted acid centers of 12MgAl–SBA-15. Accordingly, the optimal 20 wt% K_2O/12MgAl–SBA-15 catalyst was capable of converting 83% FF into condensation products with 46% selectivity (FAc, F_2Ac = 23%) after

3 h of reaction at 50 °C. Regardless of a moderate leaching of potassium metal, the spent K_2O/MgAl–SBA-15 after thermal treatment (550 °C, 2 h) could maintain fair-efficiency reactivity throughout four recycling tests.

Another promising basic solid candidate to promote the FF condensation with acetone came up with the employment of activated dolomite (O'Neill et al. 2014). The authors stated that the activation procedure (calcination-hydration) came in handy for resulting dolomite, which not only improved the textural properties (surface area and porosity) but also generated a number of accessible Brønsted basic centers, i.e. hydroxyl groups from magnesium and calcium hydroxides. As a result, the activated dolomite with a notable enhancement of surface basicity efficiently converted FF (>90%) into aldol products under biphasic conditions (140 °C, 1 h, water-methanol mixture as a solvent), in which F_2Ac was dominantly yielded (72.8%). Compared to acetone, the preferential delivery of F_2Ac was illustrated by a higher adsorption enthalpy of FAc over the surface of activated dolomite, which enabled a facile interaction of FF with FAc to produce F_2Ac. Rather than metal leaching, the deposition of carbonaceous materials was culpable for the deactivation of activated dolomite during the transformation.

In line with these batchwise processes, the continuous-flow condensation of FF with acetone was also attempted over different solid catalysts including Al_2O_3, $MgAl_2O_4$, 9.3 wt% CaO/$MgAl_2O_4$ and 9.8 wt% MgO/$MgAl_2O_4$ (Huang et al. 2022). It is found that reaction rates of these catalysts at a low conversion of FF (< 20%) followed the order of Al_2O_3 < $MgAl_2O_4$ < MgO/$MgAl_2O_4$ < CaO/$MgAl_2O_4$. Specifically, 74% FF was converted into FAc and F_2Ac with selectivity of 82% and 7%, respectively, when the CaO/$MgAl_2O_4$-assisted condensation was performed at 100 °C for 0.5 h. However, the activity of CaO/$MgAl_2O_4$ was considerably deteriorated over a short period of time-on stream, ascribable to the adsorption of 2-furoic acid byproduct, which was generated from the Cannizaro reaction of FF in basic condition. All the results of base-mediated condensation of FF with acetone are summarized in Table 10.2.

TABLE 10.2
The Aldol Condensation of FF with Acetone over Different Basic Catalysts

Catalyst	Reaction Condition	Conv. (%)	Yield (%)	Ref.
0.92Mg–Zr oxide	FF (10 mmol), acetone (10 mmol), cat. (0.5 g), 50 °C, 24 h.	80	14.3[a],60[b]	(Faba, Díaz, and Ordóñez 2012)
0.5Mg–Zr oxide	FF (12.1 mmol), acetone (136.4 mmol), cat. (0.1 g), H_2O (40 mL), 60 °C, 3 h.	70	25.9[a],7.2[b]	(Sádaba et al. 2011)
0.2Mg–Zr oxide	FF (12.1 mmol), acetone (136.4 mmol), cat. (0.1 g), H_2O (40 mL), 60 °C, 3 h.	70	28[a], 3.5[b]	(Sadaba et al. 2012)

TABLE 10.2 (Continued)
The Aldol Condensation of FF with Acetone over Different Basic Catalysts

Catalyst	Reaction Condition	Conv. (%)	Yield (%)	Ref.
0.75Mg–Al oxide	FF (67.7 mmol), acetone (680.1 mmol), cat. (2 g), 100 °C, 2 h.	98	69.8[a], 20.8[b]	(Hora et al. 2014)
0.67Mg–Al oxide	FF (10 mmol), acetone (20 mmol), cat. (0.4 g), water-toluene (1:1, 20 mL), 100 °C, 3 h.	80.4	70.8[a], 9.6[b]	(Parejas et al. 2019)
0.91Mg–Fe oxide	FF (68 mmol), acetone (680 mmol), cat. (2 g), 50 °C, 4 h.	88	32.8[a],30[b]	(Kocík et al. 2018)
0.6Co–Al oxide	FF (68 mmol), acetone (680 mmol), cat. (1 g), 140 °C, 5 h.	96.4	64.9[a], 30.5[b]	(Xu et al. 2011)
0.66Zn–Al oxide	FF (68 mmol), acetone (680 mmol), cat. (2 g), 50 °C, 6 h.	81	53.5[a], 21.1[b]	(Smoláková et al. 2018)
$La_2O_2CO_3$/ $ZnO–Al_2O_3$	FF (10 mmol), acetone (10 mmol), cat. (0.2 g), 80 °C, 4 h.	91	56.1[a], 34.9[b]	(Kong et al. 2021)
$La_2O_2CO_3/Al_2O_3$	FF (10.5 mmol), acetone (10.5 mmol), cat. (0.2 g), water (20 mL), 80 °C, 4 h.	92.3	25.8[a], 67.8[b]	(Li et al. 2021)
20 wt% MgO/NaY	FF (50 mmol), acetone (100 mmol), cat. (1.06 g), water-ethanol (20 mL,1:1), 85 °C, 8 h.	100	42.2[a], 57.1[b]	(Huang et al. 2012)
20 wt% MgO/CN	FF (3.2 mmol), acetone (29.1 mmol), cat. (0.02 g), water (40 mL), 80 °C, 3 h.	>99	87[a], 12[b]	(Li et al. 2014)
Mg–Zr/HSAG100	FF (10 mmol), acetone (10 mmol), cat. (0.5 g), 50 °C, 3 h.	>99	37[a], 51[b]	(Faba, Diaz, and Ordonez 2013)
30 wt% CaO/ZSM-5	FF (50 mmol), acetone (50 mmol), cat. (1 g), water (3 mL), 80 °C, 2 h.	100	38.4[a], 60.2[b]	(Fang et al. 2020)
10 wt% K_2O/C-PVA	FF (6.2 mmol), acetone (104 mmol), cat. (0.15 g), 100 °C, 2 h.	96.4	85.4[a],4.9[b]	(Zhan et al. 2019)
20 wt% K_2O/ 12MgAl–SBA-15	FF (34 mmol), acetone (170 mmol), cat. (0.25 g), 50 °C, 3 h.	83	19.1[a], 19.1[b]	(Arumugam et al. 2023)

(*continued*)

TABLE 10.2 (Continued)
The Aldol Condensation of FF with Acetone over Different Basic Catalysts

Catalyst	Reaction Condition	Conv. (%)	Yield (%)	Ref.
Activated dolomite	FF (57 mmol), acetone (57 mmol), cat. (0.5 g), water-methanol (60 g, 1:1.5 w/w), 140 °C, 1 h.	>90	17.2[a],72.8[b]	(O"Neill et al. 2014)
9.3 wt% CaO/ $MgAl_2O_4$	FF (1.22 mol/L in toluene, 0.5 mL/min), acetone (12.2 mol/L), cat. (0.1 g), 100 °C, 0.5 h.	74	60.7[a], 5.2[b]	(Huang et al. 2022)

Notes: [a] FAc.
[b] F_2Ac.

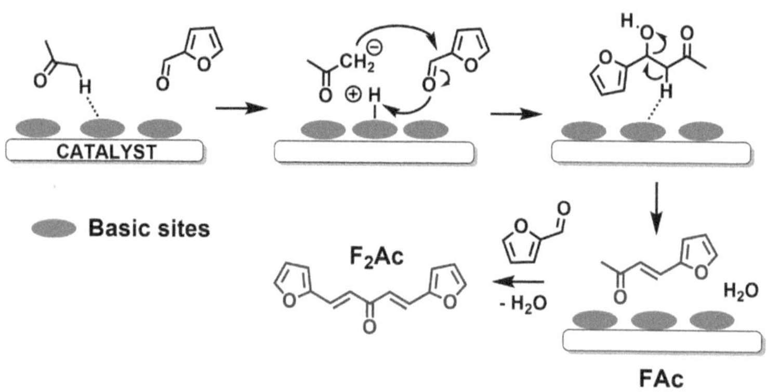

SCHEME 10.3 The proposed mechanism for the aldol condensation of FF with acetone over basic catalyst.

Contrary to the acid catalysis, the aldol condensation between FF and acetone in the presence of basic catalyst will proceed in a different manner (Scheme 10.3) (Faba, Díaz, and Ordóñez 2012; O'Neill et al. 2014). Initially, the basic sites will abstract one α-H atom of acetone to result in enolate species. In fact, this proton abstraction, which is largely influenced by the basicity of active centers and acidity of α-proton, is determined as the rate-determining step. The nucleophilic addition of this enolate into the C=O bond of FF instantly occurs to bring in β–hydroxyl ketone intermediate. In basic conditions at high temperature, this unstable intermediate is rapidly dehydrated into FAc along with regeneration of active sites. In the same manner, the enolate derived from the α-H abstraction of FAc will react with another molecule of FF to bring in F_2Ac.

10.2.3 Bifunctional Acid-Base Catalysts

Generally, the wide utility of acid-base bifunctional catalysts in several organic transformations is embraced by the cooperative catalysis of acid-base pair sites (Li et al. 2013). In detail, the interaction of acid and base centers with reactants can take place in unison or in succession, synergistically enhancing the reaction rate, product selectivity, and catalyst lifespan (Setoyama 2006). Considering such advantages, Thanh et al. adopted the amphoteric titania (TiO_2) to catalyze the aldol reaction between FF and acetone (Nguyen Thanh et al. 2016). In this study, nanocrystalline TiO_2 with pure anatase phase was easily obtained from the thermal hydrolysis of aqueous solution containing titanium (IV) oxysulfate ($TiOSO_4$) and urea. Interestingly, the calcination of the resulting TiO_2 sample was deemed to trigger several troubles of dehydration, surface dehydroxylation, and changes of textural properties, ultimately leading to a reduction of active sites. Accordingly, only uncalcined TiO_2 could offer the maximum FF conversion at 25% with 72% selectivity to FAc under testing conditions (50 °C, 4 h, acetone/FF molarity = 5). By contrast, the concerted participation of surface basic and acidic centers in 0.5Al-ZrO_2 composite (Al/Al+Zr ratio = 0.5) could upgrade the yield of FAc to 90.8% at 110 °C and 8 h (Yuan et al. 2022). Appreciably, the designed catalyst was almost impervious to metal leaching and could be reused for multiple cycles without any demand of regeneration. By the same token, the availability of surface acidity (0.14 $mmolg^{-1}$) and basicity (0.65 $mmolg^{-1}$) was rationalized for the exceptional catalysis of 0.75Mg-La oxide composite (Mg/Mg+La ratio = 0.75) in the title condensation, furnishing excellent yield (94%) and selectivity (98%) to FAc after reaction of 7 h at 150 °C (Desai and Yadav 2019). Additionally, the reactivated 0.75Mg-La catalyst after calcination (650 °C) could be reused for four cycles with no loss of activity.

Another example of acid-base bifunctional solid catalyst for the FF-acetone condensation was engaged with sol-gel-derived magnesium hydroxide fluoride [$MgF_{1.5}(OH)_{0.5}$] (Xu et al. 2019). The structural-textural characterizations confirmed that the $MgF_{1.5}(OH)_{0.5}$ sample encompassed an intimate phase mixture of MgF_2 and $Mg(OH)_{0.5}(OCH_3)_{1.5}$, a high specific area (460 m^2g^{-1}), and acid-base properties. Specifically, the proximity of Lewis acid sites (uncoordinated magnesium species in MgF_2 phase) to basic sites (moderate-strength hydroxyl groups on $Mg(OH)_{0.5}(OCH_3)_{1.5}$ phase) was pivotal for the high performance of $MgF_{1.5}(OH)_{0.5}$. Consequently, 28.5% yield of condensation products (FAc: 12.2%, F_2Ac: 16.3%) was obtained at 34% conversion of FF under mild conditions (50 °C, 6 h, acetone/FF molarity = 2).

Recently, MgO/biochar obtained from the pyrolysis of crushed soursop seed powder impregnated with $MgCl_2$ at 700 °C appeared as an active bifunctional catalyst for the aldol reaction of FF with acetone (Palermo et al. 2023). The optimal MgO/biochar catalyst containing a surface MgO content of 0.34% (w/w) offered a 52% conversion of FF and 100% selectivity to condensation products (FAc: 77%, F_2Ac: 23%) upon executing the reaction at 120 °C for 4 h, attributable to the simultaneous existence of basic MgO species and acidic centers on the oxidized surface of carbon. Besides, the aldol condensation between FF and other aromatic ketones such as 2-acetylthiophene and acetophenone was also attempted, yielding (*E*)-3-(furan-2-yl)-1-(thiophen-2-yl) prop-2-en-1-one in 79% and (*E*)-3-(furan-2-yl)-1-phenylprop-2-en-1-one in 85 %.

TABLE 10.3
The Aldol Condensation of FF with Acetone over Different Solid Acid-Base Bifunctional Catalysts

Catalyst	Reaction Condition	Conv. (%)	Yield (%)	Ref.
TiO_2	FF (68 mmol), acetone (340 mmol), cat. (1 g), 50 °C, 4 h.	25	18[a], 5[b]	(Nguyen Thanh et al. 2016)
0.5Al-Zr oxide	FF (12 mmol), acetone (270 mmol), cat. (0.2 g), 110 °C, 8 h.	92	90.8[a]	(Yuan et al. 2022)
0.75Mg-La oxide	FF (10 mmol), acetone (400 mmol), cat. (0.15 g), 150 °C, 7 h.	96	94[a]	(Desai and Yadav 2019)
$MgF_{1.5}(OH)_{0.5}$	FF (5 mmol), acetone (10 mmol), cat. (0.2 g), 50 °C, 6 h.	34	12.2[a], 16.3[b]	(Xu et al. 2019)
MgO/biochar	FF (0.5 mmol), acetone (5 mmol), cat. (0.5 g), 120 °C, 4 h.	52	40[a], 12[b]	(Palermo et al. 2023)

Notes: [a] FAc.
[b] F_2Ac.

Despite such achievements, the reactivity of spent MgO/biochar catalyst began to wind down after two trials of FAc synthesis. All the performance of heterogeneous acid-base bifunctional catalysts for the aldolization of FF with acetone are summarized in Table 10.3.

Pursuant to the cooperative catalysis of acid-base pairs, the mechanistic pathway leading to Fac and F_2Ac over acid-base bifunctional catalyst can be briefly described through their specific roles as follows: (i) the acid sites facilitate the adsorption and activation of reactants (FF and acetone) on the catalyst surface, (ii) the basic sites induce the generation of enolate species by α-H abstraction for the nucleophilic addition to FF (Faba, Díaz, and Ordóñez 2012; O"Neill et al. 2014; Desai and Yadav 2019).

10.3 ALDOL CONDENSATION OF FF WITH LEVULINIC ACID/A-ANGELICA LACTONE

In order to produce a collection of condensation compounds with higher carbon numbers, the exploitation of other biomass-derived carbonyl compounds for the aldol condensation with FF has been attempted. In this context, levulinic acid (LVA), a key C5 platform molecule derived from sugars (glucose and fructose), emerges as a promising candidate (Sajid et al. 2021). Contrary to acetone, the FF-LVA condensation taking place in the presence of either acidic or basic catalysts (Scheme 10.4) can

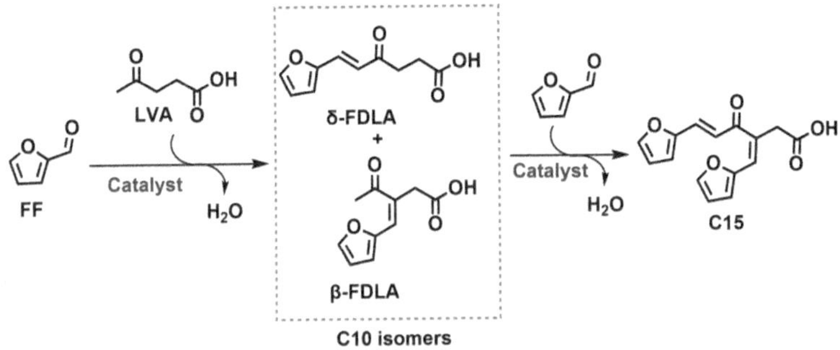

SCHEME 10.4 The catalytic aldol condensation between FF and LVA.

deliver two C10 isomers, i.e. δ- and β-furfurylidenelevulinic acid (δ- and β-FDLA) after acidification, due to the existence of two types of active α-H in the structure of LVA. Moreover, the resulting C10 adducts can condensate with another molecule of FF to provide C15 adduct, namely (3Z, 5E)-6-(furan-2-yl)-3-(furan-2-ylmethylene)-4-oxohex-5-enoic acid. No matter what kind of catalyst is applied for this condensation, sodium levulinate (Na-LVA) in place of LVA is commonly employed as the starting material in order to increase the interaction between the levulinate reactant and catalyst surface. In this section, the utility of various solid acid and base catalysts for the aldol condensation of FF with LVA is outlined.

Starting with different acidic/basic metal oxides, Liang and coworkers demonstrated that the selectivity of target products could be remarkably regulated depending on the nature of catalyst (Liang et al. 2016). For instance, basic MgO predominantly induced δ-FDLA in 70.6% yield after at 85 °C and 1.5 h, which was assignable to the simultaneous participation of active sites, i.e. moderate basic sites (Mg^{2+}–O^{2-} pairs) and surface hydroxyl (-OH) groups. Conversely, β-FDLA of 75.5 % yield was majorly constituted over ZnO calcined at 450 °C (ZnO-450) under harsh conditions (95 °C, 16 h), creditable to the attendance of catalytic surface -OH groups. Indeed, this disparity corroborated the fact that the aldol condensation of FF with LVA could proceed in either acid- or base-mediated pathway.

Following this study, the practicality of potassium-promoted MFI zeolites as basic catalysts for the FF-LVA condensation was examined (Su et al. 2019). While potassium-promoted conventional MFI zeolite (K/MFI) delivered a moderate conversion of FF (55.1%) with a low yield of C10 adducts (27.4%), potassium-modified, hierarchical MFI zeolite (K/H-MFI-0.6) showed a remarkable outcome with 70.6% yield of C10 products (δ-FDLA: 54.9%, β-FDLA: 15.7%) at 100 % FF conversion under identical conditions (100 °C, 9 h, Na-LVA/FF molarity = 1.5). Herein, the prominence of K/H-MFI-0.6 in terms of catalytic efficiency and δ-FDLA productivity was likely accredited to two reasons of: (i) a preponderance of moderate and strong basic sites, and (ii) alleviation of external mass transfer limitation by high BET surface area and high volume of mesopores. Pertaining to the reusability, it is realized that the reactivity of K/H-MFI-0.6 catalyst was acceptably preserved during five cycles by means of thermal regeneration (550 °C in air).

Obviously, the establishment of long chain carbon molecule (i.e. C15 adduct) from the condensation of FF with LVA is deemed challenging due to competitive side reactions in strong basic conditions. To overcome this obstacle, Zhao and coworkers introduced the facile fabrication of chitosan (CT) modified with NaOH 4M (4NaOH/CT) as a promising basic solid catalyst for the room-temperature synthesis of C15 compound (Zhao et al. 2022). Impressively, the 4NaOH/CT catalyst showcased an outstanding activity in the aldolization, leading to 97.8% FF conversion with 16.4% yield of C10 adducts and 78% yield of C15 adduct under mild conditions (25 °C, 3 h, FF/Na-LVA molarity = 2). The salience of 4NaOH/CT catalyst in this study was ascribed to the presence of primary amino groups in chitosan, which served to form a Schiff base with FF, circumventing the polycondensation of FF during the reaction. Aside from such accomplishment, reloading of NaOH was mandatory for the regeneration of the 4NaOH/CT catalyst.

In opposite approach, the condensation of FF with Na-LVA was also attempted in the presence of acidic catalysts. Testing with a variety of acidic zeolites (i.e. H-ZSM-5(23), H-ZSM-5(80), H-MOR, and H-BEA), Cueto and coworkers found that the surface acidity, i.e. the density of moderate acid sites, in preference to the structural properties of zeolites crucially presided over both yield and selectivity of C10 adducts (Cueto et al. 2020). In accordance with this finding, H-ZSM-5(23) guaranteed the highest FF conversion of 67% with 48.7% yield of C10 products (δ-FDLA: 7.2%, β-FDLA: 41.5%) under operating conditions (150 °C, 24 h, Na-LVA/FF molarity = 1). Interestingly, the regenerated H-ZSM-5(23) after thermal treatment could garner acceptable results of FF conversion and product yields with a negligible change of β-/δ-C10 ratio during four cycles. Afterwards, other commercial solid acids, i.e. γ-Al_2O_3, MCM-41, Amberlite IR-120, and Amberlite 36 were also applied for the title condensation (Cueto et al. 2022). The best results were acquired with γ-Al_2O_3, where the yield of C10 products amounted to 44% (δ-FDLA: 7.2%, β-FDLA: 36.8%) at 61.2 % FF conversion under milder conditions of 150 °C, 6 h, and Na-LVA/FF ratio 1:1.

Another attractive synthetic route to access C10 oxygenate compound is associated with the aldol condensation of FF with LVA-derived angelica lactone (Xu et al. 2017). The catalyst screening revealed that amphoteric Mn_2O_3 was superior in catalytic activity, resulting in a quantitative FF conversion with 96% yield of 3-(furan-2-ylmethylene)-5-methylfuran-2(3H)-one (FMFO) under specific conditions of 80 °C and 4 h. Weak acid-base strength was responsible for the exceptional catalysis of Mn_2O_3, competently negating other side reactions (polymerization or Cannizzaro reaction) of FF. Moreover, the simultaneous existence of acid and base sites on the surface of Mn_2O_3 was deemed conducive to the activation of FF and angelica lactone during the condensation. Attractively, robust Mn_2O_3 catalyst could maintain the initial activity during multiple recycles without any requirement of thermal treatment. The reactivity of heterogeneous catalysts for the condensation of FF with Na-LVA/α-angelica lactone is summarized in Table 10.4.

Based on several examples, the condensation of FF with LVA over acidic or basic catalyst is mechanistically proposed in Scheme 10.5. Initially, both FF and Na-LVA are absorbed on the surface of catalyst through the coordination of C=O groups with acidic sites (Liang et al. 2016; Cueto et al. 2020). Afterwards, the acidic sites will assist the tautomerization of Na-LVA molecule through the liberation of α-H and

TABLE 10.4
The Aldol Condensation of FF with Na-LVA/α-Angelica Lactone over Different Catalysts

Catalyst	Reaction Condition	Conv. (%)	Yield (%)	Ref.
MgO	FF (4 mmol), Na-LVA (6 mmol), cat. (0.1 g), 85 °C, 1 h.	100	70.6[a]	(Liang et al. 2016)
ZnO	FF (4 mmol), Na-LVA (6 mmol), cat. (0.1 g), 95 °C, 16 h.	95.8	75.5[b]	
K/H-MFI-0.6	FF (4 mmol), Na-LVA (6 mmol), cat. (0.2 g), 100 °C, 9 h.	100	54.9[a], 15.7[b]	(Su et al. 2019)
4NaOH/CT	FF (2 mmol), Na-LVA (1 mmol), cat. (0.15 g), water (4 mL), 25 °C, 3 h.	97.9	9.36[a], 7.04[b], 78[c]	(Zhao et al. 2022)
H-ZSM-5(23)	FF (10 mmol), Na-LVA (10 mmol), cat. (1.8 g), 150 °C, 24 h.	67	7.2[a], 41.5[b]	(Cueto et al. 2020)
γ-Al_2O_3	FF (10 mmol), Na-LVA (10 mmol), cat. (1.8 g), 150 °C, 6 h.	59.5	7.2[a], 36.8[b]	(Cueto et al. 2022)
Mn_2O_3	FF (10mmol), α-angelica lactone (10 mmol), cat. (0.16 g), 80 °C, 4 h.	>99	96[d]	(Xu et al. 2017)

Notes: [a] δ-FDLA.
[b] β-FDLA.
c C15 adduct.
[d] FMFO.

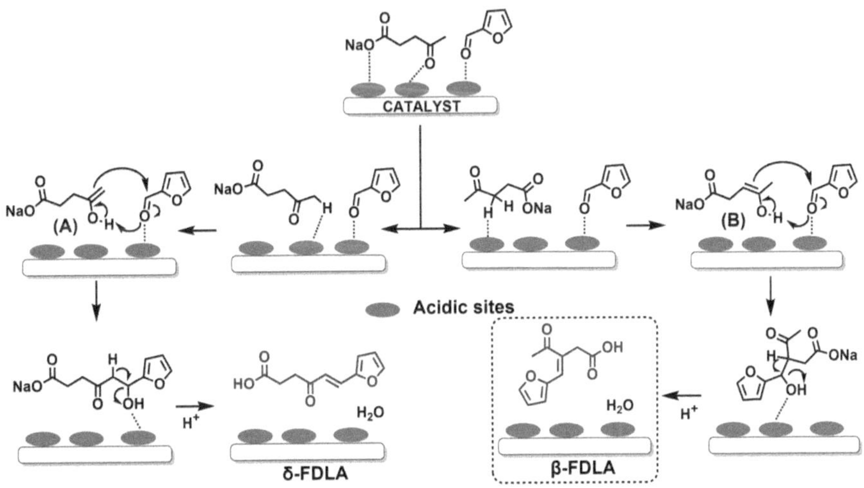

SCHEME 10.5 The proposed mechanism for the aldol condensation of FF with Na-LVA over acidic catalyst.

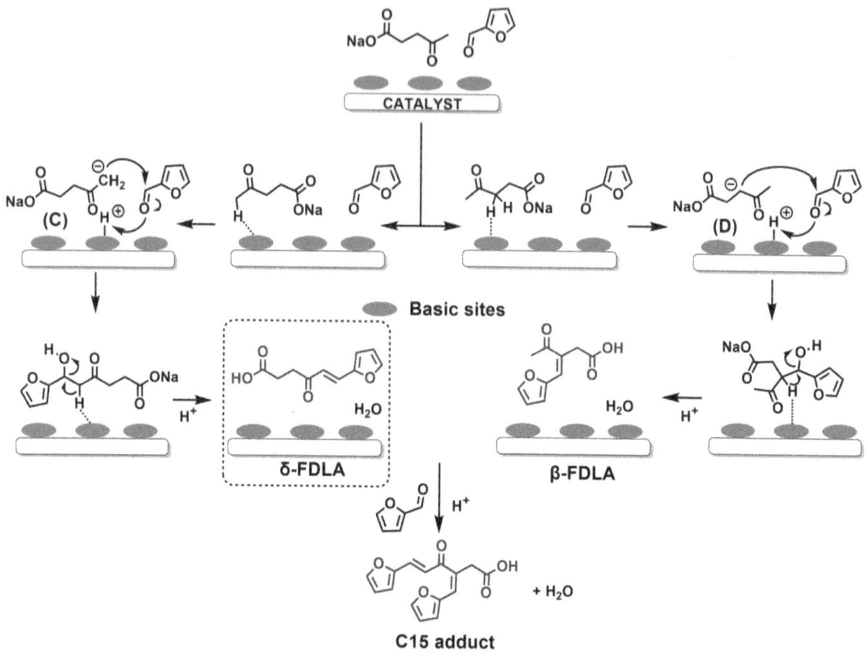

SCHEME 10.6 The proposed mechanism for the aldol condensation of FF with Na-LVA over basic catalyst.

β-H atom, giving two kinds of nucleophilic enol (A and B). Promptly, these enol species will attack the activated C=O group of FF to render two β-hydroxy ketone intermediates. Under the assistance of acidic catalyst at high temperature, the dehydration of these intermediates favorably proceeds, leading to the formation of final α,β-unsaturated ketones (δ-FDLA and β-FDLA). In the acid-catalyzed aldol condensation of FF and Na-LVA, it is worth mentioning that β-FDLA is dominantly constituted due to the high stability of enol B.

On the other hand, the base-assisted mechanism takes place through the generation of enolates (C and D) with the assembly of δ-FDLA as major final product (Scheme 10.6) (Liang et al. 2016; Su et al. 2019). Following the same base-mediated pathway, both δ-FDLA and β-FDLA can further condensate with another molecule of FF to engender the C15 adduct.

10.4 CONCLUSION AND PERSPECTIVES

It is no doubt that the aldol condensation plays an important role in the domain of biomass valorization, delivering a variety of various oxygenate compounds with elongated carbon chains. In this chapter, the state-of-the-art for the synthesis of oxygenate precursors of biofuel, from the aldol condensation of biomass-derived FF with typical carbonyl compounds (acetone, levulinic acid, and angelica lactone) have been systematically summarized. The discussion came up with an elaboration on the utility of heterogeneous catalysts with regards to their structure-activity relationship, durability, reaction parameters, and catalytic pathway. Consequently, several condensation products (i.e. C8, C10, C13, and C15 adduct) were successfully generated over different types of solid acid, base, and acid-base bifunctional catalyst. Notwithstanding promising results, the following efforts should be taken into consideration to upgrade the efficiency and feasibility of this biomass valorization strategy:

(i) It is highly concerned that acidic impurities directly originated from commercial supply or autooxidation of FF in air could severely reduce the activity of basic catalysts (e.g. metal oxides) owing to the strong adsorption on these contaminants on basic centers. Therefore, the usage of newly purchased or freshly distilled FF is recommended to address this hurdle.
(ii) Several reasons, such as surface disposition of organic compounds, leaching or structural changes of active sites critically contributed to the deactivation of designed catalysts. In response to these circumstances, the development of novel heterogenous catalysts with a special focus on optimizing structural-textural characteristics, engineering surface properties, tunable pore size/acid-base strength, water resistance, and thermal stability for post-treatment, is extensively required. Other issues such as low cost, facile preparation, and long lifespan of catalyst in both batchwise and continuous-flow operation are of great importance as well.
(iii) Convenient, economical, and high-efficiency protocols for mass-production and purification of furanic condensation products as well as starting materials should be taken into consideration to accommodate the industrial application of this synthetic strategy.
(iv) The integration of aldol condensation-hydrodeoxygenation for the synthesis of high-grade alkanes with extended carbon chains/branches from furanic derivatives or raw lignocellulosic biomass over solid catalysts is highly desired because additional steps of pretreatment or separation of intermediates can be circumvented during the operation.

With the advancement of this synthetic strategy, it is believed that resulting condensation products can be upgraded into valuable biofuels/biolubricants and petroleum-free chemicals, enhancing the market competitiveness for biomass valorization in the future.

REFERENCES

Arumugam, Mahashanon, Oleg Kikhtyanin, Amin Osatiashtiani, Veronika Kyselová, Vlastimil Fila, Iva Paterova, Ka-Lun Wong, and David Kubička. 2023. "Potassium-modified bifunctional MgAl-SBA-15 for aldol condensation of furfural and acetone." *Sustainable Energy & Fuels* 7. The Royal Society of Chemistry. 3047–3059. doi: 10.1039/D3SE00444A.

Balaga, Ravi, Peifang Yan, Kishore Ramineni, Hong Du, Zhi Xia, Mahender Reddy Marri, and Zongchao Conrad Zhang. 2022. "The role and performance of isolated zirconia sites on mesoporous silica for aldol condensation of furfural with acetone." *Applied Catalysis A: General* 648. Elsevier B.V. 118901–118909. doi: 10.1016/j.apcata.2022.118901.

Biswas, S., A. Pal, and T. Pal. 2020. "Supported metal and metal oxide particles with proximity effect for catalysis." *RSC Advances* 10 (58). The Royal Society of Chemistry. 35449–35472. doi: 10.1039/D0RA06168A.

Bohre, Ashish, Saikat Dutta, Basudeb Saha, and Mahdi M. Abu-Omar. 2015. "Upgrading furfurals to drop-in biofuels: an overview." *ACS Sustainable Chemistry & Engineering* 3 (7). American Chemical Society. 1263–1277. doi: 10.1021/acssuschemeng.5b00271.

Cueto, Jennifer, Laura Faba, Eva Díaz, and Salvador Ordóñez. 2022. "From lignocellulosic biomass to chemical precursors: simultaneous valorization of furfural and levulinic acid over mesoporous acid catalysts." *Industrial Crops and Products* 188. Elsevier B.V. 115692–115700. doi: 10.1016/j.indcrop.2022.115692.

Cueto, Jennifer, Valeria Korobka, Laura Faba, Eva Díaz, and Salvador Ordóñez. 2020. "Aldol condensation of biomass-derived levulinic acid and furfural over acid zeolites." *ACS Sustainable Chemistry & Engineering* 8 (11). American Chemical Society. 4371–4383. doi: 10.1021/acssuschemeng.9b06636.

Desai, Deepak S., and Ganapati D. Yadav. 2019. "Green synthesis of furfural acetone by solvent-free aldol condensation of furfural with acetone over La_2O_3–MgO mixed oxide catalyst." *Industrial & Engineering Chemistry Research* 58 (35). American Chemical Society. 16096–16105. doi: 10.1021/acs.iecr.9b01138.

Dubnová, Lada, Lucie Smoláková, Oleg Kikhtyanin, Jaroslav Kocík, David Kubička, Magdalena Zvolská, Miloslav Pouzar, and Libor Čapek. 2021. "The role of ZnO in the catalytic behaviour of Zn-Al mixed oxides in aldol condensation of furfural with acetone." *Catalysis Today* 379. Elsevier B.V. 181–191. doi: 10.1016/j.cattod.2020.09.011.

Faba, L., E. Diaz, and S. Ordonez. 2013. "Improvement on the catalytic performance of Mg-Zr mixed oxides for furfural-acetone aldol condensation by supporting on mesoporous carbons." *ChemSusChem* 6 (3). John Wiley & Sons, Ltd. 463–473. doi: 10.1002/cssc.201200710.

Faba, Laura, Eva Díaz, and Salvador Ordóñez. 2012. "Aqueous-phase furfural-acetone aldol condensation over basic mixed oxides." *Applied Catalysis B: Environmental* 113–114. Elsevier B.V. 201–211. doi: 10.1016/j.apcatb.2011.11.039.

Fang, Xiaowei, Ze Wang, Wenli Song, and Songgeng Li. 2020. "Aldol condensation of furfural with acetone over Ca/ZSM-5 catalyst with lower dosages of water and acetone." *Journal of the Taiwan Institute of Chemical Engineers* 108. Elsevier B.V. 16–22. doi: 10.1016/j.jtice.2020.01.004.

Gandhi, Priyanga, Biswajit Saha, Sundaramurthy Vedachalam, and Ajay K. Dalai. 2023. "Renewable fuel intermediates from furfural over copper-loaded mesoporous aldol condensation catalysts." *Sustainable Energy & Fuels* 7 (17). The Royal Society of Chemistry. 4260–4272. doi: 10.1039/d3se00504f.

Gao, Meiqi, Lili Wang, Yang Yang, Yafei Sun, Xiaorui Zhao, and Ying Wan. 2023. "Metal and metal oxide supported on ordered mesoporous carbon as heterogeneous catalysts." *ACS Catalysis* 13 (7). American Chemical Society. 4060–4090. doi: 10.1021/acscatal.2c05894.

Hora, Lukáš, Vendula Kelbichová, Oleg Kikhtyanin, Oleg Bortnovskiy, and David Kubička. 2014. "Aldol condensation of furfural and acetone over MgAl layered double hydroxides and mixed oxides." *Catalysis Today* 223. Elsevier B.V. 138–147. doi: 10.1016/j.cattod.2013.09.022.

Huang, Renjing, Jian Chang, Hyuck Choi, John M. Vohs, and Raymond J. Gorte. 2022. "Furfural upgrading by aldol condensation with ketones over solid-base catalysts." *Catalysis Letters* 152 (12). Springer Netherlands. 3833–3842. doi: 10.1007/s10562-022-03960-1.

Huang, Xiao-ming, Qing Zhang, Tie-jun Wang, Qi-ying Liu, Long-long Ma, and Qi Zhang. 2012. "Production of jet fuel intermediates from furfural and acetone by aldol condensation over MgO/NaY." *Journal of Fuel Chemistry and Technology* 40 (8). Elsevier B.V. 973–978. doi: 10.1016/S1872-5813(12)60035-8.

Jing, Yaxuan, Yu Xin, Yong Guo, Xiaohui Liu, and Yanqin Wang. 2019. "Highly efficient Nb_2O_5 catalyst for aldol condensation of biomass-derived carbonyl molecules to fuel precursors." *Chinese Journal of Catalysis* 40 (8). Elsevier B.V. 1168–1177. doi: 10.1016/s1872–2067(19)63371-1.

Ketabchi, Elham, Laura Pastor-Pérez, Tomas Ramirez Reina, and Harvey Arellano-García. 2020. "Catalytic upgrading of acetone, butanol and ethanol (ABE): a step ahead for the production of added value chemicals in bio-refineries." *Renewable Energy* 156. Elsevier B.V. 1065–1075. doi: 10.1016/j.renene.2020.04.152.

Kikhtyanin, O., P. Chlubna, T. Jindrova, and D. Kubicka. 2014. "Peculiar behavior of MWW materials in aldol condensation of furfural and acetone." *Dalton Transactions* 43 (27). The Royal Society of Chemistry. 10628–10641. doi: 10.1039/c4dt00184b.

Kikhtyanin, Oleg, Vendula Kelbichová, Dana Vitvarová, Martin Kubů, and David Kubička. 2014. "Aldol condensation of furfural and acetone on zeolites." *Catalysis Today* 227. Elsevier B.V. 154–162. doi: 10.1016/j.cattod.2013.10.059.

Kikhtyanin, Oleg, David Kubička, and Jiří Čejka. 2015. "Toward understanding of the role of Lewis acidity in aldol condensation of acetone and furfural using MOF and zeolite catalysts." *Catalysis Today* 243. Elsevier B.V. 158–162. doi: 10.1016/j.cattod.2014.08.016.

Kocík, Jaroslav, Karel Frolich, Ilona Perková, and Jan Horáček. 2018. "Pyroaurite-based Mg•Fe mixed oxides and their activity in aldol condensation of furfural with acetone: effect of oxide composition." *Journal of Chemical Technology & Biotechnology* 94 (2). John Wiley & Sons, Ltd. 435–445. doi: 10.1002/jctb.5787.

Kong, Xiao, Xiao-jie Wei, Lu-ping Li, Zhen Fang, and Hanwu Lei. 2021. "Production of liquid fuel intermediates from furfural via aldol condensation over $La_2O_2CO_3$-ZnO-Al_2O_3 catalyst." *Catalysis Communications* 149. Elsevier B.V. doi: 10.1016/j.catcom.2020.106207.

Lange, J. P., E. van der Heide, J. van Buijtenen, and R. Price. 2012. "Furfural—a promising platform for lignocellulosic biofuels." *ChemSusChem* 5 (1). John Wiley & Sons, Ltd. 150–166. doi: 10.1002/cssc.201100648.

Lewis, J. D., S. Van de Vyver, and Y. Roman-Leshkov. 2015. "Acid-base pairs in Lewis acidic zeolites promote direct aldol reactions by soft enolization." *Angewandte Chemie International Edition* 54 (34). 9835–9838. doi: 10.1002/anie.201502939.

Li, Hu, Anders Riisager, Shunmugavel Saravanamurugan, Ashok Pandey, Rajender S. Sangwan, Song Yang, and Rafael Luque. 2017. "Carbon-increasing catalytic strategies for upgrading biomass into energy-intensive fuels and chemicals." *ACS Catalysis* 8 (1). American Chemical Society. 148–187. doi: 10.1021/acscatal.7b02577.

Li, Hu, Hongguo Wu, Qiuyun Zhang, Jian Liu, Xiaofang Liu, Yancai Liu, and Song Yang. 2013. "Solid acid-base bifunctional catalysts in organic transformations." *Current Catalysis* 2 (3). Bentham Science. 173–212. doi: 10.2174/2211544711302999 0014.

Li, Lu-ping, Zhen Fang, Xiao Kong, and Wen-Jie Cong. 2021. "Synthesis of jet fuel intermediates via aldol condensation of biomass-derived furfural with lanthanide catalyst." *Molecular Catalysis* 515. Elsevier B.V. 111893–111905. doi: 10.1016/j.mcat.2021.111893.

Li, Mingming, Xuan Xu, Yutong Gong, Zhongzhe Wei, Zhaoyin Hou, Haoran Li, and Yong Wang. 2014. "Ultrafinely dispersed Pd nanoparticles on a CN@MgO hybrid as a bifunctional catalyst for upgrading bioderived compounds." *Green Chemistry* 16 (9). The Royal Society of Chemistry. 4371–4377. doi: 10.1039/c4gc00850b.

Li, W., M. Su, T. Yang, T. Zhang, Q. Ma, S. Li, and Q. Huang. 2019. "Preparation of two different crystal structures of cerous phosphate as solid acid catalysts: their different catalytic performance in the aldol condensation reaction between furfural and acetone." *RSC Advances* 9 (30). The Royal Society of Chemistry. 16919–16928. doi: 10.1039/c9ra02937k.

Liang, Guanfeng, Aiqin Wang, Xiaochen Zhao, Nian Lei, and Tao Zhang. 2016. "Selective aldol condensation of biomass-derived levulinic acid and furfural in aqueous-phase over MgO and ZnO." *Green Chemistry* 18 (11). The Royal Society of Chemistry. 3430–3438. doi: 10.1039/c6gc00118a.

Liu, Ning, Zhimin Wu, Meng Li, Shanshan Li, Yongfei Li, Rongdong Yu, Langsheng Pan, and Yuejin Liu. 2017. "A novel strategy for constructing mesoporous solid superbase catalysts: bimetallic Al–La oxides supported on SBA-15 modified with KF." *Catalysis Science & Technology* 7 (3). The Royal Society of Chemistry. 725–733. doi: 10.1039/c6cy02334g.

Luterbacher, J. S., D. Martin Alonso, and J. A. Dumesic. 2014. "Targeted chemical upgrading of lignocellulosic biomass to platform molecules." *Green Chemistry* 16 (12). The Royal Society of Chemistry. 4816–4838. doi: 10.1039/c4gc01160k.

Mariscal, R., P. Maireles-Torres, M. Ojeda, I. Sádaba, and M. López Granados. 2016. "Furfural: a renewable and versatile platform molecule for the synthesis of chemicals and fuels." *Energy & Environmental Science* 9 (4). The Royal Society of Chemistry. 1144–1189. doi: 10.1039/c5ee02666k.

Nguyen Thanh, Dong, Oleg Kikhtyanin, Ruben Ramos, Maadhav Kothari, Pavel Ulbrich, Tasnim Munshi, and David Kubička. 2016. "Nanosized TiO_2—a promising catalyst for the aldol condensation of furfural with acetone in biomass upgrading." *Catalysis Today* 277. Elsevier B.V. 97–107. doi: 10.1016/j.cattod.2015.11.027.

O"Neill, Rebecca E., Laurent Vanoye, Claude De Bellefon, and Farid Aiouache. 2014. "Aldol-condensation of furfural by activated dolomite catalyst." *Applied Catalysis B: Environmental* 144. Elsevier B.V. 46–56. doi: 10.1016/j.apcatb.2013.07.006.

Palermo, V., J. M. Camargo Lopez, M. H. Brijaldo, S. Acevedo, S. Mancipe, J. C. Castillo, H. A. Rojas, F. B. Passos, G. P. Romanelli, and J. J. Martinez. 2023. "Biochar-MgO from Soursop Seeds in the Production of Biofuel Additive Intermediates." *Chempluschem* 88. John Wiley & Sons, Ltd. e202300401. doi: 10.1002/cplu.202300401.

Parejas, Almudena, Daniel Cosano, Jesús Hidalgo-Carrillo, José Ruiz, Alberto Marinas, César Jiménez-Sanchidrián, and Francisco Urbano. 2019. "Aldol condensation of furfural with acetone over Mg/Al mixed oxides. Influence of water and synthesis method." *Catalysts* 9 (2). MDPI. 203–214. doi: 10.3390/catal9020203.

Perego, Carlo, Aldo Bosetti, Marco Ricci, and Roberto Millini. 2017. "Zeolite materials for biomass conversion to biofuel." *Energy & Fuels* 31 (8). American Chemical Society. 7721–7733. doi: 10.1021/acs.energyfuels.7b01057.

Sádaba, I., M. Ojeda, R. Mariscal, J. L. G. Fierro, and M. López Granados. 2011. "Catalytic and structural properties of co-precipitated Mg–Zr mixed oxides for furfural valorization via aqueous aldol condensation with acetone." *Applied Catalysis B: Environmental* 101 (3–4). Elsevier B.V. 638–648. doi: 10.1016/j.apcatb.2010.11.005.

Sadaba, I., M. Ojeda, R. Mariscal, R. Richards, and M. Lopez Granados. 2012. "Preparation and characterization of Mg-Zr mixed oxide aerogels and their application as aldol condensation catalysts." *Chemphyschem* 13 (14). John Wiley & Sons, Ltd. 3282–3292. doi: 10.1002/cphc.201200440.

Sajid, Muhammad, Usman Farooq, Ghulam Bary, Muhammad Mohsin Azim, and Xuebing Zhao. 2021. "Sustainable production of levulinic acid and its derivatives for fuel additives and chemicals: progress, challenges, and prospects." *Green Chemistry* 23 (23). The Royal Society of Chemistry. 9198–9238. doi: 10.1039/d1gc02919c.

Setoyama, Tohru. 2006. "Acid–base bifunctional catalysis: an industrial viewpoint." *Catalysis Today* 116. Elsevier B.V. 250–262. doi: 10.1016/j.cattod.2006.01.031.

Skrodczky, K., M. M. Antunes, Q. Zhu, A. A. Valente, N. Pinna, and P. A. Russo. 2023. "Single-step formation of metal oxide nanostructures wrapped in mesoporous silica and silica-niobia catalysts for the condensation of furfural with acetone." *Nanomaterials (Basel)* 13 (23). MDPI. 3046–3059. doi: 10.3390/nano13233046.

Smoláková, Lucie, Lada Dubnová, Jaroslav Kocík, Jan Endres, Stanislav Daniš, Peter Priecel, and Libor Čapek. 2018. "In-situ characterization of the thermal treatment of Zn-Al hydrotalcites with respect to the formation of Zn/Al mixed oxide active in aldol condensation of furfural." *Applied Clay Science* 157. Elsevier B.V. 8–18. doi: 10.1016/j.clay.2018.02.024.

Su, Mingxue, Wenzhi Li, Qiaozhi Ma, Song Li, Tao Yang, and Xiaomeng Dou. 2019. "Efficient synthesis of liquid fuel intermediates from furfural and levulinic acid via aldol condensation over hierarchical MFI zeolite catalyst." *Energy & Fuels* 33 (12). American Chemical Society. 12518–12526. doi: 10.1021/acs.energyfuels.9b03307.

Su, Mingxue, Wenzhi Li, Qiaozhi Ma, and Bowen Zhu. 2020. "Production of jet fuel intermediates from biomass platform compounds via aldol condensation reaction over iron-modified MCM-41 lewis acid zeolite." *Journal of Bioresources and Bioproducts* 5 (4). Elsevier B.V. 256–65. doi: 10.1016/j.jobab.2020.10.004.

Su, Mingxue, Wenzhi Li, Tingwei Zhang, HaoSheng Xin, Song Li, Wei Fan, and Longlong Ma. 2017. "Production of liquid fuel intermediates from furfural via aldol condensation over Lewis acid zeolite catalysts." *Catalysis Science & Technology* 7 (16). The Royal Society of Chemistry. 3555–3561. doi: 10.1039/c7cy01028a.

Taarning, Esben, Christian M. Osmundsen, Xiaobo Yang, Bodil Voss, Simon I. Andersen, and Claus H. Christensen. 2011. "Zeolite-catalyzed biomass conversion to fuels and chemicals." *Energy & Environmental Science* 4 (3). The Royal Society of Chemistry. 793–804. doi: 10.1039/c004518g.

Talanova, M. Yu, V. A. Yarchak, and E. A. Karakhanov. 2019. "Acid catalysts based on mesoporous aromatic frameworks in aldol condensation of furfural with some carbonyl compounds." *Russian Journal of Applied Chemistry* 92 (6). Springer Netherlands. 857–864. doi: 10.1134/s107042721906017x.

Tichit, Didier, and Mayra G. Álvarez. 2022. "Layered double hydroxide/nanocarbon composites as heterogeneous catalysts: a review." *ChemEngineering* 6 (4). MDPI. 45–89. doi: 10.3390/chemengineering6040045.

Tolborg, S., S. Meier, S. Saravanamurugan, P. Fristrup, E. Taarning, and I. Sadaba. 2016. "Shape-selective valorization of biomass-derived glycolaldehyde using tin-containing zeolites." *ChemSusChem* 9 (21). John Wiley & Sons, Ltd. 3054–3061. doi: 10.1002/cssc.201600757.

Wang, Fangqian, Denghao Ouyang, Ziyuan Zhou, Samuel J. Page, Dehua Liu, and Xuebing Zhao. 2021. "Lignocellulosic biomass as sustainable feedstock and materials for power generation and energy storage." *Journal of Energy Chemistry* 57. Elsevier B.V. 247–280. doi: 10.1016/j.jechem.2020.08.060.

Wang, Shenyu, Aohua Cheng, Fanhua Liu, Junjie Zhang, Tao Xia, Xiang Zeng, Wei Fan, and Ying Zhang. 2023. "Catalytic conversion network for lignocellulosic biomass valorization: a panoramic view." *Industrial Chemistry & Materials* 1 (2). The Royal Society of Chemistry. 188–206. doi: 10.1039/d2im00054g.

Wu, Zheng Ying, Qi Jiang, Yi Meng Wang, Hong Ji Wang, Lin Bing Sun, Li Ying Shi, Jia Hui Xu, Ying Wang, Yuan Chun, and Jian Hua Zhu. 2006. "Generating superbasic sites on mesoporous silica SBA-15." *Chemistry of Materials* 18 (19). American Chemical Society. 4600–4608. doi: 10.1021/cm0608138.

Xu, Jilei, Ning Li, Xiaofeng Yang, Guangyi Li, Aiqin Wang, Yu Cong, Xiaodong Wang, and Tao Zhang. 2017. "Synthesis of diesel and jet fuel range zlkanes with furfural and Angelica lactone." *ACS Catalysis* 7 (9). American Chemical Society. 5880–5086. doi: 10.1021/acscatal.7b01992.

Xu, Minrui, Stéphane Célérier, Jean-Dominique Comparot, Julie Rousseau, Matthieu Corbet, Frédéric Richard, and Jean-Marc Clacens. 2019. "Upgrading of furfural to biofuel precursors via aldol condensation with acetone over magnesium hydroxide fluorides $MgF_{2-x}(OH)_x$." *Catalysis Science & Technology* 9 (20). The Royal Society of Chemistry. 5793–5802. doi: 10.1039/c9cy01259a.

Xu, Wenjie, Xiaohui Liu, Jiawen Ren, Huihui Liu, Yichun Ma, Yanqin Wang, and Guanzhong Lu. 2011. "Synthesis of nanosized mesoporous Co–Al spinel and its application as solid base catalyst." *Microporous and Mesoporous Materials* 142 (1). Elsevier B.V. 251–257. doi: 10.1016/j.micromeso.2010.12.008.

Yuan, Enhui, Weili Dai, Guangjun Wu, Naijia Guan, and Landong Li. 2020. "Entrapped NbO_x clusters in MFI zeolite for sustainable acid catalysis." *Microporous and Mesoporous Materials* 305. Elsevier B.V. 110361–110395. doi: 10.1016/j.micromeso.2020.110361.

Yuan, Liyuan, Mengran Liu, Shanshan Liu, Guoli Fan, Lan Yang, and Feng Li. 2022. "Fabrication of Al_2O_3-ZrO_2 composite catalysts with tunable acid-base properties for highly efficient aldol condensation of furfural with acetone." *Catalysis Communications* 166. Elsevier B.V. 106451–106455.doi: 10.1016/j.catcom.2022.106451.

Zhan, Tong, Shichao Wu, Hao Ma, Chaochao Yue, Zhenquan Huang, Weibing Liu, Junjiang Teng, Dehao Li, Suhua Wang, and Hua Tan. 2019. "Production of biofuel intermediates from furfural via aldol condensation over K2O clusters containing N-doped porous carbon materials with shape selectivity." *Microporous and Mesoporous Materials* 281. Elsevier B.V. 101–109. doi: 10.1016/j.micromeso.2019.03.005.

Zhao, Xuelai, Song Li, Yuzhen Hu, Xinghua Zhang, Lungang Chen, Chenguang Wang, Longlong Ma, and Qi Zhang. 2022. "Synthesis of long chain alkanes via aldol condensation over modified chitosan catalyst and subsequent hydrodeoxygenation." *Chemical Engineering Journal* 428 (7). Elsevier B.V. doi: 10.1016/j.cej.2021.131368.

Zhou, C. H., X. Xia, C. X. Lin, D. S. Tong, and J. Beltramini. 2011. "Catalytic conversion of lignocellulosic biomass to fine chemicals and fuels." *Chemical Society Reviews* 40 (11). The Royal Society of Chemistry. 5588–5617. doi: 10.1039/c1cs15124j.

11 Domino Upgrading of Bio-Based Furfural to γ-Valerolactone via Bifunctional Catalysis

Mingrui Li, Yixuan Liu, Hu Li, and Hongguo Wu

11.1 INTRODUCTION

The environment has been seriously harmed by the massive consumption of fossil energy (Luterbacher et al. 2014, Roman-Leshkov et al. 2007). Biomass, which can be obtained from a large amount of agricultural and forestry waste, has attracted extensive attention because of its green and economical characteristics (Li et al. 2014). There are many high-value chemicals derived from biomass, such as xylose, glucose, furfural (FF), 5-hydroxymethylfurfural levulinate, and γ-valerolactone (GVL) (Li and Fukun 2017, Luo et al. 2019, Liu et al. 2015). Among them, GVL is often used as a green solvent, biofuel intermediate, plasticizer, and edible flavor due to its low melting point (31 °C), low toxicity (LD_{50} = 8800 mg/kg), and water miscibility (Yan et al. 2015, Ye et al. 2020a, Yu et al. 2019).

Most of the existing GVL synthesis methods use levulinic acid and its esters as substrates (Molleti et al. 2018, Yuan et al. 2013, Luo et al. 2014), which requires complex steps such as acid hydrolysis, dehydration, further acid hydrolysis, and ring-opening to obtain biomass raw materials (Figure 11.1). Biomass raw materials containing polysaccharides and cellulose compounds are converted into monosaccharides (e.g., glucose) via an acid hydrolysis reaction, which breaks down the complex polysaccharides into simpler monosaccharide units under acidic conditions. The monosaccharide units then undergo dehydration to form ketones or other intermediates needed for subsequent reactions, and this step typically requires high temperatures and suitable catalysts. In some cases, the intermediates produced from dehydration may undergo further acid hydrolysis to generate compounds closer to the target product. Finally, through ring-opening reactions, the ring structures of the intermediates or precursor molecules are opened to form levulinic acid and its esters, such as GVL. These steps involve precise control of reaction conditions, including temperature, pressure, and catalyst selection, to ensure high yields and efficient synthesis of the desired products. The complexity of these transformations is necessary due to the diversity and complexity of biomass raw materials and the specific requirements for the synthesis of levulinic acid and its esters. Alternatively,

DOI: 10.1201/9781003464839-11

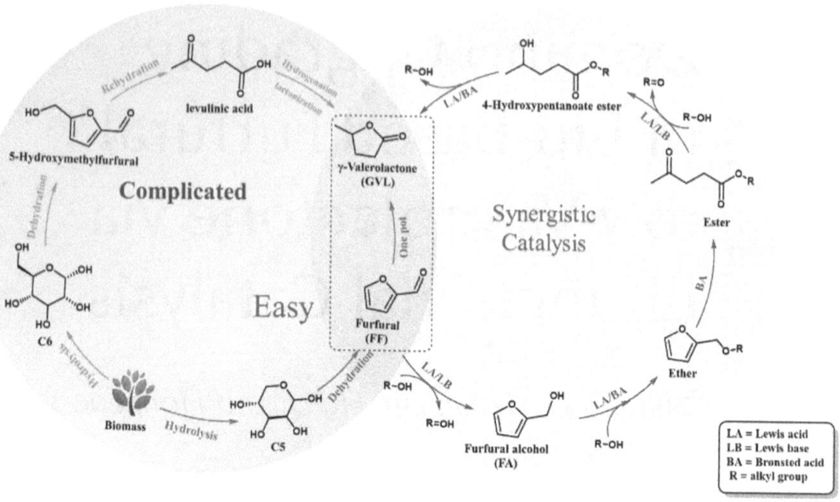

FIGURE 11.1 Schematic for cascade conversion of biomass-derived FF to GVL.

using FF to produce GVL is much closer to biomass raw materials, which means that it has the advantages of saving costs and reducing reaction operations (Bui et al. 2013). The conversion of FF to GVL can be accomplished in five different reaction steps by a one-pot cascade conversion process. Higher requirements for catalyst performance are proposed due to the complex reaction process. For example, catalytic transfer hydrogenation (CTH) and partial hydrogenation require the coordinated catalysis of Lewis acid-base sites, while ring-opening reactions need to be completed in the presence of Brønsted acid (Li et al. 2018, Zhou et al. 2019a, Li et al. 2019). This means that catalysts must simultaneously satisfy the active sites required for multiple reactions. Therefore, the effects of structure and active sites on the catalysts must be fully considered, which is a huge challenge for the efficient transformation of FF to GVL (Zhou et al. 2019b, Li et al. 2016, Zhu 2016, Li et al. 2017).

Most of the previous reviews focus on the production of GVL from levulinic acid and its esters over metal catalysts (Ye et al. 2020a, Dutta et al. 2019, Wu et al. 2018), but rarely mention bifunctional catalysts. In this chapter, the latest progress in the design of bifunctional catalysts for cascade conversion of FF to GVL and the regulation of acid-base sites were summarized. In addition, the effect of the catalyst's physical/chemical properties and the influence of reaction parameters in adjusting reaction performance were disclosed. It provides a new indication for future research on sustainable catalysts with economic and environmental advantages.

11.2 SYNERGISTIC CATALYTIC CONVERSION OF FF TO GVL

11.2.1 Dual-Component Catalysts

Many physical mixed dual-component catalysts have been reported for the conversion of FF to GVL, as shown in Table 11.1. A physical mixture of Zr-Beta with

TABLE 11.1
Mixed Catalysts for the Synthesis of GVL from FF

Entry	Catalysts	H-Donor	Adjustment of Active Sites	Reaction Conditions	GVL Yield (%)	Ref.
1	Zr-Beta + Al-MFI-ns	2-butanol	Lewis acid site and Brønsted acid site are independent of each other, which can adjust the content and strength of Lewis acid and Brønsted acid in the catalyst respectively	120 °C, 48 h	78	(Bui et al. 2013)
2	Au/ZrO_2+ ZSM-5	IPA		120 °C, 30 h	80.4	(Zhu et al. 2016)
3	Hf-MOF 808 + Al-β zeolite	IPA		120 °C, 48 h	75	(Rojas-Buzo et al. 2018)
4	chitosanRu/ PPh_3 + ZSM-5	ethanol		170 °C, 30 h	79	(Wang et al. 2020)

mesoporous Al-MFI zeolite as Lewis acid (LA) and Brønsted acid (BA) catalysts (Bui et al. 2013), or Au/ZrO_2 (Lewis acid-base sites) with ZSM-5 (BA sites) (Li et al. 2019) was developed to be efficient for direct FF-to-GVL conversion. When there were only LA sites, FF could be transformed into FA (99.0% yield) instead of GVL. Besides, when only BA was present in the catalyst, there was almost no conversion of FF due to the lack of active sites to promote hydrogenation. Likewise, Hf-MOF-808 was efficient for the CTH of FF to FA and levulinic acid to give GVL, but could not get GVL from FF (Rojas-Buzo et al. 2018). However, when the BA site was added to the catalyst, FF could be successfully converted to GVL (75%). This implies the necessity of BA for furan ring-opening of FA in this reaction process. Wang et al. reported the preparation of the CuAl catalyst by the coprecipitation method, which was the key to the transfer hydrogenation process, while the BA sites provided by H-ZSM-5 were able to catalyze the ring-opening reaction (Shao et al. 2021). It was found that the excessive addition of BA catalyst to the reaction system was not conducive to the CTH of ethyl levulinate (EL) and reduced the obtained GVL yield.

11.2.2 Bifunctional Catalysts

Although good GVL yields can be obtained using two different catalysts, the traditional FF-GVL conversion uses a two-component catalyst to construct two different active species separately, and the preparation process is complicated and costly. Conventional catalysts used in FF to GVL conversion may deactivate over time due to some factors like poisoning or aggregation, leading to decreased catalytic activity and the deactivated catalysts are difficult to regenerate. To solve this problem while efficiently converting FF to GVL, the development of bifunctional catalytic materials containing both LA and BA sites has drawn increasing attention (Table 11.2). The

TABLE 11.2
Catalysts for the Synthesis of GVL from FF

Entry	Catalysts	Acidity (mmol/g)	LA/BA	H-Source	Control of Acid Sites	Reaction Conditions	GVL Yield (%)	Ref.
1	ZrO_2-SBA-15(2)	0.32	0.08	IPA	Controlling layers number of ZrO_2 on the surface of SBA-15	170 °C, 7 h	37	(Iglesias et al. 2018)
2	Zr-KIT-5	1.86	6.5	IPA	Change the catalyst Zr loading	180 °C, 6 h	40.1	(He et al. 2020)
3	HZ-ZrP 1-5	0.87	4.1	IPA	Adjust the ratio of zeolite/ $NH_4H_2PO_4$	185 °C, 18 h	64.2	(Ye et al. 2020b)
4	HPW/Zr-Beta	0.78	3.2	IPA	Use different acid treatment catalysts	160 °C, 24 h	68	(Winoto et al. 2019)
5	20%Zr-5%T-zeolite	1.67	1.53	IPA	Adjust the ratio of TPA/Zr	170 °C, 10 h	90	(Srinivasa et al. 2019)
6	$(Sn)_{SSIE}$-beta1	--	15.9	2-butanol	Adjust Sn loading by changing the dealumination degree	120 °C, 48 h	73	(Antunes et al. 2015)
7	DUT-67(Hf)-0.06	1.28	--	IPA	Treat DUT-67-(Hf) with H_2SO_4 in different concentrations	180 °C, 8 h	70.7	(Li et al. 2019)
8	FM-Zr-ARS	0.55	0.23	IPA	Modify the catalyst with HCOOH	160 °C, 8 h	72.4	(Peng et al. 2021)
9	ZPS-1.0	--	3.25	IPA	Adjust the Zr amount	150 °C, 18 h	80.4	(Li et al. 2021)

bifunctional catalyst simplifies the catalyst preparation process and reduces the development cost. Bifunctional catalysts leverage the synergistic effects of multiple active sites, enhancing catalytic performance and stability, leading to more efficient and sustainable conversion processes. Iglesias et al. reported the synthesis of a bifunctional catalyst by loading ZrO_2 onto SBA-15 (Iglesias et al. 2018). The strength of LA and BA can be controlled by varying the content of ZrO_2 layers. Kinetic studies indicated that stronger Lewis acid sites could promote the etherification of FA, but not the CTH process. More BA sites and fewer LA sites of the catalysts prepared by impregnation of ZrO_2 and phosphotungstic acid (TPA) on β-zeolite were found to be more favorable for producing levulinic acid ester rather than GVL (Srinivasa et al. 2021a). Bifunctional catalysts with suitable strength and ratio of Lewis/Brønsted acid sites are the key to complete etherification and ring-opening reactions. For the CTH process, the corresponding aldehydes are reduced to alcohols by Meerwein–Ponndorf–Verley (MPV) reduction. Although this process can be catalyzed by Lewis acids, the presence of Lewis base sites facilitates the process by activating the hydroxyl groups of alcohols (Peng et al. 2020, Chai et al. 2021, Wang et al. 2017). Therefore, the synergistic effect of Lewis acid-base sites in the catalyst is an indispensable condition for the efficient conversion of FF to GVL.

Zeolite has a variable structure and unique pore system, and its large specific surface area facilitates the adsorption of substances. Its abundant LA and BA sites can provide active species for catalytic reactions, and the acid content can also be adjusted by dealumination (Peng et al. 2020, Chai et al. 2021, Wang et al. 2017). Thus, the frequently used catalysts for FF-to-GVL conversion are prepared on the basis of zeolites. The content of LA and LB sites in the zeolitic materials can be adjusted by changing the metal loading, or different kinds and concentrations of acids (He et al. 2020, Ye et al. 2020b, Winoto et al. 2019, Rao et al. 2019). Antunes et al. controlled the dealumination degree of zeolite by changing the conditions of acid treatment, thereby changing the Sn loading and the catalyst acid amount (Antunes et al. 2015). He et al. found that the amount of acid sites is positively correlated with the Zr loading in the catalyst (He et al. 2020). But Zr in excess loading leads to Zr clusters, which would counteract the catalyst performance. Upon soaking UT-67 (Hf) in the solution of H_2SO_4, the BA content could be adjusted accordingly by increasing the solution concentration (Li et al. 2019). However, the presence of excess BA sites caused side reactions, producing undesired products. After loading TPA and ZrO_2 in-/outside the SBA-15 pores, the catalyst's total LA content was kept unchanged, while its LA/BA molar ratio was adjusted by controlling the amount of ZrO_2 and TPA (Rao et al. 2021b). The optimized catalyst was found to exhibit good performance in the synthesis of GVL (up to 90% yield) at 170 °C for 10 h.

Except for zeolitic materials, bifunctional catalysts derived from biomass are also developed. Employing alizarin red S (ARS) as an organic ligand, a renewable FM-Zr-ARS catalyst was prepared under hydrothermal conditions (Peng et al. 2021). The SO_3H-functionalized ARS catalyst could provide BA sites for ring-opening to avoid the formation of additional LA. As a crucial step in the process of FF-to-GVL transformation, the CTH reaction typically underwent a six-membered ring transition state. Theoretical calculations showed that Hf-UiO-66 had relatively lower activation energy (13.5 kcal/mol), compared to Zr-UiO-66 (14.9 kcal/mol) (Sittiwong et al. 2021). In fact, Hf-based catalysts generally have superior performance to Zr-based counterparts in CTH reactions due to the stronger Lewis acidity of Hf-containing catalytic materials (Luo et al. 2014, Injongkol et al. 2017, Li et al. 2020, Xie et al. 2016). A series of coordination organic phosphate-Hf catalysts were prepared from vinyl phosphoric acid, *p*-toluenesulfonic acid, and $HfCl_4$ in different molar ratios (Tan et al. 2022). The contents of LA/BA sites could be accurately controlled with a relatively E factor value (0.19), suggesting that the catalyst-mediated conversion process was eco-friendly.

Bifunctional catalysts can promote specific reaction pathways while suppressing undesired side reactions, leading to higher selectivity for the desired products. The synergistic interactions between different active sites or functionalities in bifunctional catalysts can enhance catalytic performance, resulting in improved reaction kinetics, higher activity, and increased efficiency, and making them valuable for complex reaction pathways or multi-step processes. Bifunctional catalysts can be more easily regenerated or rejuvenated due to the presence of multiple active sites, extending the catalyst's lifetime and reducing the need for frequent replacement. However, designing and synthesizing bifunctional catalysts can be complex, requiring careful engineering and optimization, which can be challenging and time-consuming. Bifunctional catalysts are tailored for specific reactions, limiting their versatility compared to more general catalysts, necessitating the development of new catalysts for different applications.

11.3 EFFECT OF REACTION PARAMETERS

11.3.1 Reaction Temperature and Time

The LA/BA-bifunctionalized catalyst preparation methods are generally simple with reduced cost, but a relatively high reaction temperature (e.g., 150–180 °C) is often needed to ensure the sufficient proceeding of the overall conversion. In contrast, the same conversion process can be realized under benign conditions (e.g. 120 °C) in the presence of using the two monofunctional catalysts containing sole LA or BA sites (Table 11.1). It was proposed that the two independent active sites of mixed catalysts were able to effectively increase the mass transfer by providing more separated acid species. It is worth noting that a low reaction temperature is often accompanied by several disadvantages, including, e.g. extended time and the need for specific methods to prepare at least two catalysts with different acid types, contents, or strengths.

Bifunctionalized catalysts incorporating both LA and BA sites often require high reaction temperatures (Table 11.2). High temperatures are necessary to activate the LA and BA sites on the catalyst surface. The LA sites, typically metal oxides or metal centers, may require elevated temperatures to form active coordination sites and react with the substrate. Similarly, the BA sites, such as sulfonic acid groups, may need high temperatures to protonate reactants effectively. The LA and BA functionalities work synergistically to catalyze complex reactions. High temperatures can facilitate stronger interactions between the two acid sites and the substrate molecules, leading to enhanced catalytic activity and selectivity. Certain reactions may require high temperatures to overcome energy barriers and promote the desired chemical transformations. The combination of LA and BA sites in bifunctional catalysts can provide a unique catalytic environment that benefits from elevated temperatures to drive the reaction forward efficiently. In some cases, high-temperature treatment can help regenerate bifunctional catalysts by removing adsorbed species or byproducts that could inhibit catalytic activity. This regeneration process can restore the catalyst's performance and prolong its operational lifetime. Overall, the use of high reaction temperatures in the preparation and operation of biofunctionalized catalysts with LA and BA functionalities is aimed at optimizing catalytic performance, enhancing catalytic activity and selectivity, improving catalyst stability, and enabling efficient catalytic reactions under specific conditions.

11.3.2 Hydrogen Donor

The CTH as the key process in FF-to-GVL conversion is realized by the MPV reduction reaction. Generally, alcohols are used as both H-donors and green solvents. This also avoids using high-pressure H_2 and corrosive HCOOH. Normally, the β-H of secondary alcohols is more prone to being removed, thus showing stronger hydrogen supply capacity, compared to primary alcohols (Elsayed et al. 2020, Li et al. 2021). However, the steric hindrance of secondary alcohols gradually increases with the extension of the carbon chain. As such, its hydrogen supply capacity may be reduced by suppressing the formation of stable transition states (Li et al. 2021). Therefore, due to the small steric hindrance, IPA was used as the H-donor to prepare GVL in most cases. In addition, Cho et al. found that different components of the solvent were found to produce different FF conversion products, suggesting that it can be used to control the reaction process (Cho et al. 2020).

11.4 REUSABILITY OF CATALYSTS

The reusability of the catalyst is also an aspect that must be considered. For the two catalysts containing Lewis acid-base sites and BA sites, respectively, how to separate them after deactivation remains a challenge. This problem does not exist in LA/BA-bifunctionalized catalysts. In addition, humins were possibly generated under thermal conditions, which not only lowered the overall carbon balance, but also reduced the catalyst activity during the recovery process. It is highly desirable to remove humins attached to the catalyst and restore its activity by using some post-treatment approaches, such as temperature-programmed calcination, and microwave-assisted washing procedures (Iglesias et al. 2018, Ye et al. 2020b, Tang et al. 2021). In addition, the catalyst active site leaching is always encountered during recycling, which has to be regenerated by treatment with acid prior to the next cycle (Li et al. 2019).

11.5 CONCLUSION AND OUTLOOK

GVL has great application potential due to its unique position in organic synthesis and biofuels. However, the cascade process of FF-to-GVL conversion is complicated and requires the used catalysts with suitable active sites, which limits its further development somehow. In this chapter, the structural characteristics and synergistic catalytic processes of bifunctional catalysts were summarized. Also, the effect of involved reaction parameters (e.g. temperature and H-donor type) on the overall conversion processes was disclosed. The acid-base sites of the catalyst play a unique role in different reactions, in which LA/LB sites are active for CTH, while BA sites are crucial for the ring-opening process. In general, higher acid content means more abundant active sites in the catalyst, but the unsatisfactory ratio of LA and BA sites leads to undesirable side reactions, carbon imbalance, and GVL yield reduction. In addition, the formation of humins that are attached to the solid catalysts would reduce their reusability.

The existing bifunctional catalysts are mostly prepared from zeolites, which usually need to be further processed before using for producing GVL directly from FF. Organic hybrid materials derived from biomass have been demonstrated to be active for the CTH process, while BA sites seem not sufficient to promote the ring-opening process in some cases. Hence, it is highly required to enhance the BA site strength while ensuring catalyst stability. The accurate control of acid site strength and content of the solid catalysts would favor the overall conversion process to yield GVL. Last but not least, on the basis of laboratory research, amplification experiments should be investigated to realize the industrial production of GVL from FF.

LIST OF ABBREVIATIONS

GVL = γ-valerolactone
CTH = catalytic transfer hydrogenation
FF = furfural
FA = furfuryl alcohol
LA = Lewis acid
LB = Lewis base
BA = Brønsted acid
EL = ethyl levulinate
IPA = isopropyl alcohol
TPA = phosphotungstic acid
MPV = Meerwein–Ponndorf–Verley

REFERENCES

Antunes Margarida M., Lima Sérgio, Neves Patrícia, Magalhães Ana L., Fazio Enza, Fernandes Auguste, Neri Fortunato, Silva Carlos M., Rocha Silvia M., Ribeiro Maria F., Pillinger Martyn, Urakawa Atsushi, and Anabela A. Valente. 2015. "One-pot conversion of furfural to useful bio-products in the presence of a SnAl-containing zeolite beta catalyst prepared via post-synthesis routes." *J. Catal.* 329. Academic Press Inc.: 522–537. doi: 10.1016/j.jcat.2015.05.022.

Bui Linh, Luo Helen, Gunther William R., and Román-Leshkov Yuriy. 2013. "Domino reaction catalyzed by zeolites with Brønsted and Lewis acid sites for the production of γ-valerolactone from furfural." *Angew. Chem. Int. Ed.* 52. John Wiley & Sons Ltd.: 8022–8025. doi: 10.1002/anie.201302575.

Cho Hong Je, Kim Doyoung, and Xu Bingjun. 2020. "Selectivity control in Tandem catalytic furfural upgrading on zeolite-encapsulated Pt nanoparticles through site and solvent wngineering." *ACS Catal.* 10. ACS Publications: 4770–4779. doi: 10.1021/acscatal.0c00472.

Chai Yuchao, Dai Weili, Wu Guangjun, Guan Naijia, and Li Landong 2021. "Confinement in a zeolite and zeolite catalysis." *Acc Chem Res.* 54. American Chemical Society: 2894–2904. doi: 10.1021/acs.accounts.1c00274.

Dutta Shanta, Yu Iris K. M., Tsang Daniel C. W., Ng Yun Hau, Ok Yong Sik, Sherwood James, and Clark James H. 2019. "Green synthesis of gamma-valerolactone (GVL) through hydrogenation of biomass-derived levulinic acid using non-noble metal catalysts: A critical review." *Chem. Eng. J.* 372. Elsevier: 992–1006. doi: 10.1016/j.cej.2019.04.199.

Elsayed Islam, Jackson Michael A., and Hassan El Barbary. 2020. "Hydrogen-free catalytic reduction of biomass-derived 5-hydroxymethylfurfural into 25-bis(hydroxymethyl)furan using copper–iron oxides bimetallic nanocatalyst." *ACS Sustain. Chem. Eng.* 8. ACS Publications: 1774–1785. doi: 10.1021/acssuschemeng.9b05575.

He Jian, Li Hu, Xu Yufei, and Yang Song. 2020. "Dual acidic mesoporous KIT silicates enable one-pot production of γ-valerolactone from biomass derivatives via cascade reactions." *Renew. Energ.* 146. Elsevier BV: 359–370. doi: 10.1016/j.renene.2019.06.105.

Injongkol Yuwanda, Maihom Thana, Treesukul Piti, Sirijaraensre Jakkapan, Boekfa Bundet, and Limtrakul Jumeas. 2017. "Theoretical study on the reaction mechanism of hydrogenation of furfural to furfuryl alcohol on Lewis acidic BEA zeolites: Effects of defect structure and tetravalent metals substitution." *Phys. Chem. Chem. Phys.* 19. The Royal Society of Chemistry: 24042–24048. doi: 10.1039/c7cp04229a.

Iglesias J., Melero J. A., Morales G., Paniagua M., Hernández B., Osatiashtiani A., Lee A. F.,and Wilson K. 2018. "ZrO_2-SBA-15 catalysts for the one-pot cascade synthesis of GVL from furfural." *Catal. Sci. Technol.* 8. The Royal Society of Chemistry: 4485–4493. doi: 10.1039/c8cy01121d.

Luterbacher Jeremy S., Rand Jacqeline M., Alonso David Martin, Han Jeehoon, Youngquist J. Tyler, Maravelias Christos T., Pfleger Brian. F., and Dumesic James A. 2014. "Nonenzymatic sugar production from biomass using biomass-derived gamma-valerolactone." *Science.* 343. American Association for the Advancement of Science: 277–80. doi: 10.1126/science.1246748.

Li Hu, Bhadury Pinaki S., Riisager Anders, and Yang Song. 2014. "One-pot transformation of polysaccharides via multi-catalytic processes." *Catal. Sci. Technol.* 4. The Royal Society of Chemistry: 4138–4168. doi: 10.1039/c4cy00711e.

Luo Helen Y., Consoli Daniel F., Gunther William R., and Román-Leshkov Yuriy. 2014. "Investigation of the reaction kinetics of isolated Lewis acid sites in Beta zeolites for the Meerwein–Ponndorf–Verley reduction of methyl levulinate to γ-valerolactone." *J. Catal.* 320. Academic Press Inc.: 198–207. doi: 10.1016/j.jcat.2014.10.010.

Liu Jian, Li Hu, Liu Yan-Cai, Lu Ye-Min, He Jian, Liu Xiao-Fang, Wu Zhi-Bing, and Yang Song. 2015. "Catalytic conversion of glucose to 5-hydroxymethylfurfural over nano-sized mesoporous Al_2O_3–B_2O_3 solid acids." *Catal. Commun.* 62. Elsevier: 19–23. doi: 10.1016/j.catcom.2015.01.008.

Li Hu, He Jian, Riisager Anders, Saravanamurugan Shunmugavel, Song Baoan, and Yang Song. 2016. "Acid–base bifunctional zirconium N-alkyltriphosphate nanohybrid for hydrogen transfer of biomass-derived carboxides." *ACS Catal.* 6. ACS Publications: 7722–7727. doi: 10.1021/acscatal.6b02431.

Li Hu, Zhao Wenfeng, and Fang Zhen. 2017. "Hydrophobic Pd nanocatalysts for one-pot and high-yield production of liquid furanic biofuels at low temperatures." *Appl. Catal. B.* 215. Elsevier: 18–27. doi: 10.1016/j.apcatb.2017.05.039.

Li Fukun, France Liam John, Cai Zhenping, Li Yingwen, Liu Sijie, Lou Hongming, Long Jinxing, and Li Xuehui. 2017. "Catalytic transfer hydrogenation of butyl levulinate to γ-valerolactone over zirconium phosphates with adjustable Lewis and Brønsted acid sites." *Appl. Catal. B.* 214. Elsevier: 67–77. doi: 10.1016/j.apcatb.2017.05.013.

Li Hu, Yang Tingting, and Fang Zhen. 2018. "Biomass-derived mesoporous Hf-containing hybrid for efficient Meerwein-Ponndorf-Verley reduction at low temperatures." *Appl. Catal. B.* 227. Elsevier: 79–89. doi: 10.1016/j.apcatb.2018.01.017.

Li Hu, Li Yan, Fang Zhen, and Smith Richard L. 2019. "Efficient catalytic transfer hydrogenation of biomass-based furfural to furfuryl alcohol with recycable Hf-phenylphosphonate nanohybrids." *Catal. Today.* 319. Elsevier: 84–92. doi: 10.1016/j.cattod.2018.04.056.

Li Wenke, Cai Zhe, Li Hang, Shen Yu, Zhu Yongqiang, Li Haichao, Zhang Xubin, and Wang Fumin. 2019. "Hf-based metal organic frameworks as bifunctional catalysts for the one-pot conversion of furfural to γ-valerolactone." *Mol. Catal.* 472. Elsevier: 17–26. doi: 10.1016/j.mcat.2019.04.010.

Luo Yiping, Li Zheng, Li Xiaoling, Liu Xiaofeng, Fan Jiajun, Clark James H., and Hu Changwei. 2019. "The production of furfural directly from hemicellulose in lignocellulosic biomass: A review." *Catal. Today.* 319. Elsevier: 14–24. doi: 10.1016/j.cattod.2018.06.042.

Li Xiaomin, Du Zhengjiang, Wu Yi, Zhen Yadong, Shao Rixin, Li Bingqi, Chen Chengmeng, Liu Quansheng, and Zhou Huacong. 2020. "A novel hafnium–graphite oxide catalyst for the Meerwein–Ponndorf–Verley reaction and the activation effect of the solvent." *RSC Adv.* 10. The Royal Society of Chemistry: 9985–9995. doi: 10.1039/c9ra10795a.

Li Weile, Li Mengzhu, Liu Huai, Jia Wenlong, Yu Xin, Wang Shuai, Zeng Xianhai, Sun Yong, Wei Junnan, Tang Xing, and Lin Lu. 2021. "Domino transformation of furfural to γ-valerolactone over SAPO-34 zeolite supported zirconium phosphate catalysts with tunable Lewis and Brønsted acid sites." *Mol. Catal.* 506. Elsevier: doi: 10.1016/j.mcat.2021.111538.

Li Jie, Zhao Shuaiheng, Li Zhen, Liu Dan, Chi Yingnan, and Hu Changwen. 2021. "Efficient conversion of biomass-derived levulinic acid to gamma-valerolactone over polyoxometalate@Zr-based metal-organic Fframeworks: The synergistic effect of Bronsted and Lewis acidic sites." *Inorg Chem.* 60. ACS Publications: 7785–7793. doi: 10.1021/acs.inorgchem.1c00185.

Li Mingrui, Liu Yixuan, Lin Xialing, Tan Jinyu, Yang Song, and Li Hu. 2021. "One-step upgrading of bio-based furfural to γ-valerolactone via Hf-mediated bifunctional catalysis." *RSC Adv.* 11. The Royal Society of Chemistry: 35415–35424. doi: 10.1039/d1ra05637a.

Molleti Jayaram, Tiwari Manishkumar S., and Yadav Ganapati D. 2018. "Novel synthesis of Ru/OMS catalyst by solvent-free method: Selective hydrogenation of levulinic acid to γ-valerolactone in aqueous medium and kinetic modelling." *Chem. Eng. J.* 334. Elsevier: 2488–2499. doi: 10.1016/j.cej.2017.11.125.

Peng Peng, Gao Xiong-Hou, Yan Zi-Feng, and Mintova Svetlana. 2020. "Diffusion and catalyst efficiency in hierarchical zeolite catalysts." *National Science Review.* 7. Science Press: 1726–1742. doi: 10.1093/nsr/nwaa184.

Peng Qingrui, Wang Haijun, Xia Yongmei, and Liu Xiang. 2021. "One-pot conversion of furfural to gamma-valerolactone in the presence of multifunctional zirconium alizarin red S hybrid." *Appl. Catal. A: Gen.* 621. Elsevier: doi: 10.1016/j.apcata.2021.118203.

Roman-Leshkov Yuriy, Barrett Christopher J., Liu Zhen Y..Dumesic James A. 2007. "Production of dimethylfuran for liquid fuels from biomass-derived carbohydrates." *Nature*. 447. Spriger Nature: 982–985. doi: 10.1038/nature05923.

Rojas-Buzo Sergio, García-García Pilar, and Corma Avelino. 2018. "Catalytic transfer hydrogenation of biomass-derived carbonyls over hafnium-based metal-organic frameworks." *ChemSusChem*. 11. Wiley-VCH verlag gmbh: 432–438. doi: 10.1002/cssc.201701708.

Srinivasa Rao B., Krishna Kumari P., Koley Paramita, Tardio J., and Lingaiah N. 2019. "One pot selective conversion of furfural to γ-valerolactone over zirconia containing heteropoly tungstate supported on β-zeolite catalyst." *Mol. Catal.* 466. Elsevier: 52–59. doi: 10.1016/j.mcat.2018.12.024.

Srinivasa Rao B., Yogita, Dhana Lakshmi D., Kumari P. Krishna, and Lingaiah N. 2021a. "Influence of metal oxide and heteropoly tungstate location in mesoporous silica towards catalytic transfer hydrogenation of furfural to γ-valerolactone." *Sustain. Energy Fuels*. 5. The Royal Society of Chemistry: 3719–3728. doi: 10.1039/d1se00340b.

Srinivasa Rao B., Yogita, Dhana Lakshmi D., Kumari P. Krishna, and Lingaiah N. 2021b. "Influence of metal oxide and heteropoly tungstate location in mesoporous silica towards catalytic transfer hydrogenation of furfural to γ-valerolactone." *Sustain. Energ. Fuels*. 5. The Royal Society of Chemistry: 3719–3728. doi: 10.1039/d1se00340b.

Sittiwong Jarinya, Boonmark Sininat, Nunthakitgoson Watinee, Maihom hana, Wattanakit Chularat, and Limtrakul Jumras 2021. "Density functional investigation of the conversion of furfural to furfuryl alcohol by reaction with i-propanol over UiO-66 metal-organic framework." *Inorg Chem*. 60. ACS Publications: 4860–4868. doi: 10.1021/acs.inorgchem.0c03764.

Shao Yuewen, Li Qingyin, Dong Xinyi, Wang Junzhe, Sun Kai, Zhang Lijun, Zhang Shu, Xu Leilei, Yuan Xiangzhou, and Hu Xun. 2021. "Cooperation between hydrogenation and acidic sites in Cu-based catalyst for selective conversion of furfural to γ-valerolactone." *Fuel*. 293. Elsevier: doi: 10.1016/j.fuel.2021.120457.

Tang Bo, Li Shuang, Song Wei-Chao, Li Yan, and Yang, En-Cui. 2021. "One-pot transformation of furfural into γ-valerolactone catalyzed by a hierarchical Hf-Al-USY zeolite with balanced Lewis and Brønsted acid sites." *Sustain. Energy Fuels*. 5. The Royal Society of Chemistry: 4724–4735. doi: 10.1039/d1se00942g.

Tan Jinyu, Liu Yixuan, Li Mingrui, Li Hu, and Yang Song. 2022. "One-step catalytic upgrading of bio-based furfural to γ-valerolactone actuated by coordination organophosphate–Hf polymers." *Sustain. Energ. Fuels*. 6. The Royal Society of Chemistry: 484–501. doi: 10.1039/d1se01476e.

Wang Yeqing, Wu Qinming, Meng Xiangju, and Xiao Feng-Shou. 2017. "Insights into the organotemplate-free synthesis of zeolite catalysts." *Engineering*. 3. Strategic Study of CAE: 567–574. doi: 10.1016/j.Eng.2017.03.029.

Wu Weibo, Li Yan, Li Hu, Zhao Wenfeng, and Yang Song. 2018. "Acid–base bifunctional Hf nanohybrids enable high selectivity in the catalytic conversion of ethyl levulinate to γ-valerolactone." *Catalysts*. 8. DPI: 264. doi: 10.3390/catal8070264.

Winoto Haryo Pandu, Fikri Zuhroni Ali, Ha, Jeong-Myeong, Park, Young-Kwon, Lee, Hyunjoo, Suh, Dong Jin, and Jae, Jungho. 2019. "Heteropolyacid supported on Zr-Beta zeolite as an active catalyst for one-pot transformation of furfural to γ-valerolactone." *Appl. Catal. B*. 241. Elsevier: 588–597. doi: 10.1016/j.apcatb.2018.09.031.

Wang Tianlong, He Jianghua, and Zhang Yuetao. 2020. "Production of γ-valerolactone from one-pot transformation of biomass-derived carbohydrates over Chitosan-supported ruthenium catalyst combined with zeolite ZSM-5." *Eur. J. Org. Chem*. 2020. Elsevier: 1611–1619. doi: 10.1002/ejoc.201901704.

Xie Chao, Song Jinliang, Zhou Baowen, Hu Jiayin, Zhang Zhanrong, Zhang Pei, Jiang Zhiwei, and Han Buxing. 2016. "Porous hafnium phosphonate: Novel heterogeneous catalyst for conversion of levulinic acid and esters into γ-valerolactone." *ACS Sustain. Chem. Eng.* 4. ACS Publications: 6231–6236. doi: 10.1021/acssuschemeng.6b02230.

Yuan Jing, Li Shu-Shuang, Yu Lei, Liu Yong-Mei, Cao Yong, He He-Yong, and Fan Kang-Nian. 2013. "Copper-based catalysts for the efficient conversion of carbohydrate biomass into γ-valerolactone in the absence of externally added hydrogen." *Energy Environ. Sci.* 6. The Royal Society of Chemistry: doi: 10.1039/c3ee40857d.

Yan Kai, Yang Yiyi, Chai Jiajue, and Lu Yiran. 2015. "Catalytic reactions of gamma-valerolactone: A platform to fuels and value-added chemicals." *Appl. Catal. B.* 179. Elsevier: 292–304. doi: 10.1016/j.apcatb.2015.04.030.

Yu Zhihao, Lu Xuebin, Liu Chen, Han Yiwen, and Ji Na. 2019. "Synthesis of γ-valerolactone from different biomass-derived feedstocks: Recent advances on reaction mechanisms and catalytic systems." *Renew. Sust. Energ. Rev.* Elsevier: 140–157. doi: 10.1016/j.rser.2019.05.039.

Ye Lei, Han Yiwen, Bai Hui, and Lu Xuebin. 2020b. "HZ-ZrP catalysts with adjustable ratio of Brønsted and Lewis acids for the one-pot value-added conversion of biomass-berived furfural." *ACS Sustain. Chem. Eng.* 8. ACS Publications: 7403–7413. doi: 10.1021/acssuschemeng.0c01259.

Ye Lei, Han Yiwen, Feng Jing, and Lu Xuebin. 2020a. "A review about GVL production from lignocellulose: Focusing on the full components utilization." *Ind. Crops Prod.* 144. Elsevier. doi: 10.1016/j.indcrop.2019.112031.

Zhu Shanhui, Xue Yanfeng, Guo Jing, Cen Youliang, Wang Jianguo, and Fan Weibin. 2016. "Integrated conversion of hemicellulose and furfural into γ-valerolactone over Au/ZrO_2 catalyst combined with ZSM-5." *ACS Catal.* 6. ACS Publications: 2035–2042. doi: 10.1021/acscatal.5b02882.

Zhou Shenghui, Dai Fanglin, Xiang Zhouyang, Song Tao, Liu Detao, Lu Fachuang, and Qi Haisong. 2019a. "Zirconium–lignosulfonate polyphenolic polymer for highly efficient hydrogen transfer of biomass-derived oxygenates under mild conditions." *Appl. Catal. B.* 248. Elsevier: 31–43. doi: 10.1016/j.apcatb.2019.02.011.

Zhou Shenghui, Dai Fanglin, Chen Yian, Dang Chao, Zhang Cunzhi, Liu Detao, and Qi Haisong. 2019b. "Sustainable hydrothermal self-assembly of hafnium–lignosulfonate nanohybrids for highly efficient reductive upgrading of 5-hydroxymethylfurfural." *Green Chem.* 21. The Royal Society of Chemistry: 1421–1431. doi: 10.1039/c8gc03710h.

Index

D

E

F

G

H

I

K

L

M

T

U

V

FSC
www.fsc.org
MIX
Papier | Fördert
gute Waldnutzung
FSC® C083411

Kolibri 360 GmbH
Ferdinand-Jühlke-Straße 7
99095 Erfurt, Deutschland
produktsicherheit@kolibri360.de